T0210530

Lecture Notes in Artificial Intelligence   9166

Subseries of Lecture Notes in Computer Science

More information about this series at http://www.springer.com/series/1244

Petra Perner (Ed.)

Machine Learning and Data Mining in Pattern Recognition

11th International Conference, MLDM 2015
Hamburg, Germany, July 20–21, 2015
Proceedings

 Springer

Editor
Petra Perner
IBaI
Leipzig
Germany

ISSN 0302-9743 ISSN 1611-3349 (electronic)
Lecture Notes in Artificial Intelligence
ISBN 978-3-319-21023-0 ISBN 978-3-319-21024-7 (eBook)
DOI 10.1007/978-3-319-21024-7

Library of Congress Control Number: 2015942804

LNCS Sublibrary: SL7 – Artificial Intelligence

Springer Cham Heidelberg New York Dordrecht London

Printed on acid-free paper

Springer International Publishing AG Switzerland is part of Springer Science+Business Media
(www.springer.com)

Preface

The tenth event of the International Conference on Machine Learning and Data Mining MLDM 2015 was held in Hamburg (www.mldm.de) running under the umbrella of the World Congress "The Frontiers in Intelligent Data and Signal Analysis, DSA2015" (www.worldcongressdsa.com).

For this edition the Program Committee received 123 submissions. After the peer-review process, we accepted 41 high-quality papers for oral presentation; from these, 40 are included in this proceedings volume. The topics range from theoretical topics for classification, clustering, association rule and pattern mining to specific data-mining methods for the different multimedia data types such as image mining, text mining, video mining, and Web mining. Extended versions of selected papers will appear in the international journal *Transactions on Machine Learning and Data Mining* (www.ibai-publishing.org/journal/mldm).

A tutorial on Data Mining, a tutorial on Case-Based Reasoning, a tutorial on Intelligent Image Interpretation and Computer Vision in Medicine, Biotechnology, Chemistry and Food Industry, a tutorial on Standardization in Immunofluorescence, and a tutorial on Big Data and Text Analysis were held before the conference.

We were pleased to give out the best paper award for MLDM for the fourth time this year. There are four announcements mentioned at www.mldm.de. The final decision was made by the Best Paper Award Committee based on the presentation by the authors and the discussion with the auditorium. The ceremony took place during the banquet. This prize is sponsored by ibai solutions (www.ibai-solutions.de), one of the leading companies in data mining for marketing, Web mining, and e-commerce.

The conference was rounded up by an outlook of new challenging topics in machine learning and data mining before the Best Paper Award ceremony.

We would like to thank all reviewers for their highly professional work and their effort in reviewing the papers. We would also thank members of the Institute of Applied Computer Sciences, Leipzig, Germany (www.ibai-institut.de), who handled the conference as secretariat. We appreciate the help and understanding of the editorial staff at Springer, and in particular Alfred Hofmann, who supported the publication of these proceedings in the LNAI series.

Last, but not least, we wish to thank all the speakers and participants who contributed to the success of the conference. See you in 2016 in New York at the next World Congress (www.worldcongressdsa.com) on "The Frontiers in Intelligent Data and Signal Analysis, DSA2016," which combines under its roof the following three events: International Conference on Machine Learning and Data Mining, MLDM, the Industrial Conference on Data Mining, ICDM, and the International Conference on Mass Data Analysis of Signals and Images in Medicine, Biotechnology, Chemistry and Food Industry, MDA.

July 2015   Petra Perner

Organization

Chair

Petra Perner   IBaI Leipzig, Germany

Program Committee

Sergey V. Ablameyko	Belarus State University, Belarus
Patrick Bouthemy	Inria-VISTA, France
Michelangelo Ceci	University of Bari, Italy
Xiaoqing Ding	Tsinghua University, China
Christoph F. Eick	University of Houston, USA
Ana Fred	Technical University of Lisbon, Portugal
Giorgio Giacinto	University of Cagliari, Italy
Makato Haraguchi	Hokkaido University Sapporo, Japan
Dimitrios A. Karras	Chalkis Institute of Technology, Greece
Adam Krzyzak	Concordia University, Montreal, Canada
Thang V. Pham	Intelligent Systems Lab Amsterdam (ISLA), The Netherlands
Gabriella Sanniti di Baja	CNR, Italy
Linda Shapiro	University of Washington, USA
Tamas Sziranyi	MTA-SZTAKI, Hungary
Alexander Ulanov	HP Labs, Russian Federation
Patrick Wang	Northeastern University, USA

Additional Reviewers

Jeril Kuriakose	Manipal University Jaipur, India
Goce Ristanoski	NICTA, Australia
Hamed Bolandi	Islamic Azad University (IAU), Iran

Contents

Association and Sequential Rule Mining

Support Vector Machines

Frequent Item Set Mining and Time Series Analysis

Clustering

Text Mining

Applications of Data Mining

Data Mining in System Biology, Drug Discovery, and Medicine

Graph Mining

Greedy Graph Edit Distance

Kaspar Riesen[1,2]([✉]), Miquel Ferrer[1], Rolf Dornberger[1], and Horst Bunke[2]

[1] Institute for Information Systems, University of Applied Sciences and Arts
Northwestern Switzerland, Riggenbachstrasse 16, 4600 Olten, Switzerland
{kaspar.riesen,rolf.dornberger,miquel.ferrer}@fhnw.ch
[2] Institute of Computer Science and Applied Mathematics, University of Bern,
Neubrückstrasse 10, 3012 Bern, Switzerland
{riesen,bunke}@iam.unibe.ch

Abstract. In pattern recognition and data mining applications, where
the underlying data is characterized by complex structural relationships,
graphs are often used as a formalism for object representation. Yet, the
high representational power and flexibility of graphs is accompanied by a
significant increase of the complexity of many algorithms. For instance,
exact computation of pairwise graph dissimilarity, i.e. distance, can be
accomplished in exponential time complexity only. A previously introduced approximation framework reduces the problem of graph comparison to an instance of a linear sum assignment problem which allows graph
dissimilarity computation in cubic time. The present paper introduces an
extension of this approximation framework that runs in quadratic time.
We empirically confirm the scalability of our extension with respect to
the run time, and moreover show that the quadratic approximation leads
to graph dissimilarities which are sufficiently accurate for graph based
pattern classification.

1 Introduction

Classification is a common task in the areas of pattern recognition and data
mining [1–3]. Due to the mathematical wealth of operations available in a vector
space, a huge amount of algorithms for classification and analysis of objects given
in terms of feature vectors have been developed in recent years [4,5]. Yet, the use
of feature vectors implicates two limitations. First, as vectors always represent
a predefined set of features, all vectors in a particular application have to preserve the same length regardless of the size or complexity of the corresponding
objects. Second, there is no direct possibility to describe binary relationships
among different parts of an object. Both constraints can be overcome by graph
based representations [6,7]. That is, graphs are not only able to describe unary
properties of an object but also binary relationships among different parts. Furthermore, the number of nodes and edges of a graph is not limited a priori but can
be adapted to the size or the complexity of each individual object. Due to these
substantial advantages, a growing interest in graph-based object representation
in machine learning and data mining can be observed in recent years [8–11].

© Springer International Publishing Switzerland 2015
P. Perner (Ed.): MLDM 2015, LNAI 9166, pp. 3–16, 2015.
DOI: 10.1007/978-3-319-21024-7_1

Given that the basic objects in an application are formalized by means of graphs, one has to define an appropriate graph dissimilarity model. The process of evaluating the dissimilarity of graphs is referred to as graph matching [6,7]. Among a vast number of graph matching methods available, the concept of graph edit distance [12,13] is in particular interesting because it is able to cope with directed and undirected, as well as with labeled and unlabeled graphs. If there are labels on nodes, edges, or both, no constraints on the respective label alphabets have to be considered.

Yet, a major drawback of graph edit distance is its computational complexity. In fact, the problem of graph edit distance can be reformulated as an instance of a *Quadratic Assignment Problem (QAP)* [14] for which only exponential run time algorithms are available to date[1]. In recent years, a number of methods addressing the high complexity of graph edit distance computation have been proposed [15–19]. In [20] the authors of the present paper also introduced an algorithmic framework for the approximation of graph edit distance. The basic idea of this approach is to reduce the difficult QAP of graph edit distance computation to a *linear sum assignment problem* (LSAP). For LSAPs quite an arsenal of efficient (i.e. cubic time) algorithms exist [21].

The algorithmic procedure described in [20] consists of three major steps. In a first step the graphs to be matched are subdivided into individual nodes including local structural information. Next, in step 2, an LSAP solving algorithm is employed in order to find an optimal assignment of the nodes (plus local structures) of both graphs. Finally, in step 3, an approximate graph edit distance, which is globally consistent with the underlying edge structures of both graphs, is derived from the assignment of step 2.

The major goal of the present paper is to speed up the approximation framework presented in [20]. In particular, we aim at substantially speeding up step 2 of the algorithmic procedure. In the current framework a state-of-the art algorithm with cubic time complexity is employed in order to optimally solve the underlying LSAP. In the present paper we aim at replacing this optimal algorithm with a suboptimal greedy algorithm which runs in quadratic time. From the theoretical point of view this approach is very appealing as it makes use of an approximation algorithm for an LSAP which in turn approximates the underlying QAP. The greedy algorithm for LSAP runs in quadratic time. Hence, a substantial speed up of the approximation can be expected. Yet, the overall question to be answered is whether this faster approximation is able to keep up with the existing framework with respect to distance accuracy.

The remainder of this paper is organized as follows. Next, in Sect. 2 the concept and computation of graph edit distance as well as the original framework for graph edit distance approximation [20] are summarized. In Sect. 3 the greedy component to be employed in step 2 of our framework is introduced. An experimental evaluation on diverse data sets is carried out in Sect. 4, and in Sect. 5 we draw some conclusions.

---

[1] Note that QAPs are known to be $\mathcal{NP}$-*complete*, and therefore, an exact and efficient algorithm for the graph edit distance problem can not be developed unless $\mathcal{P} = \mathcal{NP}$.

2 Bipartite Graph Edit Distance (BP-GED)

2.1 Graph Edit Distance

A graph g is a four-tuple $g = (V, E, \mu, \nu)$, where V is the finite set of nodes, $E \subseteq V \times V$ is the set of edges, $\mu : V \to L_V$ is the node labeling function, and $\nu : E \to L_E$ is the edge labeling function. The labels for both nodes and edges can be given by the set of integers $L = \{1, 2, 3, \ldots\}$, the vector space $L = \mathbb{R}^n$, a set of symbolic labels $L = \{\alpha, \beta, \gamma, \ldots\}$, or a combination of various label alphabets from different domains. Unlabeled graphs are obtained as a special case by assigning the same (empty) label $\varnothing$ to all nodes and edges, i.e. $L_V = L_E = \{\varnothing\}$.

Given two graphs, the source graph $g_1 = (V_1, E_1, \mu_1, \nu_1)$ and the target graph $g_2 = (V_2, E_2, \mu_2, \nu_2)$, the basic idea of graph edit distance [12,13] is to transform g_1 into g_2 using some edit operations. A standard set of edit operations is given by *insertions*, *deletions*, and *substitutions* of both nodes and edges. We denote the substitution of two nodes $u \in V_1$ and $v \in V_2$ by $(u \to v)$, the deletion of node $u \in V_1$ by $(u \to \varepsilon)$, and the insertion of node $v \in V_2$ by $(\varepsilon \to v)$, where ε refers to the empty node. For edge edit operations we use a similar notation.

A sequence $(e_1, \ldots, e_k)$ of k edit operations e_i that transform g_1 completely into g_2 is called *edit path* $\lambda(g_1, g_2)$ between g_1 and g_2. Note that in an edit path $\lambda(g_1, g_2)$ each node of g_1 is either deleted or uniquely substituted with a node in g_2, and analogously, each node in g_2 is either inserted or matched with a unique node in g_1. The same applies for the edges. Yet, edit operations on edges are always defined by the edit operations on their adjacent nodes. That is, whether an edge (u, v) is substituted, deleted, or inserted, depends on the edit operations actually performed on both adjacent nodes u and v.

Let $\Upsilon(g_1, g_2)$ denote the set of all admissible and complete edit paths between two graphs g_1 and g_2. To find the most suitable edit path out of $\Upsilon(g_1, g_2)$, one introduces a cost $c(e)$ for every edit operation e, measuring the strength of the corresponding operation. The idea of such a cost is to define whether or not an edit operation e represents a strong modification of the graph. By means of cost functions for elementary edit operations, graph edit distance allows the integration of domain specific knowledge about object similarity. Furthermore, if in a particular case prior knowledge about the labels and their meaning is not available, automatic procedures for learning the edit costs from a set of sample graphs are available as well [22].

Clearly, between two similar graphs, there should exist an inexpensive edit path, representing low cost operations, while for dissimilar graphs an edit path with high cost is needed. Consequently, the exact edit distance $d_{\lambda_{\min}}(g_1, g_2)$, or $d_{\lambda_{\min}}$ for short, of two graphs g_1 and g_2 is defined as

$$d_{\lambda_{\min}}(g_1, g_2) = \min_{\lambda \in \Upsilon(g_1, g_2)} \sum_{e_i \in \lambda} c(e_i), \tag{1}$$

where $\Upsilon(g_1, g_2)$ denotes the set of all edit paths transforming g_1 into g_2, c denotes the cost function measuring the strength $c(e_i)$ of node edit operation

e_i (including the cost of the implied edge edit operations), and $\lambda_{\min}$ refers to the minimal cost edit path found in $\Upsilon(g_1, g_2)$.

Algorithms for computing the exact edit distance $d_{\lambda_{\min}}(g_1, g_2)$ are typically based on combinatorial search procedures that explore the space of all possible mappings of the nodes and edges of g_1 to the nodes and edges of g_2 (i.e. the search space corresponds to $\Upsilon(g_1, g_2)$). Such an exploration is often conducted by means of A* based search techniques [23].

2.2   Approximation of Graph Edit Distance

The computational complexity of exact graph edit distance is exponential in the number of nodes of the involved graphs. That is, considering m nodes in g_1 and n nodes in g_2, $\Upsilon(g_1, g_2)$ contains $O(m^n)$ edit paths to be explored. This means that for large graphs the computation of edit distance is intractable. The graph edit distance approximation framework introduced in [20] reduces the difficult *Quadratic Assignment Problem (QAP)* of graph edit distance computation to an instance of a *Linear Sum Assignment Problem (LSAP)* for which a large number of efficient algorithms exist [21]. The LSAP is defined as follows.

Definition 1. *Given two disjoint sets $S = \{s_1, \ldots, s_n\}$ and $Q = \{q_1, \ldots, q_n\}$ and an $n \times n$ cost matrix $\mathbf{C} = (c_{ij})$, where c_{ij} measures the cost of assigning the i-th element of the first set to the j-th element of the second set, the Linear Sum Assignment Problem (LSAP) consists in finding the minimum cost permutation*

$$(\varphi_1, \ldots, \varphi_n) = \underset{(\varphi_1, \ldots, \varphi_n) \in \mathcal{S}_n}{\arg \min} \sum_{i=1}^{n} c_{i\varphi_i},$$

where $\mathcal{S}_n$ refers to the set of all $n!$ possible permutations of n integers, and permutation $(\varphi_1, \ldots, \varphi_n)$ refers to the assignment where the first entity $s_1 \in S$ is mapped to entity $q_{\varphi_1} \in Q$, the second entity $s_2 \in S$ is assigned to entity $q_{\varphi_2} \in Q$, and so on.

By reformulating the graph edit distance problem to an instance of an LSAP, three major issues have to be resolved. First, LSAPs are generally stated on independent sets with equal cardinality. Yet, in our case the elements to be assigned to each other are given by the sets of nodes (and edges) with unequal cardinality in general. Second, solutions to LSAPs refer to assignments of elements in which every element of the first set is assigned to exactly one element of the second set and vice versa (i.e. a solution to an LSAP corresponds to a bijective assignment of the underlying entities). Yet, graph edit distance is a more general assignment problem as it explicitly allows both deletions and insertions to occur on the basic entities (rather than only substitutions). Third, graphs do not only consist of independent sets of entities (i.e. nodes) but also of structural relationships between these entities (i.e. edges that connect pairs of nodes). LSAPs are not able to consider these relationships in a global and consistent way. The first two issues are perfectly – and the third issue partially – resolvable by means of the following definition of a square cost matrix whereon the LSAP is eventually solved.

Definition 2. *Based on the node sets $V_1 = \{u_1, \ldots, u_n\}$ and $V_2 = \{v_1, \ldots, v_m\}$ of g_1 and g_2, respectively, a cost matrix $\mathbf{C}$ is established as follows.*

$$
\mathbf{C} = \left[
\begin{array}{cccc|cccc}
c_{11} & c_{12} & \cdots & c_{1m} & c_{1\varepsilon} & \infty & \cdots & \infty \\
c_{21} & c_{22} & \cdots & c_{2m} & \infty & c_{2\varepsilon} & \ddots & \vdots \\
\vdots & \vdots & \ddots & \vdots & \vdots & \ddots & \ddots & \infty \\
c_{n1} & c_{n2} & \cdots & c_{nm} & \infty & \cdots & \infty & c_{n\varepsilon} \\
\hline
c_{\varepsilon 1} & \infty & \cdots & \infty & 0 & 0 & \cdots & 0 \\
\infty & c_{\varepsilon 2} & \ddots & \vdots & 0 & 0 & \ddots & \vdots \\
\vdots & \ddots & \ddots & \infty & \vdots & \ddots & \ddots & 0 \\
\infty & \cdots & \infty & c_{\varepsilon m} & 0 & \cdots & 0 & 0
\end{array}
\right]
\tag{2}
$$

Entry c_{ij} thereby denotes the cost of a node substitution $(u_i \to v_j)$, $c_{i\varepsilon}$ denotes the cost of a node deletion $(u_i \to \varepsilon)$, and $c_{\varepsilon j}$ denotes the cost of a node insertion $(\varepsilon \to v_j)$.

Note that matrix $\mathbf{C} = (c_{ij})$ is by definition quadratic. Hence, the first issue (sets of unequal size) is instantly eliminated. Obviously, the left upper corner of the cost matrix $\mathbf{C} = (c_{ij})$ represents the costs of all possible node substitutions, the diagonal of the right upper corner the costs of all possible node deletions, and the diagonal of the bottom left corner the costs of all possible node insertions. Note that every node can be deleted or inserted at most once. Therefore any non-diagonal element of the right-upper and left-lower part is set to ∞. The bottom right corner of the cost matrix is set to zero since substitutions of the form $(\varepsilon \to \varepsilon)$ should not cause any cost.

Given the cost matrix $\mathbf{C} = (c_{ij})$, the LSAP optimization consists in finding a permutation $(\varphi_1, \ldots, \varphi_{n+m})$ of the integers $(1, 2, \ldots, (n + m))$ that minimizes the overall assignment cost $\sum_{i=1}^{(n+m)} c_{i\varphi_i}$. This permutation corresponds to the assignment

$$
\psi = ((u_1 \to v_{\varphi_1}), (u_2 \to v_{\varphi_2}), \ldots, (u_{m+n} \to v_{\varphi_{m+n}}))
$$

of the nodes of g_1 to the nodes of g_2. Note that assignment ψ includes node assignments of the form $(u_i \to v_j)$, $(u_i \to \varepsilon)$, $(\varepsilon \to v_j)$, and $(\varepsilon \to \varepsilon)$ (the latter can be dismissed, of course). Hence, the definition of the cost matrix in Eq. 2 also resolves the second issue stated above and allows insertions and/or deletions to occur in an optimal assignment.

The third issue is about the edge structure of both graphs, and it cannot be entirely considered by LSAPs. In fact, so far the cost matrix $\mathbf{C} = (c_{ij})$ considers the nodes of both graphs only, and thus mapping ψ does not take any structural constraints into account. In order to integrate knowledge about the graph structure, to each entry c_{ij}, i.e. to each cost of a node edit operation $(u_i \to v_j)$, the minimum sum of edge edit operation costs, implied by the corresponding node operation, is added. That is, we encode the minimum matching cost arising from the local edge structure in the individual entries $c_{ij} \in \mathbf{C}$. This particular encoding of the minimal edge edit operation cost enables the LSAP to consider information about the local, yet not global, edge structure of a graph. Hence, this heuristic procedure partially resolves the third issue.

Given the node assignment ψ a distance value approximating the exact graph edit distance $d_{\lambda_{\min}}(g_1, g_2)$ can be directly inferred. As stated above, the LSAP optimization finds an assignment ψ in which every node of g_1 is either assigned to a unique node of g_2 or deleted. Likewise, every node of g_2 is either assigned to a unique node of g_1 or inserted. Hence, mapping ψ refers to an admissible and complete edit path between the graphs under consideration, i.e. $\psi \in \Upsilon(g_1, g_2)$. Therefore, the edge operations, which are implied by edit operations on their adjacent nodes, can be completely inferred from ψ. Hence, we get an edit distance $d_\psi(g_1, g_2)$, or d_ψ for short, which corresponds to edit path $\psi \in \Upsilon(g_1, g_2)$.

Algorithm 1. BP-GED(g_1, g_2)

1: Build cost matrix $\mathbf{C} = (c_{ij})$ according to the input graphs g_1 and g_2
2: Compute optimal node assignment $\psi = \{u_1 \to v_{\varphi_1}, u_2 \to v_{\varphi_2}, \ldots, u_{m+n} \to v_{\varphi_{m+n}}\}$ on $\mathbf{C}$
3: Complete edit path according to ψ and **return** $d_\psi(g_1, g_2)$

Note that the edit path ψ corresponding to $d_\psi(g_1, g_2)$ considers the edge structure of g_1 and g_2 in a global and consistent way while the optimal node mapping ψ from Step 2 is able to consider the structural information in an isolated way only (single nodes and their adjacent edges). This is due to the fact that during the optimization process of the specific LSAP no information about neighboring node mappings is available. Hence, in comparison with optimal search methods for graph edit distance, our algorithmic framework might cause additional edge operations in the third step, which would not be necessary in a globally optimal graph matching. Hence, the distances found by this specific framework are – in the best case – equal to, or – in general – larger than the exact graph edit distance[2]. Yet, the proposed reduction of graph edit distance to an LSAP allows the approximate graph edit distance computation in polynomial time.

For the remainder of this paper we denote the algorithmic procedure for deriving $d_\psi(g_1, g_2)$ with *BP-GED* (*Bipartite Graph Edit Distance*[3]). In Algorithm 1 the three major steps of BP-GED are summarized.

3 Greedy Graph Edit Distance (Greedy-GED)

The primary goal of the present paper is to speed-up the approximation framework described in the section above under the condition that the level of distance accuracy provided by the current approximation framework is maintained

---

[2] In [24] it is formally proven that this approximation scheme builds an upper bound of the exact graph edit distance.

[3] The assignment problem can also be formulated as finding a matching in a *complete bipartite graph* and is therefore also referred to as *bipartite graph matching problem*.

as closely as possible. In order to approach this goal we propose to speed up step 2 of the current approximation framework.

In step 2 of Algorithm 1 an optimal assignment of the nodes (plus local structures) of both graphs has to be found. For this task a large number of algorithms exists (see [21] for an exhaustive survey). They range from primal-dual combinatorial algorithms [25–29], to simplex-like methods [30–33], cost operation algorithms [34], forest algorithms [35], and other approaches [36]. For optimally solving the LSAP in the existing framework, Munkres' algorithm [25] also referred to as Kuhn-Munkres, or Hungarian algorithm, is deployed. The time complexity of this particular algorithm (as well as the best performing other algorithms for LSAPs) is cubic in the size of the problem. According to our definition of the cost matrix $\mathbf{C}$ the size of the LSAP amounts to $(n + m)$ in our case.

Algorithm 2. Greedy-Assignment$(g_1, g_2, \mathbf{C} = (c_{ij}))$

```
 1: ψ' = {}
 2: for i = 1, ..., (m + n) do
 3:     min = ∞
 4:     for j = 1, ..., (m + n) do
 5:         if (x → v_j) ∉ ψ' then
 6:             if c_ij < min then
 7:                 min = c_ij
 8:                 φ'_i = j
 9:             end if
10:         end if
11:     end for
12:     ψ' = ψ' ∪ {(u_i → v_{φ'_i})}
13: end for
14: return ψ'
```

While for the optimal solution of LSAPs quite an arsenal of algorithms is available, only a few works are concerned with the suboptimal solution of general LSAPs (see [37] for a survey). In an early publication [38] several approximation methods for the LSAP are formulated. One of these methods is a greedy algorithm that iterates through the cost matrix from top to bottom through all rows and assign every element to the minimum unused element of the current row.

This idea is formalized in Algorithm 2. The node assignment ψ' is first initialized as empty set. Next, for each row i in the cost matrix $\mathbf{C} = (c_{ij})$ the minimum cost assignment, i.e. the column index

$$\varphi'_i = \underset{j=1,\dots,(m+n)}{\arg\min} \; c_{ij}$$

is determined. The corresponding node edit operation $(u_i \to v_{\varphi'_i})$ is eventually added to ψ' (lines 3 to 12). Note that every column of the cost matrix can be considered at most once since otherwise the assignment ψ would not be admissible in the sense of possibly assigning two nodes from g_1 to the same node of g_2 (this is verified on line 5).

In the first row we have to consider $(n + m)$ elements in order to find the minimum. In the second row we have to consider $(n + m - 1)$ elements, a.s.o. In the last row only one element remains. Hence, the complexity of this approximate assignment algorithm is $O((n + m)^2)$.

In contrast with the optimal permutation returned by the original framework in step 2, the permutation $(\varphi'_1, \varphi'_2, \ldots, \varphi'_{(n+m)})$ is suboptimal. That is, the sum of assignments costs of our greedy approach is greater than, or equal to, the minimal assignment cost provided by optimal LSAP solving algorithms:

$$\sum_{i=1}^{(n+m)} c_{i\varphi'_i} \geq \sum_{i=1}^{(n+m)} c_{i\varphi_i}$$

Yet, note that for the corresponding distance values d_ψ and $d_{\psi'}$ no globally valid order relation exists. That is, the approximate graph edit distance $d_{\psi'}$ derived from ψ' can be greater than, equal to, or smaller than d_ψ.

We substitute the optimal assignment algorithm of step 2 of Algorithm 1 with the greedy procedure described in this section. That is, we aim at solving the assignment of local graph structures in our current framework with an approximate rather than an exact algorithm. For the remainder of this paper we denote this adapted algorithmic procedure with *Greedy-GED*.

4   Experimental Evaluation

The goal of the experimental evaluation is twofold. First, we aim at empirically confirming the faster matching time of Greedy-GED compared to BP-GED. Second, we aim at answering the question whether or not the greedy graph edit distances remain sufficiently accurate for graph based pattern classification.

In order to approach the first goal we construct random graphs with a different number of nodes. For every random graph we first define a set of n nodes $V = \{v_1, \ldots, v_n\}$, where n is a random integer selected from an interval $[V_{min}, V_{max}]$. We construct 15 sets with 20 random graphs each by iteratively increasing both V_{min} and V_{max} from 50 and 60 to 190 and 200, respectively. Next, an undirected edge is inserted between any pair of nodes $(u, v) \in V \times V$ with probability p_E (the edge probability p_E is fixed to 0.5 for every graph size in our experiments). Every node $v_i \in V$ and every edge $(u_i, v_j) \in E$ is finally labeled by a random integer from $L = \{1, \ldots, V_{max}\}$ (i.e. the size of the label alphabet equals the maximum number of nodes of the corresponding graph set).

In Fig. 1 the total run time in ms for all 400 graph matchings per graph set using BP-GED (black bars) and Greedy-GED (grey bars) is shown (note that the upper part of the chart is shown on a logarithmic scale). The superiority of Greedy-GED compared to BP-GED concerning the run time is clearly observable. For instance, on graphs with 50 to 60 nodes, the total run time amounts to 358 ms and 10,426 ms for Greedy-GED and BP-GED, respectively. This corresponds to a speed up of graph edit distance computation by factor 29. Moreover, the scalability of graph edit distance computation is also assured much better by

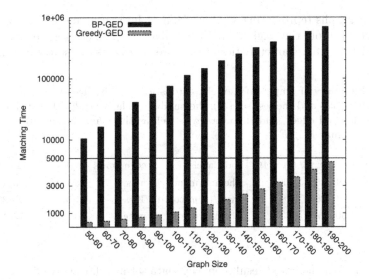

Fig. 1. Run time in ms for 400 graph matchings using BP-GED (black bars) and Greedy-GED (grey bars) for differently sized graphs. Note that the upper part of the chart is shown on a logarithmic scale.

Greedy-GED (with quadratic run time) rather than with BP-GED (with cubic run time). That is, the larger the graphs are, the bigger becomes the speed-up factor. For graphs with 120 to 130 nodes, for instance, Greedy-GED is about 90 times faster than BP-GED (1,608 ms vs. 144,563 ms in total), while for graphs with 190 to 200 nodes the speed-up factor is increased to more than 147 (4,726 ms vs. 697,812 ms in total).

For the second goal of the experimental evaluation, viz. assessing the quality of the resulting distance approximations, we use the distances returned by BP-GED and Greedy-GED in a pattern classification scenario. As basic classifier a nearest-neighbor classifier (NN) is employed. Note that there are various other approaches to graph classification that make use of graph edit distance in some form, including vector space embedding classifiers [39] and graph kernels [40]. Yet, the nearest neighbor paradigm is particularly interesting for the present evaluation because it directly uses the distances without any additional classifier training. Obviously, if an approximation algorithm leads to graph edit distances with poor quality, the classification accuracy of an NN classifier is expected to be negatively affected.

As underlying data we use three real world data sets from the IAM graph database repository [41][4]. Two graph data sets involve graphs that represent molecular compounds (AIDS and MUTA). Both data set consists of two classes, which represent molecules with activity against HIV or not (AIDS), and molecules with and without the *mutagen* property (MUTA). The third data set

---

[4] www.iam.unibe.ch/fki/databases/iam-graph-database.

consists of graphs representing symbols from architectural and electronic drawings (GREC). From the original GREC database [42], 22 classes are considered. In Table 1 the main characteristics of the three data sets are summarized.

Table 1. Summary of graph data set characteristics, viz. the size of the training (tr), the validation (va) and the test set (te), the number of classes ($|\Omega|$), the label alphabet of both nodes and edges, and the average as well as the maximum number of nodes and edges ($\varnothing/\max |V|/|E|$).

Data Set	size (tr, va, te)	$|\Omega|$	node labels	edge labels	$\varnothing$ $|V|/|E|$	max $|V|/|E|$
AIDS	250, 250, 1,500	2	Chem. symbol	none	15.7/16.2	95/103
GREC	286, 286, 528	22	Type/(x,y) coord.	Type	11.5/12.2	25/30
MUTA	1,500, 500, 2,337	2	Chem. symbol	none	30.3/30.8	417/112

In Table 2 the achieved results on each data set are shown. We first focus on the mean run time for one matching in ms ($\varnothing t$). On the relatively small graphs of the GREC data set, the speed-up by Greedy-GED compared to BP-GED is rather small (Greedy-GED is approximately 1.5 times faster than BP-GED). Yet, on the other two data sets with bigger graphs substantial speed-ups can be observed. That is, on the AIDS data set the mean matching time is decreased from 3.61 ms to 1.21 ms (three times faster) and on the MUTA data the greedy approach is more than ten times faster than the original approximation (2.53 ms vs. 26.06 ms).

Table 2. The mean run time for one matching in ms ($\varnothing t$), the mean relative deviation of Greedy-GED from BP-GED in percentage ($\varnothing e$) and the accuracy of a qNN classifier with $q = \{1, 3, 5\}$.

	Data Set					
	AIDS		GREC		MUTA	
	BP-GED	Greedy-GED	BP-GED	Greedy-GED	BP-GED	Greedy-GED
$\varnothing t$ [ms]	3.61	1.21	1.15	0.77	26.06	2.53
$\varnothing e$ [%]	–	4.89	–	7.22	–	5.13
1NN [%]	99.07	98.93	98.48	98.48	71.37	68.29
3NN [%]	97.60	97.80	98.48	98.30	72.96	70.86
5NN [%]	98.80	99.47	97.73	96.78	72.19	70.81

Next, we measure the mean relative deviation of Greedy-GED from BP-GED in percentage ($\varnothing e$). We note that the distances returned by Greedy-GED differ from the distances returned by BP-Greedy between 4.89 % and 7.22 % on the average. This deviation can also be observed in the correlation scatter plots on all data sets in Fig. 2. These scatter plots give us a visual representation of the difference between the two approximations. We plot for each pair of graphs its distance returned by BP-GED (horizontal axis) and Greedy-GED (vertical axis). Note that approximation distances derived from BP-GED as well as from Greedy-GED constitute upper bounds of the true graph edit distance. Hence, the smaller the approximation value is, the nearer it is to the exact graph edit distance. While on the GREC data set most (yet not all) of the deviation is based on an increase of distances by Greedy-GED, on the other two data sets both reductions and increases of the approximate distance values are observable in equal parts. That is, the use of an approximate – rather than an exact – solution to the LSAP does not necessarily deteriorate the derived graph edit distance accuracy.

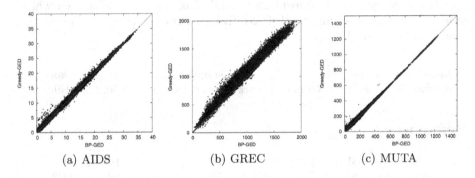

(a) AIDS   (b) GREC   (c) MUTA

Fig. 2. BP-GED (x-axis) vs. Greedy-GED (y-axis).

The question remains whether the distance values of Greedy-GED are accurate enough to serve as basis for pattern classification. We show for each data set the accuracy of a q-nearest-neighbor classifier (qNN) with $q = \{1, 3, 5\}$ using the graph edit distance approximations returned by BP-GED and Greedy-GED. On the AIDS data set the faster greedy approach results in better recognition rates than BP-GED for $q = 3$ and $q = 5$ (the improvement with $q = 5$ is statistically significant ($\alpha = 0.05$)). On the GREC data one statistically significant difference between the two approximations can be observed (with $q = 5$). Here Greedy-GED cannot keep up with the performance of BP-GED. Finally, on the MUTA data set two deteriorations provided by Greedy-GED are statistically significant ($q = 1$ and $q = 3$). Yet, the accuracies of the qNN based on BP-GED and Greedy-GED remain comparable, and the novel Greedy-GED seems to provide a reasonable trade-off between recognition accuracy and speed.

5   Conclusions and Future Work

The fast computation of accurate graph dissimilarities is still an open and active area of research in the fields of structural data mining and pattern recognition. In the present paper we propose an extension of our previous graph edit distance approximation algorithm. While in the original framework the nodes plus local edge structures are assigned to each other in an optimal way, the extension of the present paper uses a suboptimal algorithm for this task. This novel approach allows the graph edit distance approximation in quadratic – rather than cubic – time. With several experimental results we empirically confirm the superior run time of our novel approach when compared to the original framework (particularly for large graphs). Moreover, we show that the resulting approximation distances remain sufficiently accurate to serve as basis for a distance based classifier.

In future work we plan to develop several variants and refinements of the greedy method proposed in this paper. For instance, the nodes of the first graph are currently processed in fixed order. Clearly, this node ordering has a crucial impact on the greedy assignment result and thus a preprocessing step which orders the nodes according to their relative importance could be a rewarding avenue to be pursued.

Acknowledgements. This work has been supported by the *Hasler Foundation* Switzerland and the *Swiss National Science Foundation* project 200021_153249.

References

1. Perner, P. (ed.): MLDM 2012. LNCS, vol. 7376. Springer, Heidelberg (2012)
2. Perner, P. (ed.): MLDM 2013. LNCS, vol. 7988. Springer, Heidelberg (2013)
3. Duda, R.O., Hart, P.E., Stork, D.G.: Pattern Classification, 2nd edn. Wiley-Interscience, New York (2000)
4. Bishop, C.: Pattern Recognition and Machine Learning. Springer, New York (2008)
5. Shawe-Taylor, J., Cristianini, N.: Kernel Methods for Pattern Analysis. Cambridge University Press, Cambridge (2004)
6. Conte, D., Foggia, P., Sansone, C., Vento, M.: Thirty years of graph matching in pattern recognition. Int. J. Pattern Recogn. Artif. Intell. **18**(3), 265–298 (2004)
7. Foggia, P., Percannella, G., Vento, M.: Graph matching and learning in pattern recognition in the last 10 years. Int. J. Pattern Recogn. Artif. Intell. **28**(1) (2014). http://dx.doi.org/10.1142/S0218001414500013
8. Cook, D., Holder, L. (eds.): Mining Graph Data. Wiley-Interscience, New York (2007)
9. Schenker, A., Bunke, H., Last, M., Kandel, A.: Graph-Theoretic Techniques for Web Content Mining. World Scientific Publishing, Singapore (2005)
10. Gärtner, T.: Kernels for Structured Data. World Scientific Publishng, Singapore (2008)
11. Gärtner, T., Horvath, T., Wrobel, S.: Graph kernels. In: Smmut, C., Webb, G.I. (eds.) Encyclopedia of Machine Learning, pp. 467–469. Springer US, London (2010)
12. Bunke, H., Allermann, G.: Inexact graph matching for structural pattern recognition. Pattern Recogn. Lett. **1**, 245–253 (1983)
13. Sanfeliu, A., Fu, K.: A distance measure between attributed relational graphs for pattern recognition. IEEE Trans. Syst. Man Cybern. (Part B) **13**(3), 353–363 (1983)

14. Cortés, X., Serratosa, F., Solé, A.: Active graph matching based on pairwise probabilities between nodes. In: Gimelfarb, G., Hancock, E., Imiya, A., Kuijper, A., Kudo, M. (eds.) SSPR 2012. LNCS, vol. 7626, pp. 98–106. Springer, Heidelberg (2012)

15. Boeres, M., Ribeiro, C., Bloch, I.: A randomized heuristic for scene recognition by graph matching. In: Ribeiro, C., Martins, S. (eds.) WEA 2004. LNCS, vol. 3059, pp. 100–113. Springer, Heidelberg (2004)

16. Sorlin, S., Solnon, C.: Reactive tabu search for measuring graph similarity. In: Brun, L., Vento, M. (eds.) GbRPR 2005. LNCS, vol. 3434, pp. 172–182. Springer, Heidelberg (2005)

17. Neuhaus, M., Bunke, H.: An error-tolerant approximate matching algorithm for attributed planar graphs and its application to fingerprint classification. In: Fred, A., Caelli, T.M., Duin, R.P.W., Campilho, A.C. (eds.) SSPR 2004. LNCS, vol. 3138, pp. 180–189. Springer, Heidelberg (2004)

18. Justice, D., Hero, A.: A binary linear programming formulation of the graph edit distance. IEEE Trans. Pattern Anal. Mach. Intell. **28**(8), 1200–1214 (2006)

19. Dickinson, P., Bunke, H., Dadej, A., Kraetzl, M.: On graphs with unique node labels. In: Hancock, E., Vento, M. (eds.) GbRPR 2003. LNCS, vol. 2726, pp. 13–23. Springer, Heidelberg (2003)

20. Riesen, K., Bunke, H.: Approximate graph edit distance computation by means of bipartite graph matching. Image Vis. Comput. **27**(4), 950–959 (2009)

21. Burkard, R., Dell'Amico, M., Martello, S.: Assignment Problems. Society for Industrial and Applied Mathematics, Philadelphia (2009)

22. Caetano, T.S., McAuley, J.J., Cheng, L., Le, Q.V., Smola, A.J.: Learning graph matching. IEEE Trans. Pattern Anal. Mach. Intell. **31**(6), 1048–1058 (2009)

23. Hart, P., Nilsson, N., Raphael, B.: A formal basis for the heuristic determination of minimum cost paths. IEEE Trans. Syst. Sci. Cybern. **4**(2), 100–107 (1968)

24. Riesen, K., Fischer, A., Bunke, H.: Computing upper and lower bounds of graph edit distance in cubic time. Accepted for publication in Proceedings of the IAPR TC3 International Workshop on Artificial Neural Networks in Pattern Recognition

25. Munkres, J.: Algorithms for the assignment and transportation problems. J. Soc. Ind. Appl. Math. **5**(1), 32–38 (1957)

26. Kuhn, H.: The hungarian method for the assignment problem. Naval Res. Logistic Q. **2**, 83–97 (1955)

27. Jonker, R., Volgenant, A.: A shortest augmenting path algorithm for dense and sparse linear assignment problems. Computing **38**, 325–340 (1987)

28. Jonker, R., Volgenant, A.: Improving the hungarian assignment algorithm. Oper. Res. Lett. **5**, 171–175 (1986)

29. Bertsekas, D.: The auction algorithm: a distributed relaxation method for the assignment problem. Ann. Oper. Res. **14**, 105–123 (1988)

30. Hung, M.: A polynomial simplex method for the assignment problem. Oper. Res. **28**, 969–982 (1983)

31. Orlin, J.: On the simplex algorithm for networks and generalized networks. Math. Program. Stud. **24**, 166–178 (1985)

32. Ahuja, R., Orlin, J.: The scaling network simplex algorithm. Oper. Res. **40**(1), 5–13 (1992)

33. Akgül, M.: A sequential dual simplex algorithm for the linear assignment problem. Oper. Res. Lett. **7**, 155–518 (1988)

34. Srinivasan, V., Thompson, G.: Cost operator algorithms for the transportation problem. Math. Program. **12**, 372–391 (1977)

35. Achatz, H., Kleinschmidt, P., Paparrizos, K.: A dual forest algorithm for the assignment problem. Appl. Geom. Discret. Math. AMS **4**, 1–11 (1991)
36. Burkard, R., Ceia, E.: Linear assignment problems and extensions. Technical report 127, Karl-Franzens-Universität Graz und Technische Universität Graz (1998)
37. Avis, D.: A survey of heuristics for the weighted matching problem. Networks **13**, 475–493 (1983)
38. Kurtzberg, J.: On approximation methods for the assignment problem. J. ACM **9**(4), 419–439 (1962)
39. Riesen, K., Bunke, H.: Graph classification based on vector space embedding. Int. J. Pattern Recogn. Artif. Intell. **23**(6), 1053–1081 (2008)
40. Neuhaus, M., Bunke, H.: Bridging the Gap Between Graph Edit Distance and Kernel Machines. World Scientific Publishing, Singapore (2007)
41. Riesen, K., Bunke, H.: IAM graph database repository for graph based pattern recognition and machine learning. In: da Vitoria Lobo, N., et al. (eds.) Structural, Syntactic, and Statistical Pattern Recognition. LNCS, vol. 5342, pp. 287–297. Springer, Heidelberg (2008)
42. Dosch, P., Valveny, E.: Report on the second symbol recognition contest. In: Wenyin, L., Lladós, J. (eds.) GREC 2005. LNCS, vol. 3926, pp. 381–397. Springer, Heidelberg (2005)

Learning Heuristics to Reduce the Overestimation of Bipartite Graph Edit Distance Approximation

Miquel Ferrer[1]([⊠]), Francesc Serratosa[2], and Kaspar Riesen[1]

[1] Institute for Information Systems, University of Applied Sciences and Arts Northwestern Switzerland, Riggenbachstrasse 16, 4600 Olten, Switzerland
{miquel.ferrer,kaspar.riesen}@fhnw.ch
[2] Departament D'Enginyeria Informàtica I Matemàtiques, Universitat Rovira I Virgili, Avda. Països Catalans 26, 43007 Tarragoa, Spain
francesc.serratosa@urv.cat

Abstract. In data mining systems, which operate on complex data with structural relationships, graphs are often used to represent the basic objects under study. Yet, the high representational power of graphs is also accompanied by an increased complexity of the associated algorithms. Exact graph similarity or distance, for instance, can be computed in exponential time only. Recently, an algorithmic framework that allows graph dissimilarity computation in cubic time with respect to the number of nodes has been presented. This fast computation is at the expense, however, of generally overestimating the true distance. The present paper introduces six different post-processing algorithms that can be integrated in this suboptimal graph distance framework. These novel extensions aim at improving the overall distance quality while keeping the low computation time of the approximation. An experimental evaluation clearly shows that the proposed heuristics substantially reduce the overestimation in the existing approximation framework while the computation time remains remarkably low.

1 Introduction

One of the basic objectives in pattern recognition, data mining, and related fields is the development of systems for the analysis or classification of objects [1,2]. These objects (or patterns) can be of any kind [3,4]. Feature vectors are one of the most common and widely used data structure for object representation. That is, for each object a set of relevant properties, or features, is extracted and arranged in a vector. One of the main advantages of this representation is that a large number of algorithms for pattern analysis and classification is available for feature vectors [2]. However, some disadvantages arise from the rather simple structure of feature vectors. First, vectors have to preserve the same length regardless of the size or complexity of the object. Second, vectors are not able to represent binary relations among different parts of the object. As a consequence, for the representation of complex objects where relations

© Springer International Publishing Switzerland 2015
P. Perner (Ed.): MLDM 2015, LNAI 9166, pp. 17–31, 2015.
DOI: 10.1007/978-3-319-21024-7_2

between different subparts play an important role, graphs appear as an appealing alternative to vectorial descriptions.

In particular, graphs can explicitly model the relations between different parts of an object, whereas feature vectors are able to describe the object's properties only. Furthermore, the dimensionality of graphs, i.e., the number of nodes and edges, can be different for every object. Due to these substantial advantages a growing interest in graph-based object representation in machine learning and data mining can be observed in recent years [5,6].

Evaluating the dissimilarity between graphs, commonly referred to as graph matching, is a crucial task in many graph based classification frameworks. Extensive surveys about graph matching in pattern recognition, data mining, and related fields can be found in [7,8]. Graph edit distance [9,10] is one of the most flexible and versatile approaches to error-tolerant graph matching. In particular, graph edit distance is able to cope with directed and undirected, as well as with labeled and unlabeled graphs. In addition, no constraints have to be considered on the alphabets for node and/or edge labels. Moreover, through the concept of cost functions, graph edit distance can be adapted and tailored to diverse applications [11,12].

The major drawback of graph edit distance is, however, its high computational complexity that restricts its applicability to graphs of rather small size. In fact, graph edit distance belongs to the family of *quadratic assignment problems* (QAPs), which in turn belong to the class of $\mathcal{NP}$-*complete* problems. Therefore, exact computation of graph edit distance can be solved in exponential time complexity only. In recent years, a number of methods addressing the high computational complexity of graph edit distance have been proposed (e.g. [13,14]). Beyond these works, an algorithmic framework based on bipartite graph matching has been introduced recently [15]. The main idea behind this approach is to convert the difficult problem of graph edit distance to a *linear sum assignment problem* (LSAP). LSAPs basically constitute the problem of finding an optimal assignment between two independent sets of entities, for which a collection of polynomial algorithms exists [16]. In [15] the LSAP is formulated on the sets of nodes including local edge information. The main advantage of this approach is that it allows the approximate computation of graph edit distance in a substantially faster way than traditional methods. However, it generally overestimates the true edit distance due to some incorrectly assigned nodes. These incorrect assignments are mainly because the framework is able to consider local rather than global edge information only.

In order to overcome this problem and reduce the overestimation of the true graph edit distance, a variation of the original framework [15] has been proposed in [17]. Given the initial assignment found by the bipartite framework, the main idea is to introduce a post-processing step such that the number of incorrect assignments is decreased (which in turn reduces the overestimation). The proposed post-processing varies the original overall node assignment by systematically swapping the target nodes of two individual node assignments. In order to search the space of assignment variations a *beam search* (i.e. a tree search with pruning) is used. One of the most important observations derived from [17] is that given an initial node assignment, one can substantially reduce the overestimation using this local search method. Yet, the post-processing beam search still produces sub-optimal distances. The reason for this is that beam search possibly prunes the optimal solution in an early stage of the search process.

Now the crucial question arises, how the space of assignment variations could be explored such that promising parts of the search tree are not (or at least not too early) pruned. In [17] the initial assignment is systematically varied without using any kind of heuristic or additional information. In particular, it is not taken into account that certain nodes and/or local assignments have greater impact or are easier to be conducted than others, and should thus be considered first in the beam search process. Considering more important or more evident node assignments in an early stage of the beam search process might reduce the risk of pruning the optimal assignment.

In this paper we propose six different heuristics that modify the mapping order given by the original framework [15]. These heuristics can be used to influence the order in which the assignments are eventually varied during the beam search. With other words, prior to run the beam search strategy proposed in [17], the order of the assignments is varied according to these heuristics.

The remainder of this paper is organized as follows. Next, in Sect. 2, the original bipartite framework for graph edit distance approximation [15] as well as its recent extension [17], named *BP-Beam*, are summarized. In Sect. 3, our novel version of *BP-Beam* is described. An experimental evaluation on diverse data sets is carried out in Sect. 4. Finally, in Sect. 5 we draw conclusions and outline some possible tasks and extensions for future work.

2   Graph Edit Distance Computation

In this Section we start with our basic notation of graphs and then review the concept of graph edit distance. Eventually, the approximate graph edit distance algorithm (which builds the basis of the present work) is described.

2.1   Graph Edit Distance

A graph g is a four-tuple $g = (V, E, \mu, \nu)$, where V is the finite set of nodes, $E \subseteq V \times V$ is the set of edges, $\mu : V \longrightarrow L_V$ is the node labeling function, and $\nu : E \longrightarrow L_E$ is the edge labeling function. The labels for both nodes and edges can be given by the set of integers $L = \{1, 2, 3, \ldots\}$, the vector space $L = \mathbb{R}^n$, a set of symbolic labels $L = \{\alpha, \beta, \gamma, \ldots\}$, or a combination of various label alphabets from different domains. Unlabeled graphs are a special case by assigning the same (empty) label $\varnothing$ to all nodes and edges, i.e. $L_V = L_E = \{\varnothing\}$.

Given two graphs, $g_1 = (V_1, E_1, \mu_1, \nu_1)$ and $g_2 = (V_2, E_2, \mu_2, \nu_2)$, the basic idea of graph edit distance is to transform g_1 into g_2 using edit operations, namely, *insertions*, *deletions*, and *substitutions* of both nodes and edges. The substitution of two nodes u and v is denoted by $(u \rightarrow v)$, the deletion of node

u by $(u \rightarrow \epsilon)$, and the insertion of node v by $(\epsilon \rightarrow v)$[1]. A sequence of edit operations $e_1, \ldots, e_k$ that transform g_1 completely into g_2 is called an edit path between g_1 and g_2.

To find the most suitable edit path out of all possible edit paths between two graphs, a cost measuring the strength of the corresponding operation is commonly introduced (if applicable, one can also merely count the number of edit operations, i.e., the cost for every edit operation amounts to 1). The edit distance between two graphs g_1 and g_2 is then defined by the minimum cost edit path between them. Exact computation of graph edit distance is usually carried out by means of a tree search algorithm (e.g. A* [18]) which explores the space of all possible mappings of the nodes and edges of the first graph to the nodes and edges of the second graph.

2.2   Bipartite Graph Edit Distance Approximation

The computational complexity of exact graph edit distance is exponential in the number of nodes of the involved graphs. That is considering n nodes in g_1 and m nodes in g_2, the set of all possible edit paths contains $O(n^m)$ solutions to be explored. This means that for large graphs the computation of edit distance is intractable. In order to reduce its computational complexity, in [15], the graph edit distance problem is transformed into a linear sum assignment problem (LSAP).

To this end, based on the node sets $V_1 = \{u_1, \ldots, u_n\}$ and $V_2 = \{v_1, \ldots, v_m\}$ of g_1 and g_2, respectively, a cost matrix C is first established as follows:

$$
C = \begin{bmatrix}
c_{11} & c_{12} & \cdots & c_{1m} & c_{1\epsilon} & \infty & \cdots & \infty \\
c_{21} & c_{22} & \cdots & c_{2m} & \infty & c_{2\epsilon} & \cdots & \infty \\
\vdots & \vdots & \ddots & \vdots & \vdots & \vdots & \ddots & \vdots \\
c_{n1} & c_{n2} & \cdots & c_{nm} & \infty & \infty & \cdots & c_{n\epsilon} \\
c_{\epsilon1} & \infty & \cdots & \infty & 0 & 0 & \cdots & 0 \\
\infty & c_{\epsilon2} & \cdots & \infty & 0 & 0 & \cdots & 0 \\
\vdots & \vdots & \ddots & \infty & \vdots & \vdots & \ddots & \vdots \\
\infty & \infty & \cdots & c_{\epsilon m} & 0 & 0 & \cdots & 0
\end{bmatrix}
$$

Entry c_{ij} denotes the cost of a node substitution $(u_i \rightarrow v_j)$, $c_{i\epsilon}$ denotes the cost of a node deletion $(u_i \rightarrow \epsilon)$, and $c_{\epsilon j}$ denotes the cost of a node insertion $(\epsilon \rightarrow v_j)$. The left upper corner of the cost matrix represents the costs of all possible node substitutions, the diagonal of the right upper corner the costs of all possible node deletions, and the diagonal of the bottom left corner the costs of all possible node insertions. In every entry $c_{ij} \in C$, not only the cost of the node operation, but also the minimum sum of edge edit operation costs implied by the corresponding node operation is taken into account. That is, the matching cost of the local edge structure is encoded in the individual entries $c_{ij} \in C$.

---

[1] Similar notation is used for edges.

Note that in [19] another definition of matrix $C = (c_{ij})$ has been proposed. The major idea of [19] is to define a smaller square cost matrix in combination with some (weak) conditions on the cost function. This particular redefinition of C is able to further speed up the assignment process while not affecting the distance accuracy. In the present paper we make use of the original cost matrix definition without any constraints on the cost function.

In the second step of [15], an assignment algorithm is applied to the square cost matrix $C = (c_{ij})$ in order to find the minimum cost assignment of the nodes (and their local edge structure) of g_1 to the nodes (and their local edge structure) of g_2. Note that this task exactly corresponds to an instance of an LSAP and can thus be optimally solved in polynomial time by several algorithms [16].

Any of the LSAP algorithms will return a permutation $(\varphi_1, \ldots, \varphi_{n+m})$ of the integers $(1, 2, \ldots, (n+m))$, which minimizes the overall mapping cost $\sum_{i=1}^{(n+m)} c_{i\varphi_i}$. This permutation corresponds to the mapping

$$\psi = \{u_1 \rightarrow v_{\varphi 1}, u_2 \rightarrow v_{\varphi 2}, \ldots, u_{m+n} \rightarrow v_{\varphi_{m+n}}\}$$

of the nodes of g_1 to the nodes of g_2. Note that ψ does not only include node substitutions ($u_i \rightarrow v_j$), but also deletions and insertions ($u_i \rightarrow \epsilon$), ($\epsilon \rightarrow v_j$) and thus perfectly reflects the definition of graph edit distance (substitutions of the form ($\epsilon \rightarrow \epsilon$) can be dismissed, of course). Hence, mapping ψ can be interpreted as partial edit path between g_1 and g_2, which considers operations on nodes only.

In the third step of [15], the partial edit path ψ between g_1 and g_2 is completed with respect to the edges. This can be accomplished since edge edit operations are uniquely implied by the adjacent node operations. That is, whether an edge is substituted, deleted, or inserted, depends on the edit operations performed on its adjacent nodes. The total cost of the completed edit path between graphs g_1 and g_2 is finally returned as approximate graph edit distance $d_{\langle\psi\rangle}(g_1, g_2)$. We refer to this graph edit distance approximation algorithm as $BP(g_1, g_2)$ from now on[2]. The three major steps of BP are summarized in Algorithm 1.

Algorithm 1. $BP(g_1, g_2)$

1: Build cost matrix $C = (c_{ij})$ according to the input graphs g_1 and g_2
2: Compute optimal node assignment $\psi = \{u_1 \rightarrow v_{\varphi 1}, \ldots, u_{m+n} \rightarrow v_{\varphi_{m+n}}\}$ on C using any LSAP solver algorithm
3: **return** Complete edit path according to ψ and $d_{\langle\psi\rangle}(g_1, g_2)$

Note that the edit path corresponding to $d_{\langle\psi\rangle}(g_1, g_2)$ considers the edge structure of g_1 and g_2 in a global and consistent way while the optimal node mapping ψ is able to consider the structural information in an isolated way only (single nodes and their adjacent edges). This is due to the fact that during the optimization process of the specific LSAP no information about neighboring node

---

[2] BP stands for *Bipartite*. The assignment problem can also be formulated as finding a matching in a *complete bipartite graph* and is therefore also referred to as *bipartite graph matching problem*.

assignments is available. Hence, in comparison with optimal search methods for graph edit distance, this algorithmic framework might cause additional edge operations in the third step, which would not be necessary in a globally optimal graph matching. Hence, the distances found by this specific framework are – in the best case – equal to, or – in general – larger than the exact graph edit distance. Yet, the proposed reduction of graph edit distance to an LSAP allows the approximate graph edit distance computation in polynomial time.

2.3 Beam Search Graph Edit Distance Approximation

Several experimental evaluations indicate that the suboptimality of BP, i.e. the overestimation of the true edit distance, is very often due to a few incorrectly assigned nodes in ψ with respect to the optimal edit path [15]. An extension of BP presented in [17] ties in at this observation. In particular, the node mapping ψ is used as a starting point for a subsequent search in order to improve the quality of the distance approximation.

Algorithm 2 gives an overview of this process (named BP-$Beam$ from now on). First, BP is executed using graphs g_1 and g_2 as input. As a result, both the approximate distance $d_{\langle \psi \rangle}(g_1, g_2)$ and the node mapping ψ are available. In a second step, the swapping procedure $BeamSwap$ (Algorithm 3) is executed. $Beam$-$Swap$ takes the input graphs g_1 and g_2, distance $d_{\langle \psi \rangle}$, mapping ψ, and a meta-parameter b as parameters. The swapping procedure of Algorithm 3 basically varies mapping ψ by swapping the target nodes v_{φ_i} and v_{φ_j} of two node assignments $(u_i \rightarrow v_{\varphi_i}) \in \psi$ and $(u_j \rightarrow v_{\varphi_j}) \in \psi$, resulting in two new assignments $(u_i \rightarrow v_{\varphi_j})$ and $(u_j \rightarrow v_{\varphi_i})$. For each swap it is verified whether (and to what extent) the derived distance approximation stagnates, increases or decreases.

Algorithm 3 ($BeamSwap$) systematically processes the space of possible swappings by means of a tree search. The tree nodes in the search procedure correspond to triples $(\psi, q, d_{\langle \psi \rangle})$, where ψ is a certain node mapping, q denotes the depth of the tree node in the search tree and $d_{\langle \psi \rangle}$ is the approximate distance value corresponding to ψ. The root node of the search tree refers to the optimal node mapping ψ found by BP. Hence, the root node (with depth $= 0$) is given by the triple $(\psi, 0, d_{\langle \psi \rangle})$. Subsequent tree nodes $(\psi', q, d_{\langle \psi' \rangle})$ with depth $q = 1, \ldots, (m+n)$ contain node mappings ψ' where the individual node assignment $(u_q \rightarrow v_{\varphi_q})$ is swapped with any other node assignment of ψ.

Algorithm 2. BP-$Beam(g_1, g_2, b)$

1: $d_{\langle \psi \rangle}(g_1, g_2) = BP(g_1, g_2)$
2: $d_{Beam}(g_1, g_2) = BeamSwap(g_1, g_2, d_{\langle \psi \rangle}, \psi, b)$
3: **return** $d_{Beam}(g_1, g_2)$

As usual in tree search based methods, a set *open* is employed that holds all of the unprocessed tree nodes. Initially, *open* holds the root node $(\psi, 0, d_{\langle \psi \rangle})$ only. The tree nodes in *open* are kept sorted in ascending order according to their depth in the search tree (known as *breadth-first search*). As a second order criterion the approximate edit distance $d_{\langle \psi \rangle}$ is used. As long as *open* is not empty, we retrieve (and remove) the triple $(\psi, q, d_{\langle \psi \rangle})$ at the first position in

Algorithm 3. $BeamSwap(g_1, g_2, d_{\langle\psi\rangle}, \psi, b)$

1: $d_{best} = d_{\langle\psi\rangle}$
2: Initialize $open = \{(\psi, 0, d_{\langle\psi\rangle})\}$
3: **while** $open$ is not empty **do**
4:     Remove first tree node in $open$: $(\psi, q, d_{\langle\psi\rangle})$
5:     **for** $j = (q+1), \ldots, (m+n)$ **do**
6:         $\psi' = \psi \setminus \{u_{q+1} \to v_{\varphi_{q+1}}, u_j \to v_{\varphi_j}\} \cup \{u_{q+1} \to v_{\varphi_j}, u_j \to v_{\varphi_{q+1}}\}$
7:         Derive approximate edit distance $d_{\langle\psi'\rangle}(g_1, g_2)$
8:         $open = open \cup \{(\psi', q+1, d_{\langle\psi'\rangle})\}$
9:         **if** $d_{\langle\psi'\rangle}(g_1, g_2) < d_{best}$ **then**
10:           $d_{best} = d_{\langle\psi'\rangle}(g_1, g_2)$
11:         **end if**
12:     **end for**
13:     **while** size of $open > b$ **do**
14:         Remove tree node with highest approximation value $d_{\langle\psi\rangle}$ from $open$
15:     **end while**
16: **end while**
17: **return** d_{best}

$open$, generate the successors of this specific tree node and add them to $open$. To this end all pairs of node assignments $(u_{q+1} \to v_{\varphi_{q+1}})$ and $(u_j \to v_{\varphi_j})$ with $j = (q+1), \ldots, (n+m)$ are individually swapped resulting in two new assignments $(u_{q+1} \to v_{\varphi_j})$ and $(u_j \to v_{\varphi_{q+1}})$. In order to derive node mapping ψ' from ψ, the original node assignment pair is removed from ψ and the swapped node assignment is added to ψ'. Since index j starts at $(q+1)$ we also allow that a certain assignment $(u_{q+1} \to v_{\varphi_{q+1}})$ remains unaltered at depth $(q+1)$ in the search tree.

Note that the search space of all possible permutations of ψ contains $(n+m)!$ possibilities, making an exhaustive search (starting with ψ) both unreasonable and intractable. Therefore, only the b assignments with the lowest approximate distance values are kept in $open$ at all time (known as *beam search*). Note that parameter b can be used as trade-off parameter between run time and approximation quality. That is, it can be expected that larger values of b lead to both better approximations and increased run time (and vice versa).

Since every tree node in our search procedure corresponds to a complete solution and the cost of these solutions neither monotonically decrease nor increase with growing depth in the search tree, we need to buffer the best possible distance approximation found during the tree search in d_{best} (which is finally returned by *BeamSwap* to the main procedure *BP-Beam*).

As stated before, given a mapping ψ from *BP*, the derived edit distance $d_{\langle\psi\rangle}$ overestimates the true edit distance in general. Hence, the objective of any post-processing should be to find a variation ψ' of the original mapping ψ such that $d_{\langle\psi'\rangle} < d_{\langle\psi\rangle}$. Note that the distance d_{best} returned by *BeamSwap* (Algorithm 3) is smaller than, or equal to, the original approximation $d_{\langle\psi\rangle}$ (since $d_{\langle\psi\rangle}$ is initially taken as best distance approximation). Hence, the distance d_{Beam} (finally returned by *BP-Beam* (Algorithm 2))is in any case smaller than, or equal to $d_{\langle\psi\rangle}$.

3 Sorted BP-Beam

Note that the successors of tree node $(\psi, q, d_{\langle\psi\rangle})$ are generated in an arbitrary yet fixed order in *BP-Beam* (or rather in its subprocess *BeamSwap*). In particular, the assignments of the original node mapping ψ are processed according to the depth q of the current search tree node. That is, at depth q the assignment $(u_q \rightarrow v_{\varphi_q})$ is processed and swapped with other assignments. Note that beam search prunes quite large parts of the tree during the search process. Hence, the fixed order processing, which does not take any information about the individual node assignments into account, is a clear drawback of the procedure described in [17].

Clearly, it would be highly favorable to process important or evident node assignments as early as possible in the tree search. To this end, we propose six different sorting strategies that modify the order of mapping ψ obtained from BP and feed this reordered mapping into *BeamSwap*. Using these sorting strategies we aim at verifying whether we can learn about the strengths and weaknesses of a given assignment returned by BP before the post processing is carried out. In particular, we want to distinguish assignments from ψ that are most probably incorrect from assignments from ψ that are correct with a high probability (and should thus be considered early in the post processing step).

Algorithm 4. SBP-Beam(g_1,g_2,b)

1: $d_{\langle\psi\rangle}(g_1, g_2) = BP(g_1, g_2)$
2: $\psi' = SortMatching(\psi)$
3: $d_{SortedBeam}(g_1, g_2) = BeamSwap(g_1, g_2, d_{\langle\psi\rangle}, \psi', b)$
4: **return** $d_{SortedBeam}(g_1, g_2)$

The proposed algorithm, referred to as *SBP-Beam* (the initial S stands for *Sorted*), is given in Algorithm 4. Note that *SBP-Beam* is a slightly modified version of *BP-Beam*. That is, the sole difference to *BP-Beam* is that the original mapping ψ (returned by BP) is reordered according to a specific sorting strategy (line 2). Eventually, *BeamSwap* is called using mapping ψ' (rather than ψ) as parameter. Similar to d_{Beam}, $d_{SortedBeam}$ is always lower than, or equal to, $d_{\langle\psi\rangle}(g_1, g_2)$, and can thus be securely returned. Note that the reordering of nodes does not influence the corresponding edit distances and thus $d_{\langle\psi\rangle} = d_{\langle\psi'\rangle}$.

3.1 Sorting Strategies

On line 2 of Algorithm 4 (*SBP-Beam*), the original mapping ψ returned by BP is altered by reordering the individual node assignments according to a certain criterion. Note that the resulting mapping ψ' contains the same node assignments as ψ but in a different order. That is, the order of the assignments is varied but the individual assignments are not modified. Note that this differs from the swapping procedure given in Algorithm 3, where the original node assignments in ψ are modified, changing the target nodes of two individual assignments.

In the following, we propose six different strategies to reorder the original mapping ψ. The overall aim of these sorting strategies is to put more evident assignments (i.e. those to be supposed that are correct) at the beginning of

ψ' such that they are processed first in the tree search. More formally, for each strategy we first give a weight (or rank) to each source node u_i of the assignments in mapping ψ. Then, the order of the assignments $(u_i \rightarrow v_{\varphi i}) \in \psi'$ is set either in ascending or descending order according to the corresponding weight of u_i.

Confident: The source nodes u_i of the assignments $(u_i \rightarrow v_{\varphi i}) \in \psi$ are weighted according to $c_{i\varphi_i}$. That is, for a given assignment $(u_i \rightarrow v_{\varphi i})$, the corresponding value $c_{i\varphi_i}$ in the cost matrix C is assigned to u_i as a weight. The assignments of the new mapping ψ' are then sorted in ascending order according to the weights of u_i. Thus, with this sorting strategy assignments with low costs, i.e. assignments which are somehow evident, appear first in ψ'.

Unique: The source nodes u_i of the assignments $(u_i \rightarrow v_{\varphi i}) \in \psi$ are given a weight according to the following function

$$\max_{\forall j=1,\dots,m} c_{ij} - c_{i\varphi_i}$$

That is, the weight given to a certain source node u_i corresponds to the maximum difference between the cost $c_{i\varphi_i}$ of the actual assignment $(u_i \rightarrow v_{\varphi i})$ and the cost of a possible alternative assignment for u_i. Note that this difference can be negative, which means that the current assignment $(u_i \rightarrow v_{\varphi i}) \in \psi$ is rather suboptimal (since there is at least one other assignment for u_i with lower cost than $c_{i\varphi_i}$). Assignments in ψ' are sorted in descending order with respect to this weighting criteria. That is, assignments with a higher degree of confidence are processed first.

Divergent: The aim of this sorting strategy is to prioritize nodes u_i that have a high divergence among all possible node assignment costs. That is, for each row i we sum up the absolute values of cost differences between all pairs of assignments

$$\sum_{j=1}^{m-1} \sum_{k=j+1}^{m} |c_{ij} - c_{ik}|$$

Rows with a high divergence correspond to local assignments that are somehow easier to be conducted than rows with low sums. Hence we sort assignments $(u_i \rightarrow v_{\varphi i})$ in descending order with respect to the corresponding divergence in row i.

Leader: With this strategy, nodes u_i are weighted according to the maximum difference between the minimum cost assignment of node u_i and the second minimum cost assignment of u_i. Assume we have $\min 1_i = \min_{j=1,\dots,m} c_{ij}$ and $\min 2_i = \min_{j=1,\dots,m, j \neq k} c_{ij}$ (k refers to the column index of the minimum cost entry $\min 1_i$). The weight for node u_i amounts to (the denominator normalizes the weight)

$$\frac{\min 1_i - \min 2_i}{\min 2_i}$$

The higher this difference, the easier is the local assignment to be conducted. Hence, assignments $(u_i \rightarrow v_{\varphi i})$ are sorted in descending order with respect to the weight of u_i.

Interval: First we compute the interval for each row i and each column j of the upper left corner of C, denoted as δ_{r_i} and δ_{c_j} respectively. Given a row i (or column j), the interval is defined as the absolute difference between the maximum and the minimum entry in row i (or column j). We also compute the mean of all row and column intervals, denoted by $\overline{\delta}_r$ and $\overline{\delta}_c$ respectively. The weight assigned to a given assignment $(u_i \rightarrow v_{\varphi i})$ is then

- 1, if $\delta_{r_i} > \overline{\delta}_r$ and $\delta_{c_{\varphi_i}} > \overline{\delta}_c$
- 0, if $\delta_{r_i} < \overline{\delta}_r$ and $\delta_{c_{\varphi_i}} < \overline{\delta}_c$
- 0.5, otherwise

That is, if the intervals of both row and column are greater than the corresponding means, the weight is 1. Likewise, if both intervals are lower than the mean intervals, the weight is 0. For any other case the weight is 0.5.

  If the intervals of row i and column φi are larger than the mean intervals, the row and column of the assignment $(u_i \rightarrow v_{\varphi i})$ are in general easier to handle than others. On the other hand, if the row and column intervals of a certain assignment are below the corresponding mean intervals, the individual values in the row and column are close to each other making an assignment rather difficult. Hence, we reorder the assignments $(u_i \rightarrow v_{\varphi i})$ of the original mapping ψ in decreasing order according to these weights.

Deviation: For each row i and each column j of the left upper corner of the cost matrix C we compute the mean $\overline{\theta}_{r_i}$ and $\overline{\theta}_{c_j}$ and the deviation $\overline{\sigma}_{r_i}$ and $\overline{\sigma}_{c_j}$. Then, for each assignment $(u_i \rightarrow v_{\varphi_i}) \in \psi$ we compute its corresponding weight according to the following rule:

- Initially, the weight is 0.
- If $c_{i\varphi_i} < \overline{\theta}_{r_i} - \overline{\sigma}_{r_i}$ we add 0.25 to the weight and compute the total number p of assignments in row i that also fulfill this condition. Note that p is always greater than, or equal to, 1 ($p = 1$ when $c_{i\varphi_i}$ is the sole cost that fulfills the condition in row i). We add $0.5/p$ to the weight.
- Repeat the previous step for column $j = \varphi_i$ using $\overline{\theta}_{c_{\varphi_i}}$ and $\overline{\sigma}_{c_{\varphi_i}}$.

Given an assignment $(u_i \rightarrow v_{\varphi i}) \in \psi$ with cost $c_{i\varphi_i}$ that is lower than the mean minus the deviation, we assume the assignment cost is low enough to be considered as evident (and thus we add 0.25 to the weight). The weight increases if in the corresponding row only few (or no) other evident assignments are available (this is why we add $0.5/p$ to the the weight, being p the number of evident assignments). The same applies for the columns. So at the end, assignments with small weights correspond to rather difficult assignments, while assignments with higher weights refer to assignments which are more evident than others. Hence, we reorder the original mapping ψ in decreasing order according to this weight.

4   Experimental Evaluation

The goal of the experimental evaluation is to verify whether the proposed extension *SBP-Beam* is able to reduce the overestimation of graph edit distance
approximation returned by *BP* and in particular *BP-Beam* and how the different
sorting strategies affect the computation time. Three data sets from the IAM
graph database repository involving molecular compounds (AIDS), fingerprint
images (Fingerprint), and symbols from architectural and electronic drawings
(GREC) are used to carry out this experimental evaluation. For details about
these data sets we refer to [20]. For all data sets, small subsets of 100 graphs are
randomly selected on which 10,000 pairwise graph edit distance computations
are performed. Three algorithms will be used as reference systems, namely A^*
which computes the true edit distance, *BP* as it is the starting point of our
extension, and *BP-Beam* (the system to be further improved).

In a first experiment we aim at researching whether there is a predominant
sorting strategy that generally leads to better approximations than the others. To
this end we run *SBP-Beam* six times, each using an individual sorting strategy.
Parameter b is set to 5 for these and the following experiments.

Table 1 shows the mean relative overestimation ϕo for each dataset and for
each sorting strategy. The mean relative overestimation of a certain approximation is computed as the relative difference to the sum of exact distances returned
by A^*. The relative overestimation of A^* is therefore zero and the value of ϕo
for *BP* is taken as reference value and corresponds to 100 % (not shown in the
table). In addition, given a dataset we rank each sorting strategy in ascending
order according to the amount of relative overestimation (in brackets after ϕo).
Thus, lower ranks mean lower overestimations and therefore better results. The
last row of Table 1 shows the aggregation of ranks for a given sorting strategy,
which is a measure of how a particular sorting strategy globally behaves. Table 1
also shows the mean computation time ϕt of every sorting strategy.

Major observations can be made in Table 1. First, there are substantial differences in the overestimation among all sorting strategies for a given dataset.

Table 1. The mean relative overestimation of the exact distance (ϕo) in % together
with rank (1-6) of the sorting strategies (in brackets), and the mean computation time
ϕt. Last row sums all the ranks for a given sorting strategy.

		Confident	Unique	Divergent	Leader	Interval	Deviation
AIDS	ϕo [%]	15.81 (3)	15.91 (4)	14.03 (1)	15.38 (2)	18.31 (5)	20.94 (6)
	ϕt [ms]	1.82	1.83	1.80	1.81	1.80	1.82
Fingerprint	ϕo [%]	11.99 (1)	18.06 (6)	16.29 (5)	14.44 (4)	13.64 (3)	13.23 (2)
	ϕt [ms]	1.48	1.50	1.49	1.49	1.50	1.50
GREC	ϕo [%]	20.57 (4)	15.74 (2)	16.67 (3)	21.41 (5)	13.73 (1)	21.65 (6)
	ϕt [ms]	2.63	2.63	2.63	2.61	2.65	2.58
Sum of ranks		8	12	9	11	9	14

For instance, on the AIDS dataset, 5 % of difference between the minimum and the maximum overestimation can be observed (on GREC and Fingerprint the differences amount to about 8 % and 6 %, respectively). This variability indicates that not all of the sorting strategies are able to reduce the overestimation with the same degree. Second, focusing on the sum of ranks we observe that there is not a clear winner among the six strategies. That is, we cannot say that there is a sorting strategy that clearly outperforms the others in general. However, the sum of ranks suggests there are two different clusters, one composed by *Confident*, *Divergent* and *Interval* with a sum of ranks between 8 and 9, and a second composed by the rest of the sorting strategies with (slightly) higher sums. Third, regarding the computation time of the various sorting strategies almost no differences can be observed. That is, all sorting strategies show approximately the same computation time.

In a second experiment, we use the ranks obtained in the previous experiment to measure the impact on the overestimation by running *SBP-Beam* with subsets of sorting strategies. To this end, the following experimental setup is defined. First, for each dataset we run *SBP-Beam* using the sorting strategy with rank 1 (referenced to as *SBP-Beam(1)*). That is, for AIDS *SBP-Beam(1)* employs the *Divergent* strategy, while *Confident* and *Interval* are used for Fingerprint and GREC, respectively. Then *SBP-Beam* is carried out on every dataset using the two best sorting strategies, and return the minimum distance obtained by both algorithms (referred to as *SBP-Beam(2)*). We continue adding sorting strategies in ascending order with respect to their rank until all six strategies are employed at once. Table 2 shows the mean relative overestimation ϕo and the mean computation time ϕt for this evaluation.

Regarding the overestimation ϕo we observe a substantial improvement of the distance quality using *BP-Beam* rather than *BP*. For instance, on the AIDS data the overestimation is reduced by 85 % (similar results are obtained on the other data sets). But more important is the comparison between *BP-Beam* and *SBP-Beam(1)*. Note that *SBP-Beam(1)* is identical to *BP-Beam* but uses reordered

Table 2. The mean relative overestimation of the exact distance (ϕo) in %, and the mean run time for one matcing (ϕt) in ms. The beam size b for *BP-Beam* and *SBP-Beam* is set to 5.

Algorithm	AIDS		Fingerprint		GREC	
	ϕo	ϕt	ϕo	ϕt	ϕo	ϕt
A^*	0.00	25750.22	0.00	2535.42	0.00	7770.81
BP	100.00	0.28	100.00	0.35	100.00	0.27
BP-Beam	15.09	1.82	19.64	1.45	16.98	2.65
SBP-Beam(1)	14.03	1.81	11.99	1.49	13.73	2.64
SBP-Beam(2)	9.83	3.34	8.81	2.61	9.14	5.01
SBP-Beam(3)	8.18	4.85	5.01	3.73	7.33	7.37
SBP-Beam(4)	7.28	6.38	4.42	4.84	6.89	9.74
SBP-Beam(5)	6.06	7.86	3.81	5.95	6.55	12.05
SBP-Beam(6)	5.75	9.41	3.57	7.09	6.25	14.36

mappings ψ' according to the best individual sorting strategy. Thus, by comparing these two algorithms we can assess the true impact of the reordering of mapping ψ. *SBP-Beam(1)* always obtains lower overestimations than *BP-Beam*. The improvement on the overestimation ranges from around 1 % in the case of AIDS dataset to near 8 % in the case of the Fingerprint dataset, which refers to a substantial improvement of the distance quality. This result confirms the hypothesis that the order of the mapping ψ being fed into *BeamSwap* is important and that further improvements of the distance quality can be achieved by means of intelligently ordered assignments in ψ.

Moreover, we observe further substantial reductions of the overestimation as we increase the number of sorting strategies. Note that the overestimation monotonically decreases as the number of sorting strategies is increased, reaching very low overestimations with *SBP-Beam(6)* (between 3.57 % and 6.25 % only). This means that we are able to obtain very accurate mappings (with very few incorrect node assignments) that lead to results very close to the exact distance with our novel procedure.

These substantial reductions of the overestimation from *BP* to *BP-Beam* and *SBP-Beam(6)* can also be seen in Fig. 1 where for each pair of graphs in the Fingerprint dataset, the exact distance (x-axis) is plotted against the distance obtained by an approximation algorithm (y-axis). In fact, for *SBP-Beam(6)* the line-like scatter plot along the diagonal suggests that the approximation is in almost every case equal to the to the optimal distance.

Regarding the computation time ϕt we can report that *BP* provides the lowest computation time on all data sets (approximately 0.3ms per matching on all data sets). *BP-Beam* increases the computation time to approximately 2ms per matching. Note that there is no substantial difference in computation time between *BP-Beam* and *SBP-Beam(1)* (which differ only in the reordering of the individual node assignments). As expected, the run time of *SBP-Beam* linearly grows with the number of sorting strategies. For instance, on the AIDS data set every additional search strategy increases the run time by approximately 1.5ms (on the other data sets a similar run time increase can be observed). However, it

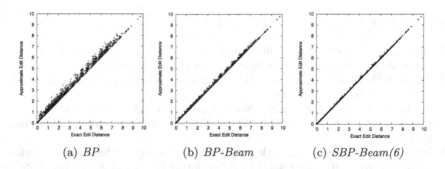

Fig. 1. Exact (x-axis) vs. approximate (y-axis) edit distance on the Fingerprint dataset computed with (a) *BP*, (b) *BP-Beam*, (c) *SBP-Beam(6)*.

is important to remark that in any case the computation time remains far below the one provided by A^*.

5 Conclusions and Future Work

In recent years a bipartite matching framework for approximating graph edit distance has been presented. In its original version this algorithm suffers from a high overestimation of the computed distance with respect to the true edit distance. Several post-processing approaches, based on node assignment variations, have been presented in order to reduce this overestimation. In this paper, we propose six different heuristics to sort the order of the individual node assignments before the mapping is post-processed. These novel strategies sort the individual node assignments with respect to a weight that measures the confidence of the assignments. The overall aim of the paper is to empirically demonstrate that the order in which the assignments are explored has a great impact on the resulting distance quality. The experimental evaluation on three different databases supports this hypothesis. Despite the results show that there is not a sorting strategy that clearly outperforms the others, we see that all of them are able to reduce the overestimation with respect to the unsorted version. Though the run times are increased when compared to our former framework (as expected), they are still far below the run times of the exact algorithm.

These results encourage pursuing several research lines in future work. First, in order to verify whether processing non-evident assignments first could also be beneficial, the proposed strategies can be used sorting in reverse order. Second, new and more elaborated sorting strategies based on complex learning algorithms can be developed. Finally, modified versions of *BeamSwap* using different criteria to sort and process the tree nodes could be tested.

Acknowledgements. This work has been supported by the Swiss National Science Foundation (SNSF) project Nr. 200021_153249, the Hasler Foundation Switzerland, and by the Spanish CICYT project DPI2013–42458–P and TIN2013–47245–C2-2–R.

References

1. Bishop, C.M.: Pattern Recognition and Machine Learning (Information Science and Statistics). Springer-Verlag New York Inc, Secaucus (2006)
2. Duda, R.O., Hart, P.E., Stork, D.G.: Pattern Classification, 2nd edn. Wiley-Interscience, New York (2000)
3. Silva, A., Antunes, C.: Finding multi-dimensional patterns in healthcare. In: Perner, P. (ed.) MLDM 2014. LNCS, vol. 8556, pp. 361–375. Springer, Heidelberg (2014)
4. Dittakan, K., Coenen, F., Christley, R.: Satellite image mining for census collection: a comparative study with respect to the ethiopian hinterland. In: Perner, P. (ed.) MLDM 2013. LNCS, vol. 7988, pp. 260–274. Springer, Heidelberg (2013)
5. Schenker, A., Bunke, H., Last, M., Kandel, A.: Graph-Theoretic Techniques for Web Content Mining. World Scientific, Singapore (2005)

6. Cook, D.J., Holder, L.B.: Mining Graph Data. John Wiley and Sons, New York (2006)
7. Foggia, P., Percannella, G., Vento, M.: Graph matching and learning in pattern recognition in the last 10 years. IJPRAI **28**(1), 1450001 (2014)
8. Conte, D., Foggia, P., Sansone, C., Vento, M.: Thirty years of graph matching in pattern recognition. IJPRAI **18**(3), 265–298 (2004)
9. Sanfeliu, A., Fu, K.-S.: A distance measure between attributed relational graphs for pattern recognition. IEEE Trans. Syst. Man Cybern. SMC-13 (3), 353–362 (1983)
10. Bunke, H., Allermann, G.: Inexact graph matching for structural pattern recognition. Pattern Recogn. Lett. **1**(4), 245–253 (1983)
11. Neuhaus, M., Bunke, H.: A graph matching based approach to fingerprint classification using directional variance. In: Kanade, T., Jain, A., Ratha, N.K. (eds.) AVBPA 2005. LNCS, vol. 3546, pp. 191–200. Springer, Heidelberg (2005)
12. Robles-Kelly, A., Hancock, E.R.: Graph edit distance from spectral seriation. IEEE Trans. Pattern Anal. Mach. Intell. **27**(3), 365–378 (2005)
13. Sorlin, S., Solnon, C.: Reactive tabu search for measuring graph similarity. In: Brun, L., Vento, M. (eds.) GbRPR 2005. LNCS, vol. 3434, pp. 172–182. Springer, Heidelberg (2005)
14. Justice, D., Hero, A.O.: A binary linear programming formulation of the graph edit distance. IEEE Trans. PAMI **28**(8), 1200–1214 (2006)
15. Riesen, K., Bunke, H.: Approximate graph edit distance computation by means of bipartite graph matching. Image Vis. Comput. **27**(7), 950–959 (2009)
16. Burkard, R.E., Dell'Amico, M., Martello, S.: Assignment problems. SIAM **157**(1), 183–190 (2009)
17. Riesen, K., Fischer, A., Bunke, H.: Combining bipartite graph matching and beam search for graph edit distance approximation. In: El Gayar, N., Schwenker, F., Suen, C. (eds.) ANNPR 2014. LNCS, vol. 8774, pp. 117–128. Springer, Heidelberg (2014)
18. Nilsson, N.J., Hart, P.E., Raphael, B.: A formal basis for the heuristic determination of minimum cost paths. IEEE Trans. Syst. Sci. Cybern. **SSC**–4(2), 100–107 (1968)
19. Serratosa, F.: Fast computation of bipartite graph matching. Pattern Recogn. Lett. **45**, 244–250 (2014)
20. Riesen, K., Bunke, H.: IAM graph database repository for graph based pattern recognition and machine learning. In: da Vitoria Lobo, N., Kasparis, T., Roli, F., Kwok, J.T.-Y., Georgiopoulos, M., Anagnostopoulos, G.C., Loog, M. (eds.) SSPR/SPR. LNCS, vol. 5342, pp. 287–297. Springer, Heidelberg (2008)

Seizure Prediction by Graph Mining, Transfer Learning, and Transformation Learning

Nimit Dhulekar[1], Srinivas Nambirajan[2], Basak Oztan[3], and Bülent Yener[1(✉)]

[1] Computer Science Department, Rensselaer Polytechnic Institute, Troy, NY, USA
yener@cs.rpi.edu
[2] Department of Mathematical Sciences, Rensselaer Polytechnic Institute,
Troy, NY, USA
[3] American Science and Engineering Inc., Billerica, MA, USA

Abstract. We present in this study a novel approach to predicting EEG epileptic seizures: we accurately model and predict non-ictal cortical activity and use prediction errors as parameters that significantly distinguish ictal from non-ictal activity. We suppress seizure-related activity by modeling EEG signal acquisition as a cocktail party problem and obtaining seizure-related activity using Independent Component Analysis. Following recent studies intricately linking seizure to increased, widespread synchrony, we construct dynamic EEG synchronization graphs in which the electrodes are represented as nodes and the pair-wise correspondences between them are represented by edges. We extract 38 intuitive features from the synchronization graph as well as the original signal. From this, we use a rigorous method of feature selection to determine minimally redundant features that can describe the non-ictal EEG signal maximally. We learn a one-step forecast operator restricted to just these features, using autoregression (AR(1)). We improve this in a novel way by cross-learning common knowledge across patients and recordings using *Transfer Learning*, and devise a novel transformation to increase the efficiency of transfer learning. We declare imminent seizure based on detecting outliers in our prediction errors using a simple and intuitive method. Our median seizure detection time is 11.04 min prior to the labeled start of the seizure compared to a benchmark of 1.25 min prior, based on previous work on the topic. To the authors' best knowledge this is the first attempt to model seizure prediction in this manner, employing efficient seizure suppression, the use of synchronization graphs and transfer learning, among other novel applications.

1 Introduction

Epilepsy is one of the most common disorders of the central nervous system characterized by recurring seizures. An epileptic seizure is described by abnormally excessive or synchronous neuronal activity in the brain [24]. Epilepsy patients show no pathological signs of the disease during inter-seizure periods, however, the uncertainty with regards to the onset of the next seizure deeply affects the lives of the patients.

© Springer International Publishing Switzerland 2015
P. Perner (Ed.): MLDM 2015, LNAI 9166, pp. 32–52, 2015.
DOI: 10.1007/978-3-319-21024-7_3

Seizure prediction refers to predicting the onset of epileptic seizures by analyzing electroencephalographic (EEG) recordings without any apriori knowledge of the exact temporal location of the seizure [22]. A method with the capacity to successfully predict the occurrence of an epileptic seizure would make it possible for the patient to be administered therapeutic treatments thereby alleviating the pain [21]. Many approaches have been suggested as possible seizure prediction algorithms, with modest levels of success. The first attempt at a seizure prediction algorithm was made by Viglione and Walsh in 1975 [71] and investigated spectral components and properties of EEG data. This was followed in the 80 s decade by different groups attempting to apply linear approaches, in particular autoregressive modeling, to seizure prediction [59, 60, 62]. Moving forward 20 years, Mormann et al. posed the question whether characteristic features can be extracted from the continuous EEG signal that are predictive of an impending seizure [48]. In 2002, the First International Workshop on Seizure Prediction [41] was conducted to bring together experts from a wide range of background with the common goal of improving the understanding of seizures, and thus advancing the current state of seizure prediction algorithms on a joint data set [41].

Most general approaches to the seizure prediction problem share several common steps including (i) processing of multichannel EEG signals, (ii) discretization of the time series into fixed-size overlapping windows called epochs, (iii) extraction of frequency bands to analyze the signal in frequency and/or time domains using techniques such as wavelength transformation [1], (iv) extraction of linear and non-linear features from the signal or its transformations; these features can be univariate, computed on each EEG channel separately, or multivariate, computed between two or more EEG channels, and (v) learning a model of the seizure statistics given the features by using supervised machine learning techniques such as Artificial Neural Networks [26, 34, 64, 69, 70], or Support Vector Machines [10, 11]. A thorough survey of the various linear and non-linear features can be found in [23, 25, 57]. These features are usually calculated over epochs of predetermined time duration (around 20 s) via a moving window analysis. It has been found that univariate features, such as Lyapunov exponents, correlation dimension, and Hjorth parameters, calculated from the EEG recordings performed poorly as compared to bivariate and multivariate features [16, 22, 27, 29, 31, 39, 50]. This is understandable given that the seizure spreads to all the electrodes, whereas not all electrical activity in the brain may result in the onset of a seizure - it might be a localized discharge at a certain electrode. Although it has been shown that univariate features are less significant for seizure prediction, the importance of non-linear features over linear features is not quite as straightforward. It has recently been observed that non-linear techniques might not enhance the performance of the seizure prediction algorithm considerably over linear techniques, and also have considerable limitations with respect to computational complexity and description of epileptic events [14, 15, 43, 49, 51].

A phase-locking bivariate measure, which captures brainwave synchronization patterns, has been shown to be important in differentiating interictal from pre-ictal states [37, 38]. In particular, it is suggested that the interictal period is characterized by moderate synchronization at large frequency bands while the pre-ictal period is marked by a decrease in the beta range synchronization between the epileptic focus and other brain areas, followed by a subsequent hypersynchronization at the time of the onset of the ictal period [47].

Many different approaches have been applied towards determining these features, such as frequency domain tools [8,53], wavelets [28,52], Markov processes [45], autoregressive models [3,11,68], and artificial neural networks [44]. If it were possible to reliably predict seizure occurrence then preventive clinical strategies would be replaced by patient specific proactive therapy such as resetting the brain by electrical or other methods of stimulation. While clinical studies show early indicators for a pre-seizure state including increased cerebral blood flow, heart rate change, the research in seizure prediction is still not reliable for clinical use.

Recently there has been an increased focus in analyzing multivariate complex systems such as EEG recordings using concepts from *network theory* [4,7,18,42,54,67], describing the topology of the multivariate time-series through *interaction networks*. The interaction networks enable characterization of the pair-wise correlations between electrodes using graph theoretical features over time [9,66]. In the spatio-temporal interaction networks, *nodes* (vertices) represent the EEG channels and the *edges* (links) represent the level of neuronal synchronization between the different regions of the brain. This approach has been exploited in the analysis of various neuropsychiatric diseases including schizophrenia, autism, dementia, and epilepsy [36,66,73]. Within epilepsy research, evolution of certain graph features over time revealed better understanding of the interactions of the brain regions and the seizures. For instance, Schindler et al. analyzed the change in path lengths and clustering coefficients to highlight the evolution of seizures on epileptic patients [61], Kramer et al. considered the evolution of local graph features including betweenness centrality to explain the coupling of brain signals at seizure onset [35], and Douw et al. recently showed epilepsy in glioma patients was attributed to the theta band activity in the brain [20]. In [46] authors independently suggest a similar approach that combines tensor decompositions with graph theory. Even with this significant body of research what remained unclear was whether the network-related-approach can adequately identify the inter-ictal to pre-ictal transition [36]. In this paper, we continue studying a form of interaction networks dubbed *synchronization graphs* [19] and introduce new features as the early indicators of a seizure onset, thereby identifying the inter-ictal to pre-ictal transition.

Summarily, the current approaches aim to develop features that are naturally characteristic of seizure activity. While these approaches are both intuitive and instructive, ictal activity is often a small portion of the available data, and statistical learning techniques, which require a large corpus of data for reliable prediction, can be expected to perform poorly as seizure-predictors. However, these techniques seem promising for accurate prediction of non-ictal activity with respect to which ictal activity may be identified as an anomaly. Provided that the only anomalous activity in the data is the seizure, or that other anomalies present with discernible signatures, this provides an equivalent method of predicting seizure. In general we operate under the paradigm that any feature or parameter that distinguishes between ictal and non-ictal activity is a mathematical characteristic of seizure, although it may not be a natural physiological indicator. We rigorously define the notion of what it means to be a good mathematical characteristic of a seizure, rate our seizure-discriminating parameter accordingly, systematically increase how well it discriminates between ictal and non-ictal activity, and qualify our predictions using such a discriminating parameter.

Furthermore, since cortical activity is continuously recorded as EEG signals, it can be represented as a time-series, and analyzed using time-series forecasting methods. The objective of time-series forecasting is to use equally spaced past and current observations to accurately predict future values. *Autoregressive* models (AR) are commonly used tools for time-series prediction, and have been used to capture the spatio-temporal properties of EEG signals [3,68]. We further improve on the AR model by using *Transfer Learning* [56] to learn the best forecast operator for a particular EEG recording from other EEG recordings. Transfer learning is a general form of learning such that there need not be any similarity in the distributions of the training and testing data. In our context, transfer learning does not require the past values and future values of the output variable to be correlated. In addition, transfer learning is particularly useful when data is only partially available or corrupted by noise, where such data can be effectively supplemented by clean data from a different experiment. We further improve on the transfer learning by modifying the transfer set into the most similar form to the dataset being investigated by means of a simple transformation (based on the Procrustes problem [72]).

The three main contributions of this work are as follows: (i) Formulating seizure prediction as a problem of anomaly detection and developing a discriminating parameter for the anomaly (ii) bridging the two concepts of AR modeling and interaction graphs by constructing an AR(1) model on the features extracted from the time-evolving EEG synchronization graphs (as well as other features obtained from the EEG signal itself), and (iii) introducing the concepts of transfer and transformation learning to improve the predictions of the AR(1) model.

The organization of the paper is as follows: in Sect. 2, we describe our methodology starting with the epileptic EEG dataset, initial noise removal, and procedure to construct EEG synchronization graphs and extract features from the graph. We then detail the working of a feature selection method based on quadratic programming, build autoregressive models on the selected features, and transfer and transformation learn on these features. Finally, we use an alarm-based detection system to signal the seizure. In Sect. 3 we present and discuss the results for seizure prediction, and comparison to a benchmark. We provide an overview and outline possible extensions to this study, in Sect. 4.

2   Methodology

Our seizure-prediction paradigm is centered around discriminating between seizure and non-seizure activity. So we attempt to learn from normal activity

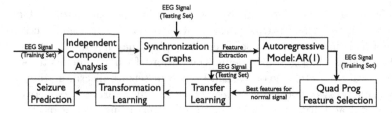

Fig. 1. Block diagram representing the entire sequence of steps involved in the methodology. We apply Independent Component Analysis (ICA) for artifact removal and noise reduction, which allows us to learn non-ictal activity. This step is carried out only on the training set of EEG recordings, and the testing set of EEG recordings is kept separate. Synchronization graphs are constructed by using Phase Lag Index as explained in Sect. 2.3. These graphs are constructed for both the training and testing sets. Based on the features extracted from the synchronization graphs and the signal itself, an autoregressive model is built Sect. 2.6, and this model allows us to identify predictive importance of features that are determined via a Quadratic Programming Feature Selection (QPFS) technique Sect. 2.5. The feature selection technique is applied only on the training set. The important features are then used for transfer and transformation learning Sect. 2.7 which improves the performance of seizure prediction Sect. 2.8.

maximally and from seizure activity minimally. We then interpret consistent and clear deviations from our understanding of normal activity as seizure. To this end: we initially suppress seizure activity; develop synchronization graphs to describe the seizure-suppressed cortical activity well; select maximally descriptive, minimally redundant features; cross-learn common attributes of seizure-suppressed activites across patients; and measure how well we predict on data from which the seizure has not been suppressed. Conditioned on this sequence of operations performing well, we reason that any prediction error on the data where the seizure has not been suppressed is due to seizure-related activity. We develop a simple way to determine when such seizure-related activity is reliable enough to declare an imminent seizure, and use it to make such a declaration. The following sections describe the steps involved from taking the EEG signal as input to predicting the seizure. The steps are illustrated in the block diagram shown in Fig. 1.

2.1   Epileptic EEG Data Set

Our dataset consists of scalp EEG recordings of 41 seizures from 14 patients. All the patients were evaluated with scalp video-EEG monitoring in the international 10–20 system (as described in [30]), magnetic resonance imaging (MRI), fMRI for language localization, and position emission tomography (PET). All the patients had *Hippocampal Sclerosis* (HS) except one patient (Patient-1) who suffered from *Cortical Dysplasia* (CD). After selective amygdalohippocampectomy, all the patients were seizure free. The patient information is provided in Table 1. For 4 patients, the seizure would onset from the right, whereas for 10 patients the seizure would onset from the left.

   The recordings include sufficient pre-ictal and post-ictal periods for the analysis. Two of the electrodes (A_1 and A_2) were unused and C_z electrode was

Table 1. Patient Types. Almost all the patients (except one) exhibited hippocampal sclerosis (HS). There are two types of lateralizations in HS: left (L) and right (R). One patient (Patient-1) exhibited cortical dysplasia (CD).

Patient	Pathology	Lateralization	Number of recordings	Length of individual recordings (in minutes)
Patient-1	CD	R	2	30
Patient-2	HS	R	2	30
Patient-3	HS	R	3	60
Patient-4	HS	R	5	60
Patient-5	HS	L	1	60
Patient-6	HS	L	1	30
Patient-7	HS	L	2	60
Patient-8	HS	L	2	60
Patient-9	HS	L	3	60
Patient-10	HS	L	3	30
Patient-11	HS	L	2	60
Patient-12	HS	L	5	41
Patient-13	HS	L	5	35
Patient-14	HS	L	5	35

used for referential montage that yielded 18-channel EEG recordings. A team of doctors diagnosed the initiation and the termination of each seizure and reported these periods as the ground truth for our analysis. An example of such a recording can be found in Fig. 2 in [63]. Seizures were 77.12 s long on average and their standard deviation was 48.94 s. The high standard deviation of the data is an indication of the vast variability in the data which makes the task of seizure prediction complicated.

2.2   Seizure Suppression

In order to suppress seizure-activity, we resort to modeling EEG signal acquisition as follows. We assume that: (1) seizure activity is statistically independent of normal activity and (2) there may be numerous statistically independent cortical activities, both seizure related and otherwise, that combine to provide the signal captured by a single electrode (3) the seizure activity is non-gaussian. Based on these two assumptions, we look to locate and discard the seizure-related activity, thereby suppressing the seizure. Under the assumptions stated above, the problem of extracting seizure-related activity is mathematically equivalent to the cocktail party problem exemplifying blind source separation, which is solved by the state of the art technique of Independent Component Analysis [12,13,33], which has thus far been used mainly to remove artifacts from EEG data [17,32,40]. Here we use ICA to locate seizure-related activity and remove it in a manner similar to artifact-removal. Formally, given that $\mathbf{X} \in \mathbb{R}^{n,d}$ is a linear mixture of k d-dimensional independent components contained in $\mathbf{S} \in \mathbb{R}^{k,d}$, we may write

$$\mathbf{X} = \mathbf{AS},$$

where $\mathbf{A} \in \mathbb{R}^{n \times k}$ is *the mixing matrix* and $\mathbf{S} \in \mathbb{R}^k$. In general, both $\mathbf{A}$ and $\mathbf{S}$ are unknown and we compute the independent components, with respect to an independence maximization measure, as $\mathbf{S} = \mathbf{WX}$, where $\mathbf{W}$ is the inverse of the mixing matrix.

Once $\mathbf{A}$ is computed, we discard seizure related activity by zeroing the columns having the lowest euclidean norms. We reason this as follows: since much of the data is normal function, the independent components corresponding to seizure-related activity do not contribute to most of the data; their contribution is concentrated in time (corresponding to concentration in row-indices of $\mathbf{A}$). Due to the inherent scaling-degeneracy in the problem of blind source separation, we obtain an $\mathbf{A}$ having unit row-norms. This leads to the coefficients corresponding to seizure-related independent components to be tightly controlled, resulting in columns corresponding to the seizure-related independent components being of low euclidean norm. We heuristically zero the lowest two columns of $\mathbf{A}$ to form $\mathbf{A}_o$ and declare

$$\mathbf{X}_o = \mathbf{A}_o \mathbf{S}$$

to be the seizure-suppressed EEG data. It is important to note that the seizure is not completely suppressed, but the independent components retrieved allow us to model the non-ictal activity more precisely.

2.3   Construction of EEG Synchronization Graphs

For the signal $\mathbf{f}X[i, m]$, we construct epochs of equal lengths with an overlap of 20 % between the preceding and following epochs. The number of epochs, n, is equal to $1.25M/L$, where L is the duration of the epoch in same time units. Since the EEG recordings contain both temporal and spatial information, we construct *time-evolving EEG Synchronization Graphs* on the EEG datasets. A synchronization graph is constructed for each epoch, giving an indication of the spatio-temporal correspondence between electrodes - these relationships can then be utilized to obtain changes in the network by identifying descriptive features. The *nodes* represent the EEG electrodes and the *edges* represent a closeness relationship between the nodes in a given epoch. We use an epoch length of 5 s.

A sample time-evolving graph on an EEG recording is shown in Fig. 2. The pair-wise relationships between the electrodes during an epoch are used to construct the graph edges. If the pair-wise distance between two nodes i and k, where $i, k \in \{1, \ldots, 18\}$, and $i \neq k$, for epoch t, given as $d_{i,k}^n$, is less than a specified threshold, τ, then an edge is inserted into the graph between the two nodes. Note that smaller threshold values seek higher correlation between the electrodes, thereby yielding sparser graphs. Similarly, higher threshold values would establish an edge even if there is small correlation between the data, thereby yielding denser graphs. For our analysis, we performed a parametric search and found the best value of τ to be 1.

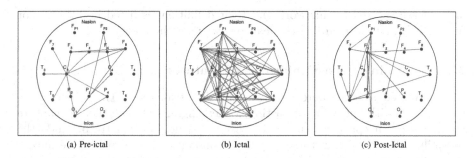

| (a) Pre-ictal | (b) Ictal | (c) Post-Ictal |

Fig. 2. Sample EEG Synchronization Graphs for pre-ictal, ictal, and post-ictal epochs. It is clearly seen that the ictal period has more coherence between different regions of the brain. The Phase Lag Index measure is used as the synchronization measure (or cost function) to set up edges between the nodes in the graph.

Several synchronization measures have been proposed as plausible options for $d_{i,k}^n$ to set up the edges in the graph. Based on earlier results presented in [19], we chose Phase Lag Index (PLI) [65] for $d_{i,k}^n$. PLI is defined as follows:

$$PLI_{i,k}(n) = \frac{1}{Lf_s} \left| \sum_{m=1}^{Lf_s} \text{sgn}\left(\phi_i^n(m) - \phi_k^n(m)\right) \right| \tag{1}$$

where $\phi_i^n = \arctan(\frac{\hat{\mathbf{fx}}_i^n}{\mathbf{fx}_i^n})$ is the angle of the Hilbert transform $\hat{\mathbf{fx}}_i^n$ of the signal $\mathbf{fx}_i^n$.

2.4 Feature Extraction from EEG Synchronization Graphs

We extract 26 features from the EEG synchronization graph for each epoch. These features quantify the compactness, clusteredness, and uniformity of the graph. Apart from these graph-based features, we compute two spectral features - the variance of the stationary distribution on an undirected markov chain on the graph, and the second largest eigenvalue of the Laplacian of the graph. In addition we compute certain natural statistics: the mean jump size between epochs and its variance, to measure the similarity to a Weiner process, and finally the *hinged mean* and *hinged variance*, defined as the mean and variance, respectively, of the signal at the current epoch centered/hinged at the mean of a strictly trailing window. These features arise naturally in change-point-detection and are motivated by the natural belief that, as a stochastic process, the EEG signal undergoes a statistical change when a seizure begins. To this feature set we also added time-domain and spectral features. The time-domain features include the Hjorth parameters - activity, mobility, and complexity, and the frequency-domain features include skewness of amplitude spectrum and spectral entropy [1]. In all, we calculated 38 features.

In subsequent text, we refer to the feature matrix as $\mathbf{D} \in \mathbb{R}^{n,d}$, with n epochs and d features. We refer to the feature vector at time t (row t of $\mathbf{D}$) as $\mathbf{d}^t$, and the time-series corresponding to feature i (column i of $\mathbf{D}$) as $\mathbf{d}_i$. A complete list of the features used in this work and their definitions is listed in Table 2. For further information regarding the features, we refer the reader to [1,6].

2.5   Determining the Significance of Features

The computed features were motivated by discussions with the subject matter experts, with the view of casting a meaningful but wide net to capture attributes of an epileptic seizure. However, this doesn't strictly preclude the possibility that certain features may be redundant or low in predictive importance. Furthermore, we wish to select features that are particularly descriptive of the non-ictal activity, of which the data is largely comprised. Therefore, we quantify the predictive significance of the features in a natural but effective way, and score the features to maximize their predictive importance for the entire data, and minimize redundancy, using the method in [58], which we summarize here. The primary advantages of using QPFS are as follows: (i) QPFS is based on efficient quadratic programming technique [5]. The quadratic term quantifies the dependence between each pair of variables, whereas the linear term quantifies the relationship between each feature and the class label. (ii) QPFS provides a considerable time complexity improvement over current methods on very large data sets with high dimensionality.

Measuring Redundancy. Our notion of redundancy arises naturally from the interpretation of brain activity as a stochastic process, whence the usual notion of linear dependence is replaced with the notion of statistical correlation. Specifically, suppose the data matrix, $\mathbf{D} \in \mathbb{R}^{n,d}$, spanning n epochs and consisting of d features. We define, the correlation matrix, $\mathbf{Q} \in \mathbb{R}^{d,d}$, element-wise, where $\mathbf{Q}(i,j)$ is the Pearson correlation coefficient between the feature vectors $\mathbf{d}_i, \mathbf{d}_j \in \mathbb{R}^n$:

$$\mathbf{Q}(i,j) = \frac{\mathbf{d}_i^\top \mathbf{d}_j}{\|\mathbf{d}_i\| \, \|\mathbf{d}_j\|}.$$

The quadratic form $\mathbf{x}^\top \mathbf{Q} \mathbf{x}$ thus has the natural interpretation of yielding the sample-covariance of a compound feature, with coefficients contained in $\mathbf{x}$, which is the notion of redundancy that we wish to minimize.

Measuring Predictive Importance. We first recall that the activity of the brain at time t is completely captured by $\mathbf{d}^t$. We define the predictive importance, f_i, of the feature i, as the r.m.s. influence of $\mathbf{d}_i^t$ on $\mathbf{d}_j^{t+1}, 1 \leq j \leq n$, measured by the coefficients in the forecast operator corresponding to i. Formally, let $\boldsymbol{\Psi} \in \mathbb{R}^{d,d+1}$ be the forecast operator. Then our best prediction of $\mathbf{d}^{t+1}$ is $\tilde{\mathbf{d}}^{t+1}$ where

$$\tilde{\mathbf{d}}^{t+1} = \mathbf{d}^t \boldsymbol{\Psi}^\top.$$

The influence, $\mathbf{p}_i(j)$, of feature i on j, contained in $\mathbf{p}_i \in \mathbb{R}^d$, may be determined by predicting via $\boldsymbol{\Psi}$ using its indicator vector, $\mathbf{e}_i$:

$$\mathbf{p}_i = \mathbf{e}_i \boldsymbol{\Psi}^\top,$$

Table 2. Names and description of EEG global graph features. Features 1–26 are computed on the synchronization graph, features 27 and 28 are signal-based. Features 29–31 are representative of change-point detection, and features 32 and 33 are spectral features. Features 34–36 and 37–38 are time-domain and frequency-domain features, respectively.

Index	Feature name	Description
1	Average Degree	Average number of edges per node
2	Clustering Coefficient C	Average of the ratio of the links a node's neighbors have in between to the total number that can possibly exist
3	Clustering Coefficient D	Same as feature 2 with node added to both numerator and denominator
4	Average Eccentricity	Average of node eccentricities, where the *eccentricity* of a node is the maximum distance from it to any other node in the graph
5	Diameter of graph	Maximum of node eccentricities
6	Radius of graph	Minimum of node eccentricities
7	Average Path Length	Average hops along the shortest paths for all possible pairs of nodes
8	Giant Connected Component Ratio	Ratio between the number of nodes in the largest connected component in the graph and total The number of nodes
9	Number of Connected Components	Number of clusters in the graph excluding the isolated nodes
10	Average Connected Component Size	Average number of nodes per connected component
11	% of Isolated Points	% of isolated nodes in the graph, where an *isolated node* has a degree 0
12	% of End Points	% of endpoints in the graph, where an *endpoint* has a degree 1
13	% of Central Points	% of nodes in the graph whose eccentricity is equal to the graph radius
14	Number of Edges	Number of edges between all nodes in the graph
15	Spectral Radius	Largest eigenvalue of the adjacency matrix
16	Adjacency Second Largest Eigenvalue	Second largest eigenvalue of the adjacency matrix
17	Adjacency Trace	Sum of the adjacency matrix eigenvalues
18	Adjacency Energy	Sum of the square of adjacency matrix eigenvalues
19	Spectral Gap	Difference between the magnitudes of the two largest eigenvalues

(Continued)

Table 2. (*Continued*)

Index	Feature name	Description
20	Laplacian Trace	Sum of the Laplacian matrix eigenvalues
21	Laplacian Energy	Sum of the square of Laplacian matrix eigenvalues
22	Normalized Laplacian Number of 0's	Number of eigenvalues of the normalized Laplacian matrix that are 0
23	Normalized Laplacian Number of 1's	Number of eigenvalues of the normalized Laplacian matrix that are 1
24	Normalized Laplacian Number of 2's	Number of eigenvalues of the normalized Laplacian matrix that are 2
25	Normalized Laplacian Upper Slope	The sorted slope of the line for the eigenvalues that are between 1 and 2
26	Normalized Laplacian Trace	Sum of the normalized Laplacian matrix eigenvalues
27	Mean of EEG recording	Mean of EEG signal for each electrode and epoch
28	Variance of EEG recording	Variance of EEG signal for each electrode and epoch
29, 30	Change-based Features	Mean and variance of jump size in EEG signal for each electrode and epoch
31	Change-based Feature 3	Variance of EEG signal for particular electrode in given epoch after subtracting the mean of up to 3 previous windows
32	Spectral Feature 1	Variance of eigenvector of the product of the adjacency matrix and the inverse of the degree matrix
33	Spectral Feature 2	Second largest eigenvalue of the Laplacian matrix
34, 35, 36	Hjorth parameters (time-domain)	Activity, Mobility, and Complexity
37, 38	Frequency-domain features	Skewness of amplitude spectrum and Spectral entropy

whence the r.m.s. influence, f_i, of i is simply

$$f_i = \|\mathbf{p}_i\| = \|\mathbf{\Psi}_i\|,$$

the column-norm of the forecast operator corresponding to column i. We define $\mathbf{f} \in \mathbb{R}^d$ such that $\mathbf{f}_i = \|\mathbf{\Psi}_i\|$, as the predictive importance vector.

Optimizing Redundancy and Predictive Importance. We obtain a significance-distribution over the features that maximizes predictive importance and minimize redundancy by solving

$$\mathbf{x}^* = \quad \arg\min q(\mathbf{x}); \\ \text{subject to } \mathbf{x} \in \mathbb{R}^d, \ \mathbf{x} \geq \mathbf{0}, \ \textstyle\sum_i \mathbf{x}_i = 1, \tag{2}$$

where the constraints arise from forcing the resulting vector to be a distribution, from which we omit an appropriately sized tail, or select just the support if it is small. To make the objective function stable under scaling of the data, we normalize $\mathbf{f}$ to obtain

$$\hat{\mathbf{f}} = \mathbf{f}/\|\mathbf{f}\|_\infty .$$

To effect a meaningful trade-off between minimizing redundancy and maximizing predictive importance, we take a convex combination of the corresponding terms:

$$q(\mathbf{x}) = (1 - \alpha)\mathbf{x}^\top \mathbf{Q}\mathbf{x} - \alpha\hat{\mathbf{f}}^\top \mathbf{x},$$

where α is chosen, as in [58] as

$$\alpha = \frac{\sum_{i,j} \mathbf{Q}(i,j)/d^2}{\sum_{i,j} \mathbf{Q}(i,j)/d^2 + \sum_k \mathbf{f}_k/d} .$$

Since both the predictive importance and the correlation matrix are statistical in nature, they are less affected by the relatively fleeting seizure. So we expect the significant features obtained via QPFS to be features that are highly predictive of non-ictal activity. We use the MATLAB utility quadprog to solve Eq. 2.

2.6   Autoregressive Modeling on Feature Data

Research has indicated that promising results regarding early detection or prediction of the seizure can be achieved by application of an autoregressive model (AR) to the EEG signal [2,11]. Also, AR models are linear and as shown in prior research are comparable to non-linear models in their predictive capability [14,15,43,49,51]. We expand on these earlier results by applying an autoregressive model to the features extracted both from the graph and the signal itself.

An autoregressive model of order 1, AR(1), is applied to the matrix $\mathbf{D}$, extracted from the time-evolving EEG synchronization graphs. For an AR(1) model the output at time t is only dependent on the values of the time-series at time $t - 1$. As a result, the implicit assumption when using AR(1) is that $\mathbf{d}^t$ is a markov chain indexed by t. Formally

$$\mathbf{d}_i^t = \rho_{i0} + \rho_{i1}\mathbf{d}_1^{t-1} + \rho_{i2}\mathbf{d}_2^{t-1} + \ldots + \rho_{im}\mathbf{d}_m^{t-1} + \epsilon_t \tag{3}$$

where ρ_{ij} are the linear coefficients computed via autoregression. In matrix form, (3) is $[\mathbf{D}]_1^t = \boldsymbol{\Psi}\cdot[1, \mathbf{D}]_0^{t-1} + \epsilon$, where the notation $[\mathbf{A}]_a^b$ denotes a matrix containing all rows from a to b of $\mathbf{A}$, including rows a, b. We compute $\boldsymbol{\Psi}$ to minimize the error ϵ in euclidean norm,

$$\boldsymbol{\Psi} = \arg\min_{\mathbf{Z}} \left\| [\mathbf{D}]_1^t - \mathbf{Z} \cdot [1, \mathbf{D}]_0^{t-1} \right\|_F^2 , \tag{4}$$

$$\boldsymbol{\Psi} = [\mathbf{D}]_1^t \cdot \left([1, \mathbf{D}]_0^{t-1}\right)^\dagger .$$

where $\mathbf{A}^\dagger$ denotes the moore-penrose pseudoinverse of $\mathbf{A}$. The role of the operator $\boldsymbol{\Psi}$ is to predict $\mathbf{D}(t)$ as a function of $\mathbf{D}(t-1)$. Any operator that does this will be called subsequently as the *forecast operator*. Thus, using an AR(1) model we arrive at a forecast operator, $\boldsymbol{\Psi}$.

2.7 Transfer Learning and Transformation Learning on Autoregressive Model

We critically improve this forecast operator obtained from AR(1) in two directions. First, we improve it under the assumption that the data obtained from different patients are not completely independent of each other; that data obtained from one patient holds some information common to all patients, along with information specific to the patient. Thus, we *transfer* knowledge from one patient to another, motivated by the existing work on Transfer Learning. Specifically, given a feature data set $\mathbf{D}$, the feature transfer set, $\hat{\mathbf{D}}$, and the corresponding forecast operators $\boldsymbol{\Psi}$ and $\hat{\boldsymbol{\Psi}}$ respectively we transfer knowledge from $\hat{\mathbf{D}}$ to $\mathbf{D}$ by regularizing (4) with

$$\lambda||\boldsymbol{\Psi} - \hat{\boldsymbol{\Psi}}||_F^2.$$

The parameter λ, playing the familiar role of the Tikhonov Regularizer, is the *transfer coefficient*, governing how much we learn from $\hat{\mathbf{D}}$ onto $\mathbf{D}$. The forecast operator obtained from this transfer learning is simply

$$\bar{\boldsymbol{\Psi}} = \arg\min \left\{ \left\| [\mathbf{D}]_{t-1}^0 * \boldsymbol{\Psi} - [\mathbf{D}]_t^1 \right\|_F^2 + \lambda \left\| \boldsymbol{\Psi} - \hat{\boldsymbol{\Psi}} \right\|_F^2 \right\}, \tag{5}$$

the analytical solution for which is:

$$\boldsymbol{\Psi} = \hat{\boldsymbol{\Psi}} + \left([\mathbf{D}]_{t-1}^{0\top} [\mathbf{D}]_{t-1}^0 + \lambda\mathbf{I} \right)^\dagger \left([\mathbf{D}]_{t-1}^0 \right)^\top \left([\mathbf{D}]_t^1 - [\mathbf{D}]_{t-1}^{0\top} \hat{\boldsymbol{\Psi}} \right) \tag{6}$$

As more data is obtained for $\mathbf{D}$, the value of λ is reduced because now the core set is getting better at predicting its own future values. To test our estimates, we use the following split between the training and testing data. First, we split $\mathbf{D}$ into training (TR) and testing (TE) sets. Within the training set, we create a further split thereby creating training prime ($\overline{TR}$) and validation (Val) sets. We then train our AR(1) model on $\overline{TR}$, and then use $\hat{\boldsymbol{\Psi}}$ to improve this estimate by testing on Val. Then, we retrain the model using the learned parameters on the entire training set TR and finally test on TE.

Next, we account for the differences in the collected data that may arise as a result of non-uniformities in the process of acquiring data. We do this under the assumption that the spectral nature of the data is minimally variant with changes across the various setups for acquiring data from multiple patients, and that the flow of time is immutable. To learn from $\hat{\mathbf{D}}$ onto $\mathbf{D}$, we find the object, $\hat{\mathbf{D}}_{\mathbf{D}}$, retaining the spectral nature of $\hat{\mathbf{D}}$, and respecting the directionality of time, that is the closest to $\mathbf{D}$. Formally, we find a rotation $\boldsymbol{\Gamma}(\hat{\mathbf{D}}, \mathbf{D})$ such that

$$\hat{\mathbf{D}}_{\mathbf{D}} = \hat{\mathbf{D}}\boldsymbol{\Gamma}(\hat{\mathbf{D}}, \mathbf{D}),$$

$$\Gamma(\hat{\mathbf{D}}, \mathbf{D}) = \arg \min_{\substack{\mathbf{U} \\ \mathbf{U}^\top \mathbf{U} = \mathbf{I}}} \left\| \mathbf{D} - \hat{\mathbf{D}} \mathbf{U} \right\|.$$

This is the Procrustes problem that has been well-studied [72], and has a closed form solution in terms of the SVD of $\mathbf{D}, \hat{\mathbf{D}}$. Let these SVDs be

$$\mathbf{D} = \mathbf{U}_{\mathbf{D}} \mathbf{\Sigma}_{\mathbf{D}} \mathbf{V}_{\mathbf{D}}^\top, \quad \hat{\mathbf{D}} = \mathbf{U}_{\hat{\mathbf{D}}} \mathbf{\Sigma}_{\hat{\mathbf{D}}} \mathbf{V}_{\hat{\mathbf{D}}}^\top.$$

Then

$$\Gamma(\hat{\mathbf{D}}, \mathbf{D}) = \mathbf{V}_{\hat{\mathbf{D}}} \mathbf{V}_{\mathbf{D}}^\top, \Rightarrow \hat{\mathbf{D}}_{\mathbf{D}} = \mathbf{U}_{\hat{\mathbf{D}}} \mathbf{\Sigma}_{\hat{\mathbf{D}}} \mathbf{V}_{\mathbf{D}}^\top.$$

We now transfer-learn using $\hat{\mathbf{D}}_{\mathbf{D}}$. In summary, we first notice that knowledge can be transferred from other but similar data and then transform such similar data sets into their *most learnable* forms using a simple transformation.

2.8 Declaration of Imminent Seizure

We use the prediction errors incurred by the use of our forecast operator as the eventual *ictal discriminator*. We compute an estimate of the probability of deviation towards seizure using these errors, and declare that a seizure is imminent when this probability is reliably high. We outline how we compute this probability and quantify the sense of reliability we use, in that order.

Probability of Deviation Towards Seizure. Let $\epsilon(t)$ be the prediction error at time, t. Using a moving window of size $\Delta = 30$, we first use a simple statistical thresholding on the errors to determine if an *alarm* has to be thrown, which signifies an outlier to normal function, and a potential seizure. Specifically, let $\kappa(t)$ be the binary variable indicating whether or not an alarm is thrown - 1 when it is thrown and 0 when it is not. Let $\mu(t, \Delta), \sigma(t, \Delta)$ denote the mean and standard deviation of the sequence $\epsilon(t), \epsilon(t+1), \cdots, \epsilon(t+\Delta)$. Then

$$\kappa(t) = \begin{cases} 1 & \text{if } \epsilon(t+\Delta) - \mu(t, \Delta) > \tau^* \sigma(t, \Delta) \\ 0 & \text{otherwise} \end{cases}$$

where τ^* is a tolerance/sensitivity parameter. Clearly, κ is an indicator of the one-sided tail of the distribution from which $\epsilon(t)$ is drawn. Under our assumption that a recorded activity is either normal function or seizure, the measure of the tail of the distribution of errors during normal function is an appropriate estimate of the probability of seizure. We estimate the size of this *tail of normality* for an interval by the ensemble average of $\kappa(t)$ for the interval. When $\kappa(t)$ indicates seizure repeatedly in a manner highly unlikely to have arisen from random sampling from the tail of normality, we declare an imminent seizure. In practice, we choose $\tau^* = 3$, and declare a seizure when we see 3 consecutive alarms. We justify this choice as follows: in the case where $\kappa(t)$ indicates the result of the high-entropy fair coin toss (i.e. 1 and 0 with equal probability), the probability of obtaining 3 consecutive alarms is 12.5 %. In practice our estimate of the size of the tail of normality is significantly below $1/5$, resulting in our three-in-a-row rule to be an even rarer occurrence than once in 125 occurrences or 0.8 %.

3 Results

We present results for the Quadratic Programming Feature Selection algorithm, determining the best forecast operator, and a comparison of the performance of our seizure prediction algorithm on basic autoregression vs. with the addition of transfer and transformation learning.

3.1 Quadratic Programming Feature Selection Results

The feature-significance vectors obtained from solving the QPFS problem in (2) were found to be highly sparse, and the features that were supported by these vectors were chosen without exception - 9 in all: (i) Average Degree, (ii) Diameter of graph, (iii) Average Path Length, (iv) Giant Connected Component Ratio, (v) Number of Connected Components, (vi) Percentage of Isolated Points, (vii) Number of eigenvalues with value 0 of the normalized Laplacian matrix, (viii) Number of eigenvalues with value 2 of the normalized Laplacian matrix, and (ix) Normalized Laplacian trace.

3.2 Baseline SVM Results

To establish a baseline to validate the efficacy of our results, we compare our algorithm to the following benchmark algorithm:
Application of Support Vector Machine to feature matrix $\mathbf{D}$:

We provide as input the features identified by QPFS from $\mathbf{D}$ to a two-class Support Vector Machine (SVM). We learn a model of the inter-ictal and ictal states based on their respective feature values. We then classify using the SVM the seizure onset in the pre-ictal region based on the feature values in that region. The intuition being that the initial part of the pre-ictal region will have features similar to the inter-ictal region, whereas the latter part will be more similar to the ictal region in the feature space. We consider the pre-ictal region to start 10 min prior to the onset of the seizure.

We found that the benchmark did not predict the seizure in 14 of the 41 analyzed recordings. The median prediction time for the recordings for which seizures were predicted was 1.25 min prior to the seizure.

3.3 Autoregression vs Transfer and Transformation Learning

One of the objectives of this study was to improve the basic autoregressive model by the application of transfer learning and transformation learning. We now show that the additional functionality makes the prediction either at least as good as that by the AR(1) model or better for a significant percentage of the dataset. We found that in 60 % of the analyzed EEG recordings, transformation learning was able to predict the seizure prior to AR(1), or transfer learning. In 52.5 % of the analyzed EEG recordings, transfer learning performed better than the AR(1) model predicting the seizure earlier. Finally, in 67.5 % of the cases, either transfer learning or transformation learning was better than the AR(1)

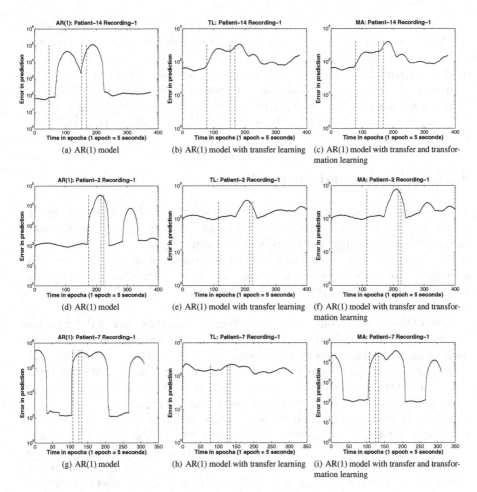

Fig. 3. Comparison of the AR(1) model ((a), (d), and (g)) with the AR(1) model improved by transfer learning ((b), (e), and (h)) and transformation learning ((c), (f), and (i)). The epoch at which the seizure is detected is shown in blue, the start and end of the seizure region are marked in red. The first row is an example of where the AR model does better than both transfer learning and transformation learning. The second row is a typical example of the AR(1) model with transfer and transformation learning outperforming both the AR(1) model and the AR(1) model enhanced by transfer learning. The third row is an example of an anomaly where the AR(1) model with transfer learning performs much better than the AR(1) model and the AR(1) model with transfer and transformation learning.

model. The median prediction times prior to the occurrence of the seizure for the three methods are 10 min, 10.96 min, and 11.04 min for AR(1), transfer learning, and transformation learning, respectively. Considering only the recordings where transformation learning or transfer learning outperformed AR(1), the median prediction times change to 9.33 min for AR(1), 11.17 min for transfer learning, and 11.92 min for transformation learning.

In Fig. 3, the first row ((a)–(c)) consists of an analyzed EEG recording where the AR(1) model was able to predict the seizure before the other two techniques. The second row ((d)–(f)) consists of an analyzed EEG recording where the AR(1) model with transfer learning and transformation learning was able to predict the seizure before the other two techniques. Finally, the third row ((g)–(i)) consists of an anomalous result where the AR(1) model with transfer learning predicted the seizure before either AR(1) model or the AR(1) model with transfer and transformation learning.

4   Conclusions and Future Work

In this study, we outline a seizure prediction algorithm designed for EEG epileptic seizure data by constructing an autoregressive model improved by the addition of transfer learning and transformation learning on features extracted by building synchronization graphs on the independent components of the EEG signal. We use a quadratic programming algorithm called Quadratic Programming Feature Selection (QPFS) to select the features with the highest predictive importance and minimal redundancy.

One of the primary concerns with the seizure prediction area is the definition of a Seizure Prediction Horizon (SPH). In the literature prediction horizons have varied from several minutes to a few hours [55]. We would like to come up with a more rigorous theoretical basis for assigning prediction horizons. Another future direction is with respect to the various thresholds used in the study. Although, well-motivated and justified from the literature, we would like to obtain these thresholds from first principles. Examples of these thresholds include epoch lengths for the synchronization graphs, sensitivity parameters for raising an alarm, and number of columns to zero out from the mixing matrix in Independent Component Analysis (ICA). Yet another important future direction is analyzing the partial contribution of each module in the pipeline to determine the effect of individual modules in improving the basic prediction. Specifically, we would like to examine the influence of ICA vs. transformation learning to determine which of the two is better used for the initial surgery to suppress seizure - to ensure that we don't use a gas-engine for the short haul and a horse for the long one. Furthermore, we would like to qualify the use of transfer learning based on the similarity of the data sets being learned across: establish a metric of closeness of data sets/learnability across patients and recordings. Finally, the problem of seizure prediction is accompanied by the problem of localizing seizure, which, apart from requiring new methods, also sets a higher standard for understanding seizure. We hope to contribute to this problem in the future as well.

References

1. Acar, E., Aykut-Bingol, C., Bingol, H., Bro, R., Yener, B.: Multiway analysis of epilepsy tensors. Bioinformatics **23**(13), i10–i18 (2007)

2. Alkan, A., Koklukaya, E., Subasi, A.: Automatic seizure detection in EEG using logistic regression and artificial neural network. J. Neurosci. Meth. **148**(2), 167–176 (2005)

3. Anderson, N.R., Wisneski, K., Eisenman, L., Moran, D.W., Leuthardt, E.C., Krusienski, D.J.: An offline evaluation of the autoregressive spectrum for electrocorticography. IEEE Trans. Biomed. Eng. **56**(3), 913–916 (2009)

4. Barrat, A., Barthelemy, M., Vespignani, A.: Dynamical Processes on Complex Networks, vol. 1. Cambridge University Press, Cambridge (2008)

5. Bertsekas, D.P.: Nonlinear Programming, 2nd edn. Athena Scientific, Belmont (1999)

6. Bilgin, C.C., Ray, S., Baydil, B., Daley, W.P., Larsen, M., Yener, B.: Multiscale feature analysis of salivary gland branching morphogenesis. PLoS ONE **7**(3), e32906 (2012)

7. Boccaletti, S., Latora, V., Moreno, Y., Chavez, M., Hwang, D.U.: Complex networks: structure and dynamics. Phys. Rep. **424**(4–5), 175–308 (2006). http://www.sciencedirect.com/science/article/pii/S037015730500462X

8. Bronzino, J.D.: Principles of electroencephalography. In: Biomedical Engineering Handbook, 3rd edn. Taylor and Francis, New York (2006)

9. Bullmore, E., Sporns, O.: Complex brain networks: graph theoretical analysis of structural and functional systems. Nat. Rev. Neurosci. **10**(3), 186–198 (2009)

10. Chandaka, S., Chatterjee, A., Munshi, S.: Cross-correlation aided support vector machine classifier for classification of EEG signals. Expert Syst. Appl. **36**(2 Part 1), 1329–1336 (2009)

11. Chisci, L., Mavino, A., Perferi, G., Sciandrone, M., Anile, C., Colicchio, G., Fuggetta, F.: Real-time epileptic seizure prediction using AR models and support vector machines. IEEE Trans. Biomed. Eng. **57**(5), 1124–1132 (2010)

12. Comon, P.: Independent component analysis - a new concept. Signal Process. **36**, 287–314 (1994)

13. Comon, P., Jutten, C.: Handbook of Blind Source Separation: Independent Component Analysis and Applications, 1st edn. Academic Press, Oxford (2010)

14. Corsini, J., Shoker, L., Sanei, S., Alarcon, G.: Epileptic seizure predictability from scalp EEG incorporating constrained blind source separation. IEEE Trans. Biomed. Eng. **53**, 790–799 (2006)

15. Cranstoun, S.D., Ombao, H.C., von Sachs, R., Guo, W., Litt, B., et al.: Time-frequency spectral estimation of multichannel EEG using the auto-slex method. IEEE Trans. Biomed. Eng. **49**, 988–996 (2002)

16. D'Alessandro, M., Vachtsevanos, G., Esteller, R., Echauz, J., Cranstoun, S., Worrell, G., et al.: A multi-feature and multi-channel univariate selection process for seizure prediction. Clin. Neurophysiol. **116**, 506–516 (2005)

17. Delorme, A., Sejnowski, T.J., Makeig, S.: Enhanced detection of artifacts in EEG data using higher-order statistics and independent component analysis. Neuroimage **34**(4), 1443–1449 (2007)

18. Demir, C., Gultekin, S.H., Yener, B.: Augmented cell-graphs for automated cancer diagnosis. Bioinformatics **21**(Suppl. 2), ii7–ii12 (2005)

19. Dhulekar, N., Oztan, B., Yener, B., Bingol, H.O., Irim, G., Aktekin, B., Aykut-Bingol, C.: Graph-theoretic analysis of epileptic seizures on scalp EEG recordings. In: Proceedings of the 5th ACM Conference on Bioinformatics, Computational Biology, and Health Informatics, BCB 2014, pp. 155–163. ACM, New York (2014). http://doi.acm.org/10.1145/2649387.2649423

20. Douw, L., van Dellen, E., de Groot, M., Heimans, J.J., Klein, M., Stam, C.J., Reijneveld, J.C.: Epilepsy is related to theta band brain connectivity and network topology in brain tumor patients. BMC Neurosci. 11(1), 103 (2010)

21. Elger, C.E.: Future trends in epileptology. Curr. Opin. Neurol. 14, 185–186 (2001)

22. Esteller, R., Echauz, J., D'Alessandro, M., Worrell, G., Cranstoun, S., Vachtsevanos, G., et al.: Continuous energy variation during the seizure cycle: towards an on-line accumulated energy. Clin. Neurophysiol. 116, 517–526 (2005)

23. Fisher, N., Talathi, S.S., Carney, P.R., Ditto, W.L.: Epilepsy detection and monitoring. In: Tong, S., Thankor, N.V. (eds.) Quantitative EEG Analysis Methods and Applications, pp. 157–183. Artech House (2008)

24. Fisher, R.S., van Emde Boas, W., Blume, W., Elger, C., Genton, P., Lee, P., Engel, J.J.: Epileptic seizures and epilepsy: definitions proposed by the international league against epilepsy (ILAE) and the international bureau for epilepsy (IBE). Epilepsia 46(4), 470–472 (2005)

25. Giannakakis, G., Sakkalis, V., Pediaditis, M., Tsiknakis, M.: Methods for seizure detection and prediction: an overview. Neuromethods, 1–27 (2014)

26. Güler, N.F., Übeyli, E.D., Güler, I.: Recurrent neural networks employing lyapunov exponents for EEG signals classification. Expert Syst. Appl. 29(3), 506–514 (2005)

27. Harrison, M.A., Frei, M.G., Osorio, I.: Accumulated energy revisited. Clin. Neurophysiol. 116, 527–531 (2005a)

28. Hazarika, N., Chen, J.Z., Tsoi, A.C., Sergejew, A.: Classification of EEG signals using the wavelet transform. Signal Process. 59, 61–72 (1997)

29. Iasemidis, L.D., Shiau, D.S., Pardalos, P.M., Chaovalitwongse, W., Narayanan, K., Prasad, A., et al.: Long-term prospective on-line real-time seizure-prediction. Clin. Neurophysiol. 116, 532–544 (2005)

30. Jasper, H.H.: The ten-twenty electrode system of the international federation. Electroencephalogr Clin. Neurophysiol. Suppl. 10, 371–375 (1958)

31. Jouny, C.C., Franaszczuk, P.J., Bergey, G.K.: Signal complexity and synchrony of epileptic seizures: is there an identifiable preictal period? Clin. Neurophysiol. 116, 552–558 (2005)

32. Jung, T.P., Makeig, S., Humphries, C., Lee, T.W., McKeown, M.J., Iragui, V., Sejnowski, T.J.: Removing electroencephalographic artifacts by blind source separation. Psychophysiology 37, 163–178 (2000)

33. Jutten, C., Herault, J.: Blind separation of sources, part I: an adaptive algorithm based on neuromimetic architecture. Signal Process. 24, 1–10 (1991)

34. Kannathal, N., Choo, M.L., Rajendra Acharya, U., Sadasivan, P.K.: Entropies for detection of epilepsy in EEG. Comput. Meth. Programs Biomed. 80(3), 187–194 (2005)

35. Kramer, M.A., Kolaczyk, E.D., Kirsch, H.E.: Emergent network topology at seizure onset in humans. Epilepsy Res. 79(2), 173–186 (2008)

36. Kuhnert, M.T., Elger, C.E., Lehnertz, K.: Long-term variability of global statistical properties of epileptic brain networks. Chaos: Interdisc. J. Nonlinear Sci. 20(4), 043126 (2010). http://scitation.aip.org/content/aip/journal/chaos/20/4/10.1063/1.3504998

37. Le Van, Q.M., Navarro, V., Martinerie, J., Baulac, M., Varela, F.J.: Toward a neurodynamical understanding of ictogenesis. Epilepsia 44(12), 30–43 (2003)

38. Le Van, Q.M., Soss, J., Navarro, V., Robertson, R., Chavez, M., Baulac, M., Martinerie, J.: Preictal state identification by synchronization changes in long-term intracranial EEG recordings. Clin. Neurophysiol. 116, 559–568 (2005)

39. Le Van Quyen, M., Soss, J., Navarro, V., Robertson, R., Chavez, M., Baulac, M., et al.: Preictal state identification by synchronization changes in long-term intracranial EEG recordings. Clin. Neurophysiol. **116**, 559–568 (2005)
40. Lee, T.W., Girolami, M., Sejnowski, T.J.: Independent component analysis using an extended infomax algorithm for mixed sub-gaussian and super-gaussian sources. Neural Comput. **11**(2), 417–441 (1999)
41. Lehnertz, K., Litt, B.: The first international collaborative workshop on seizure prediction: summary and data description. Clin. Neurophysiol. **116**, 493–505 (2005)
42. Li, G., Semerci, M., Yener, B., Zaki, M.J.: Effective graph classification based on topological and label attributes. Stat. Anal. Data Min. ASA Data Sci. J. **5**(4), 265–283 (2012)
43. Litt, B., Esteller, R., Echauz, J., D'Alessandro, M., Shor, R., et al.: Epileptic seizures may begin hours in advance of clinical onset: a report of five patients. Neuron **30**, 51–64 (2001)
44. Liu, H.S., Zhang, T., Yang, F.S.: A multistage, multimethod approach for automatic detection and classification of epileptiform EEG. IEEE Trans. Biomed. Eng. **49**(12 Pt 2), 1557–1566 (2002)
45. Lytton, W.W.: Computer modeling of Epilepsy. Nat. Rev. Neurosci. **9**(8), 626–637 (2008)
46. Mahyari, A., Aviyente, S.: Identification of dynamic functional brain network states through tensor decomposition. In: 39th IEEE International Conference on Acoustics, Speech, and Signal Processing (ICASSP 2014) (2014)
47. Mirowski, P., Madhavan, D., LeCun, Y., Kuzniecky, R.: Classification of patterns of EEG synchronization for seizure prediction. Clin. Neurophysiol. **120**, 1927–1940 (2009)
48. Mormann, F., Andrzejak, R.G., Elger, C.E., Lehnertz, K.: Seizure prediction: the long and winding road. Brain **130**, 314–333 (2007)
49. Mormann, F., Kreuz, T., Andrzejak, R., David, P., Lehnertz, K., et al.: Epileptic seizures are preceded by a decrease in synchronization. Epilepsy Res. **53**, 173–185 (2003)
50. Mormann, F., Kreuz, T., Rieke, C., Andrzejak, R.G., Kraskov, A., David, P., et al.: On the predictability of epileptic seizures. Clin. Neurophysiol. **116**, 569–587 (2005)
51. Mormann, F., Lehnertz, K., David, P., Elger, C.E.: Mean phase coherence as measure for phase synchronization and its application to the EEG of epilepsy patients. Physica D **144**, 358–369 (2000)
52. Murali, S., Kulish, V.V.: Modeling of evoked potentials of electroencephalograms: an overview. Digit. Signal Process. **17**, 665–674 (2007)
53. Muthuswamy, J., Thakor, N.V.: Spectral analysis methods for neurological signals. J. Neurosci. Meth. **83**, 1–14 (1998)
54. Newman, M.E.J.: The structure and function of complex networks. SIAM Rev. **45**, 167–256 (2003)
55. Osorio, I., Zaveri, H., Frei, M., Arthurs, S.: Epilepsy: The Intersection of Neurosciences, Biology, Mathematics, Engineering, and Physics. Taylor & Francis (2011). http://books.google.com/books?id=O97hKvyyYgsC
56. Pan, S.J., Yang, Q.: A survey on transfer learning. IEEE Trans. Knowl. Data Eng. **22**(10), 1345–1359 (2010)
57. van Putten, M.J.A.M., Kind, T., Visser, F., Lagerburg, V.: Detecting temporal lobe seizures from scalp EEG recordings: a comparison of various features. Clin. Neurophysiol. **116**(10), 2480–2489 (2005)
58. Rodriguez-Lujan, I., Huerta, R., Elkan, C., Cruz, C.S.: Quadratic programming feature selection. J. Mach. Learn. Res. **11**, 1491–1516 (2010)

59. Rogowski, Z., Gath, I., Bental, E.: On the prediction of epileptic seizures. Biol. Cybern. **42**, 9–15 (1981)
60. Salant, Y., Gath, I., Henriksen, O.: Prediction of epileptic seizures from two-channel EEG. Med. Biol. Eng. Comput. **36**, 549–556 (1998)
61. Schindler, K.A., Bialonski, S., Horstmann, M.T., Elger, C.E., Lehnertz, K.: Evolving functional network properties and synchronizability during human epileptic seizures. CHAOS: Interdisc. J. Nonlinear Sci. **18**(3), 033119 (2008)
62. Siegel, A., Grady, C.L., Mirsky, A.F.: Prediction of spike-wave bursts in absence epilepsy by EEG power-spectrum signals. Epilepsia **116**, 2266–2301 (1982)
63. Smith, S.J.M.: EEG in the diagnosis, classification, and management of patients with epilepsy. J. Neurol. Neurosurg. Psychiatry **76**, ii2–ii7 (2005)
64. Srinivasan, V., Eswaran, C., Sriraam, N.: Artificial neural network based epileptic detection using time-domain and frequency-domain features. J. Med. Syst. **29**(6), 647–660 (2005)
65. Stam, C.J., Nolte, G., Daffertshofer, A.: Phase lag index: assessment of functional connectivity from multi channel EEG and MEG with diminished bias from common sources. Hum. Brain Mapp. **28**, 1178–1193 (2007)
66. Stam, C., van Straaten, E.: The organization of physiological brain networks. Clin. Neurophysiol. **123**, 1067–1087 (2012)
67. Strogatz, S.H.: Exploring complex networks. Nature **410**(6825), 268–276 (2001). http://dx.doi.org/10.1038/35065725
68. Subasi, A., Alkan, A., Koklukaya, E., Kiymik, M.K.: Wavelet neural network classification of EEG signals by using ar models with mle processing. Neural Netw. **18**(7), 985–997 (2005)
69. Tzallas, A.T., Tsipouras, M.G., Fotiadis, D.I.: The use of time-frequency distributions for epileptic seizure detection in EEG recordings. In: 29th Annual International Conference of the IEEE Engineering in Medicine and Biology Society, pp. 1265–1268 (2007)
70. Tzallas, A.T., Tsipouras, M.G., Fotiadis, D.I.: Epileptic seizure detection in EEGs using time-frequency analysis. IEEE Trans. Inf. Technol. Biomed. **13**(5), 703–710 (2009)
71. Viglione, S.S., Walsh, G.O.: Epileptic seizure prediction. Electroencephalogr. Clin. Neurophysiol. **39**, 435–436 (1975)
72. Wang, C., Mahadevan, S.: Manifold alignment using procrustes analysis. In: Proceedings of the 25th International Conference on Machine Learning, ICML 2008, pp. 1120–1127. ACM, New York (2008). http://doi.acm.org/10.1145/1390156.1390297
73. Wu, H., Li, X., Guan, X.: Networking property during epileptic seizure with multichannel EEG recordings. In: Wang, J., Yi, Z., Żurada, J.M., Lu, B.-L., Yin, H. (eds.) ISNN 2006. LNCS, vol. 3973, pp. 573–578. Springer, Heidelberg (2006). http://dx.doi.org/10.1007/11760191_84

Classification and Regression

Local and Global Genetic Fuzzy Pattern Classifiers

Søren Atmakuri Davidsen[✉], E. Sreedevi, and M. Padmavathamma

Department of Computer Science, Sri Venkateswara University,
Tirupati, India
sorend@acm.org, sreedevi.mca15@gmail.com, prof.padma@yahoo.com

Abstract. Fuzzy pattern classifiers are a recent type of classifiers making use of fuzzy membership functions and fuzzy aggregation rules, providing a simple yet robust classifier model. The fuzzy pattern classifier is parametric giving the choice of fuzzy membership function and fuzzy aggregation operator. Several methods for estimation of appropriate fuzzy membership functions and fuzzy aggregation operators have been suggested, but considering only fuzzy membership functions with symmetric shapes found by heuristically selecting a "middle" point from the learning examples. Here, an approach for learning the fuzzy membership functions and the fuzzy aggregation operator from data is proposed, using a genetic algorithm for search. The method is experimentally evaluated on a sample of several public datasets, and performance is found to be significantly better than existing fuzzy pattern classifier methods. This is despite the simplicity of the fuzzy pattern classifier model, which makes it interesting.

Keywords: Approximate reasoning · Pattern recognition · Fuzzy classifier · Genetic algorithm

1 Introduction

Ideas from uncertain reasoning using fuzzy logic have found their way into classification, where new methods are continuously being developed, as their ability to deal with the uncertainty, or "fuzziness" of training data is intuitive and useful [3]. In particular the fuzzy methods are useful for robustness, meaning how stable they are in a testing situation as compared to training; and interpretability, meaning how easy it is to understand the learned model; while maintaining good generalization, meaning how well is dealt with unseen examples. Many conventional classifiers are designed from the idea that a training example should only belong to one class, and hence perform best in cases where classes are well-defined and separable by clear boundaries. This may not be the case in many real world problems, where classes are hard to define, and feature-value regions may often be overlapping. Fuzzy classifiers, by their fuzzy nature, assume that training examples may have gradual membership to a class, and feature-value regions

P. Perner (Ed.): MLDM 2015, LNAI 9166, pp. 55–69, 2015.
DOI: 10.1007/978-3-319-21024-7_4

may be allowed to take on "nearby" values and still be considered relevant for the class at hand.

Fuzzy pattern classifiers have recently been introduced as a new method for approaching classification problems [7, 8]. A fuzzy pattern describes a prototype, the *look*, of a class. The fuzzy pattern consists of a fuzzy membership function for each feature, and a fuzzy aggregation operator used for inference. A fuzzy pattern classifier is an ensemble of such fuzzy patterns. The ensemble classifies new instances by selecting the fuzzy pattern with highest aggregated membership. The fuzzy pattern classifier is "fuzzy" not only because of using concepts from fuzzy logic, but, due to the way fuzzy membership functions capture the uncertain knowledge from the learning examples and describe the look of the class in a way that unseen examples may still be recognized though values are outside the boundaries of what was already seen [6].

Existing work on fuzzy pattern classifiers have proposed modifications along:

(1) which fuzzy membership function is most appropriate to describe a feature;
(2) which fuzzy aggregation operator is most appropriate for inferring the class.

However, the focus has been on providing a heuristic for constructing the fuzzy membership functions [7, 8], or simply trying out new fuzzy aggregation operators [8].

The contributions of this paper are two new fuzzy pattern classifier methods using genetic algorithms, which address the issue of constructing fuzzy membership functions and selecting fuzzy aggregation operators. The first proposed method uses a global search where all parameters of the fuzzy pattern classifier are searched simultaneously and accuracy is used to measure fitness, based on the idea that global search will provide appropriate separation of the classes. The second proposed method searches fuzzy membership functions for each fuzzy pattern, and an within-class similarity measure as fitness (local search). The local search method is based on the idea that learning shapes of fuzzy membership functions (i.e. covering the training examples properly) will give a more robust classifier.

Fuzzy logic has been applied to many types of fuzzy classifiers, and the idea of using genetic algorithms to search for parameters is not new either. In the area of genetic fuzzy systems, the use of genetic algorithms in fuzzy rule based classifiers has been extensively researched. In fuzzy rule based classifiers, the model is known as the knowledge base and consists of a data base and a rule base. The data base defines how fuzzification is to be done, and the rules define the inference. In genetic fuzzy systems, genetic algorithms have been used in two contexts: (1) "tuning" the knowledge base, meaning, adjusting the membership functions used for fuzzification and adjusting weights for each of the inference rules; (2) learning the knowledge base, meaning, searching for membership functions useful for fuzzification and searching for inference rules. For a recent survey and overview, see [2] and [9].

In fuzzy pattern classifiers however, to best knowledge of the authors, this is the first example of using genetic algorithms to search for fuzzy membership functions and fuzzy aggregation operators. The rule base in a fuzzy rule based classifier can contain several rules for each class, depending on the training examples and how they cover the domain. Likewise, each rule in a fuzzy rule based classifier can consider only a subset of the features. A fuzzy pattern classifier on the other hand is analogous to a fuzzy rule based classifier, which learns only a single rule (pattern) for each class, and the single rule considers all the features of the training examples. Thus, selection of membership functions to represent the rule and aggregation operator is important. In this paper the focus is selection of membership functions, through use of genetic algorithms.

The rest of the paper is organized as follows: Sect. 2 introduces the classification setting, fuzzy pattern classifiers and genetic algorithms. In Sect. 3 the fuzzy pattern classifiers with searched membership functions are proposed, while in Sect. 4 experimental analysis of the proposed classifiers are shown for a sample of public datasets. Finally, concluding remarks and further work is given in Sect. 5.

2 Preliminaries

2.1 Classification Setting

The supervised classification problem can be defined as: Given training examples $\mathcal{D}$ with n examples, where each example contains m features $x_i = (a_{i,1}, a_{i,2}, ..., a_{i,m})$; and each example belongs to a class $y_i \in \mathcal{Y}$:

$$\mathcal{D} = \{(x_1, y_1), (x_2, y_2), ..., (x_n, y_n)\} \subset (\mathcal{X} \times \mathcal{Y})^n \tag{1}$$

select a hypothesis h^* from the classifier's hypothesis space $\mathcal{H}$ which minimizes the loss function l on the training examples:

$$h^* \in \arg\min{}_{h \in \mathcal{H}} \int_{(x,y) \in \mathcal{D}} l(h(x), y) \tag{2}$$

The loss function must be selected, and typically a measure such as accuracy is used (note, that the negation of accuracy should be used to make it measure loss). Accuracy can be calculated as follows for a binary classification problem:

$$\text{accuracy} = (TP + TN)/(TP + FP + TN + FN) \tag{3}$$

where TP and TN are true positive and true negative (correctly classified) examples, and FP and FN are false positives and false negatives (incorrectly classified) examples.

2.2 Fuzzy Pattern Classifiers

Fuzzy pattern classifiers (FPCs) are popular because they are simple to construct, their model is easy to understand, they deals well with uncertainty of unobserved examples, and the run-time evaluation of the model is very little. The fuzzy pattern classifier is an ensemble of fuzzy patterns F, one for each class, and a fuzzy aggregation operator λ used for inference. Thus, the model is denoted (λ, F).

Defuzzification is done by selecting the class belonging to the pattern with the highest inferred value. In literature, the fuzzy aggregation operator must be the same for all fuzzy patterns, and this restriction is upheld here.

A fuzzy pattern is a model of a class, and consists of a set of fuzzy membership functions (μ-functions), one for each feature of the example:

$$F_i = [\mu_{i,1}, \mu_{i,2}, ..., \mu_{i,m}] \tag{4}$$

Each μ-function is a prototype of what values belong to the specific feature of the given class. The μ-function is automatically *learned* from examples and not assigned by experts. In literature, the type of fuzzy membership function is decided by the method, and the shape is found by heuristic approaches, such as a π-function using min, max and arithmetic mean of the feature values [7]; or a gaussian function calculated from mean, width and steepness [8].

An example using the Fuzzy Reduction Rule (FRR) classifier [7]: Given the π-function's definition:

$$\pi(x) = \begin{cases} 2^{m-1}\left((x-a)/(r-a)\right)^m & a < x \leq p \\ 1 - \left(2^{m-1}\left((r-x)/(r-a)\right)^m\right) & p < x \leq r \\ 1 - \left(2^{m-1}\left((x-r)/(b-r)\right)^m\right) & r < x \leq q \\ 2^{m-1}\left((b-x)/(b-r)\right)^m & q < x \leq b \\ 0 & \text{Otherwise} \end{cases} \tag{5}$$

where $m > 0$, $a < r < b$, $p = \frac{1}{2}(r+a)$, and $q = \frac{1}{2}(r+b)$.

For the jth feature of examples belonging to class k, determine π-function parameters r, p, q as follows:

$$\begin{aligned} r_{k,j} &= \overline{a_{i,j}}, \quad y_i = k \\ \delta_{j,k} &= \max a_{i,j} - \min a_{i,j}, \quad y_i = k \\ p_{k,j} &= r_{k,j} - \delta_{k,j} \\ q_{k,j} &= r_{k,j} + \delta_{k,j} \end{aligned} \tag{6}$$

where $r_{k,j}$ is the mean value of the ith feature for values belonging to class k, specifying the center of the π function. Likewise, $p_{k,j}$ and $q_{k,j}$ are left and right points estimated from the min and max. Figure 1 exemplifies this type of learned function.

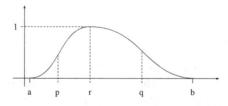

Fig. 1. Example of learned π-function.

Table 1. Common fuzzy aggregation operators

Name	Symbol	Definition
Product	$\otimes_{\text{PROD}}$	$\prod_i^n a_i$
Minimum	$\otimes_{\text{min}}$	$\min(a_1, a_2, ..., a_n)$
Arithmetic mean	h_{MEAN}	$\frac{1}{n} \sum_i^n a_i$
OWA	h_{OWA}	$\sum_i^n v_i \cdot a_{(i)}$ where $a_{(\cdot)}$ is an ordered permutation of a and v a sequence of weights.
AIWA	h_{AIWA}	$1 - \left(\sum_i^n (1 - a_i)^{(1/\rho)} \right)^\rho$, for $p > 0.5$, $\rho = (1/p) - 1$
Maximum	$\oplus_{\text{max}}$	$\max(a_1, a_2, ..., a_n)$
Algebraic sum	$\oplus_{\text{alg.sum}}$	$1 - \prod_i^n (1 - a_i)$

The inference step is done by a fuzzy aggregation operator $\lambda : [0, 1]^n \to [0, 1]$. A list of common fuzzy aggregation operators are shown in Table 1. An operator such as the product is "strict" in the sense that it requires contribution from all aggregating values, somewhat like a boolean AND. The opposite is true for e.g. the algebraic sum, which is satisfied by contribution from few of the aggregating values (like boolean OR). Reference [5] contains a complete introduction to fuzzy aggregation operators.

Once the fuzzy patterns have been learned and the fuzzy aggregation operator has been selected, classification is done by matching the unseen example's feature-values $u = (u_1, u_2, ..., u_m)$ to each of the fuzzy patterns and inferring using the fuzzy aggregation operator λ:

$$Q_k = \lambda(\mu_{k,1}(u_1), \mu_{k,2}(u_2), ..., \mu_{k,m}(u_m)), \quad k \in \mathcal{Y} \tag{7}$$

after which defuzzication is simply selecting the pattern with the highest inferred value:

$$\hat{y} = \arg \max_{k \in Y}(Q_1, Q_2, ..., Q_k) \tag{8}$$

2.3 Genetic Algorithm Optimization

A genetic algorithm (GA) is a type of search and optimization method, which mimic how living creatures evolve in nature the concepts of: Mixing the gene material of parents, random mutation, and selection of the *fittest* over generations.

The basic component in a GA is the chromosome Φ, which represents a candidate solution. Each chromosome consists of z genes $\Phi = (\phi_1, \phi_2, ..., \phi_z)$, which are roughly a representation of a parameter in the candidate solution. The chromosomes belong to a population $P_g = \{\Phi_1, ..., \Phi_Z\}$ which are the current candidate solutions. Through genetic operators crossover and mutation, new candidate solutions are generated, and the most *fit* are selected to become parents for the next generation of candidate solutions. The algorithm outline for a GA is given in Algorithm 1.

Gene values are typically boolean $\{0, 1\}$ or real-valued. The mutation operator changes the value of each gene with a specified probability (denoted the mutation probability). For boolean values, mutation simply flip the value, while real-valued mutation adds a value drawn using a uniform random distribution within an interval, for example $[-0.5, 0.5]$. The crossover operator produce a new chromosome provided two parent chromosomes. For selecting the mix of genes from each parent, one or more crossover-points are used: At each cross-over point, the child's genes will switch origin from one parent to the other. The cross-over points can be selected randomly, or, from a specified set of crossover-points (this is also called uniform crossover).

In order to use a GA, the following steps should be considered: (1) Decide coding of chromosome into candidate solution; (2) Decide fitness function to evaluate fitness of candidate solution; (3) Define which specific genetic operators to use; (4) Define a stopping criteria.

Algorithm 1. Genetic algorithm pseudocode.

1: $g \leftarrow 0$
2: $P_g \leftarrow$ Initialize population
3: $f_g \leftarrow$ Fitness P_g
4: **repeat**
5:     $g \leftarrow g + 1$
6:     $P_g \leftarrow$ Select fittest from P_{g-1}
7:     Crossover P_g
8:     Mutate P_g
9:     $f_g \leftarrow$ Fitness P_g
10: **until** $g = G$ or $\epsilon \geq f_g - f_{g-1}$

Coding means defining a function for creating the candidate solution S from the chromosome, $f : \Phi \rightarrow S$. The fitness function J, as already mentioned, evaluates how good a candidate solution is for the given problem typically measuring by measuring some kind of loss, $J : S \rightarrow [0, 1]$ where 0 means no loss (best solution) and 1 complete loss (worst). Stopping criteria is for example a fixed number of iterations, or reaching a specific minimum improvement threshold ϵ. For a complete introduction to genetic algoritms please see [10]. For simplicity, we denote use of an algorithm G : fitness $\rightarrow \Phi$, meaning, the genetic algorithm is a function G, which provided a fitness function, yields a *fittest* chromosome Φ.

3   Genetic Algorithm for Fuzzy Pattern Classifiers

As described in Sect. 2.2, the FPC model is described by (λ, F) where λ is the aggregation operator and $F = (F_1, F_2, ..., F_C)$ are the fuzzy patterns, one for each class; and $F_i = (\mu_{i,1}, \mu_{i,2}, ..., \mu_{i,m})$ contains fuzzy membership functions for each feature of the ith class.

Depending on which μ-function is used, a different number of parameters are used to decide it's shape, for example the π-function takes three parameters

p, r, q. Likewise, the fuzzy aggregation operator λ can take parameters, depending on what operator is chosen. Together, the choice of fuzzy aggregation operator, the choice of fuzzy membership function; and their combined parameters make the candidate solution for a fuzzy pattern classifier. These choices and parameters must all be present in the chromosome coding.

In this work, two approaches for chromosome coding and search are proposed:

1. FPC-GA: Search and evaluate fitness of the fuzzy pattern classifier (all classes together, global search);
2. FPC-LGA: Search and evaluate fitness for one fuzzy pattern at a time (local search).

A *priori* the FPC-GA method may perform better for a given problem, but, the learned μ-functions may not be as interpretable as optimization is done on separation of the classes. FPC-LGA is expected to learn μ-functions similar to the heuristic approaches but with better coverage of values in training examples.

For crossover, a uniform crossover method was selected, where crossover-points were selected such that a group of genes belonging to a μ-function will not be separated. In practice this means that μ-functions will be mixed from each parent, and mutation will modify the parameters of the μ-function. Figure 2 gives an overview of the general idea behind this coding.

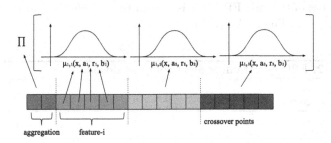

Fig. 2. Chromosome representation of candidate solution in FPC-GA

3.1   FPC-GA Classifier

In the FPC-GA classifier, the GA considers all fuzzy patterns in the fuzzy pattern classifier simultaneously. This means the fitness of the selected chromosome/solution considers the separation of the classes, and not only covering the feature-values of the training examples most effectively.

For FPC-GA, a chromosome is of length $5mC + 2$, where 5 is the number of genes used to represent a membership function and its parameters, m is the number of features and C the number of classes. The last 2 are used for coding of the aggregation operator and its parameters.

Given a set of fuzzy membership function definitions $M = \{\pi, \Delta, ...\}$ and a set of fuzzy aggregation operators $\Lambda = \{\oplus_{\text{prod}}, h_{\text{mean}}, ...\}$, the chromosome coding for FPC-GA is as follows:

$$\Phi = \begin{cases} \lambda = \Lambda_{\phi_1'} \\ F_{1,1} = M_{\phi_2'}(\phi_3, \phi_4, \phi_5) \\ F_{1,2} = M_{\phi_6'}(\phi_7, \phi_8, \phi_9) \\ \quad ... \\ F_{1,m} = M_{\phi_{1+m*3-3}'}(\phi_{1+m*3-2}, \phi_{1+m*3-1}, \phi_{1+m*3}) \\ \quad ... \\ F_{C,m} = M_{\phi_{1+Cm*3-3}'}(\phi_{1+Cm*3-2}, \phi_{1+Cm*3-1}, \phi_{1+Cm*3}) \end{cases} \tag{9}$$

where ϕ_1' is the value of ϕ_1 normalized to the number of elements in Λ corresponding to one of the fuzzy aggregation operators; and likewise for $M_{\phi_i'}$ corresponding to one of the fuzzy membership function definitions.

The fitness function for FPC-GA is the negation of accuracy, calculated from the training examples:

$$J = 1 - \text{accuracy FPC}(X, Y) \tag{10}$$

The process of training FPC-GA is left up to the genetic algorithm, as seen in the FPC-GA outline in Algorithm 2.

Algorithm 2. FPC-GA training algorithm pseudocode.

1: $\Phi \leftarrow G(J)$
2: $(\lambda, F) \leftarrow$ Decoded from Φ

3.2   FPC-LGA Classifier

For the second proposed classifier, FPC-LGA, the GA searches for fuzzy membership functions one fuzzy pattern at a time, i.e. the fitness is how well the fuzzy membership functions cover a specific class, not the separation of the classes.

The method of FPC-LGA coding is similar to the FPC-GA coding, except a chromosome represents a single fuzzy pattern, and the fuzzy aggregation operator is fixed:

$$\Phi = \begin{cases} F_1 = M_{\phi_1'}(\phi_{i_1}, \phi_{i_2}, \phi_{i_3}) \\ F_2 = M_{\phi_2'}(\phi_{i_4}, \phi_{i_5}, \phi_{i_6}) \\ \quad ... \\ F_m = M_{\phi_m'}(\phi_{i_{m*3-3}}, \phi_{i_{m*3-2}}, \phi_{i_{m*3-1}}) \end{cases} \tag{11}$$

Fitness is calculated by comparing predictions by the fuzzy pattern with an ideal target vector using standard root mean square error (RMSE). The predictions by the fuzzy pattern k as previously defined in Eq. 7 and ideal target vector $\tilde{Q}$ for fuzzy pattern k is:

$$\tilde{Q}_k = \begin{cases} 1 & \text{if } y_i = k \\ 0 & \text{otherwise} \end{cases}, \quad \text{for } i = 1, .., n \tag{12}$$

This gives the fitness function:

$$J_L = \text{RMSE}(\tilde{Q}_k, Q_k) \tag{13}$$

As FPC-LGA searches one fuzzy pattern (class) at a time, the GA need to be applied several times to learn all fuzzy patterns of the ensemble, as seen in the outline in Algorithm 3.

Algorithm 3. FPC-LGA training algorithm pseudocode.

1: $\lambda \leftarrow$ Selected aggregation operator.
2: **for** $k \in Y$ **do**
3:      $X_k \leftarrow$ Subset of X where class is k
4:      $\Phi \leftarrow G(J_L)$
5:      $F_k \leftarrow$ Decoded from Φ
6: **end for**

---

4   Experimental Analysis

The proposed classifier was implemented in the library FyLearn, which is a Python language library for fuzzy machine learning algorithms, API-compatible with SciKit-Learn. It is available for download[1].

The classifier was evaluated experimentally to get an idea of performance and pitfalls. Most interesting are the comparison with two other FPCs using heuristic membership function construction approaches, and the comparison between local and global construction methods.

4.1   Datasets

Twelve available datasets were selected, available through the UCI dataset repository[2], except for *Telugu Vowels*, which is from the Indian Statistical Institute website[3]. All datasets had examples with missing values removed and values feature-wise normalized to the unit interval $[0, 1]$. The 12 datasets were selected based on the absence of nominal data, hence, only numeric value features (real and/or integer), which are supported by the FPC. The datasets used are shown in Table 2.

---

[1] Available for download through GitHub: https://github.com/sorend/fylearn.

[2] http://archive.ics.uci.edu/ml/datasets.html.

[3] http://www.isical.ac.in/~sushmita/patterns/.

Table 2. Datasets and their basic properties.

#	Dataset	#-examples	#-features	#-classes	Majority class
1	Iris	150	4	3	0.33
2	Pima Indians Diabetes	768	8	2	0.65
3	Wisconsin Breast Cancer	683	9	2	0.65
4	Bupa Liver Disorders	345	6	2	0.58
5	Wine	178	12	3	0.40
6	Telugu Vowels	871	3	6	0.24
7	Haberman	306	3	2	0.74
8	Indian Liver Patient	579	10	2	0.72
9	Vertebral Column	310	6	2	0.68
10	Glass	214	10	6	0.36
11	Ionosphere	351	34	2	0.64
12	Balance Scale	625	4	3	0.46

4.2 Method

For comparison, the FyLearn implementation of the two FPC classifiers "FRR"[7] and "MFPC"[8] were included. The classifiers were configured with the parameters suggested by their authors. For FRR π-membership functions and two different fuzzy aggregation rules were selected; product and arithmetic mean, FRR_{PROD} and FRR_{MEAN} respectively. For MFPC two aggregation operators were selected; OWA and AIWA, $MFPC_{OWA}$ and $MFPC_{AIWA}$ respectively. MFPC further allows for parameter tuning, which was done by allowing a grid-search for the parameter values suggested in [8].

For the proposed FPC-GA and FPC-LGA configurations were created to match FRR and MFPC. Two FPC-GA variants similar to FRR's parameters were included: Both restricted to using a π-function for membership function and for fuzzy aggregation rule, one with the product and another with the arithmetic mean, $FPC\text{-}GA_{PROD}$ and $FPC\text{-}GA_{MEAN}$ respectively. For comparison with MFPC, two variants were configured with the π-membership function the fuzzy aggregation operator replaced with OWA and AIWA respectively. In FPC-GA the genetic algorithm was allowed to set parameter values for OWA/AIWA, while for FPC-LGA the parameters were fixed to those chosen by the corresponding MFPC's grid-search. For FPC-(L)GAs genetic algorithm, 100 chromosomes and 100 iterations was used, together with a mutation probability of 0.3 and tournament selection.

4.3 Visual Comparison of GA μ-functions

Figure 3 show an example of membership functions from FRR (heuristic approach), FPC-GA and FPC-LGA classifiers. Similar visualizations were created for the other datasets, with similar results, but, omitted for clarity. The GA searched fuzzy membership functions are not easily identifiable or interpretable as the ones produced by the heuristic approach, however regions appear similar.

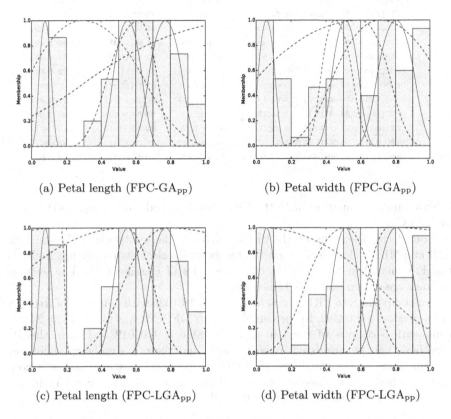

(a) Petal length (FPC-GA$_{pp}$)

(b) Petal width (FPC-GA$_{pp}$)

(c) Petal length (FPC-LGA$_{pp}$)

(d) Petal width (FPC-LGA$_{pp}$)

Fig. 3. Example of heuristic (solid) and learned (dashed) μ-functions of two features from the Iris dataset for FPC-GA$_{PROD}$ and FPC-LGA$_{PROD}$ classifiers. Each class is a separate color. (Color figure online)

According to the hypothesis that FPC-LGA produces membership functions which are more easily understood, the visualization of the membership functions supports this claim: The membership functions produced by FPC-LGA coincides better with the data of the class, than FPC-GA.

4.4 Accuracy

Reported results are mean accuracy and standard deviation from conducting the experiment using a strategy of 10-fold stratified cross-validation and repeating the classification 100 times (100 × 10 CV). The results are reported in two sets:

Table 3. Mean accuracy and standard deviation. Best classifier is marked with bold.

#	FRR_{prod}	FRR_{mean}	$FPC\text{-}GA_{PROD}$	$FPC\text{-}GA_{MEAN}$	$FPC\text{-}LGA_{PROD}$	$FPC\text{-}LGA_{MEAN}$
Avg rank	10.87500	7.58333	8.16667	7.25000	5.79167	4.33333
1	53.56(9.56)	77.93(5.98)	83.85(5.25)	84.22(7.16)	**93.19(4.65)**	92.22(2.32)
2	67.71(1.45)	70.96(1.86)	70.72(2.73)	70.89(2.34)	**72.77(2.24)**	70.24(1.41)
3	74.35(1.87)	94.06(1.58)	90.34(2.04)	92.13(2.68)	88.56(20.78)	**95.69(1.22)**
4	44.41(4.68)	56.72(2.34)	54.78(3.67)	55.14(2.20)	55.98(3.64)	**59.80(3.99)**
5	34.58(1.41)	69.92(7.71)	65.74(9.71)	69.37(8.45)	73.13(11.78)	**85.39(4.68)**
6	29.23(8.83)	49.29(3.45)	62.12(7.50)	60.93(6.00)	58.49(11.01)	**62.62(6.43)**
7	**71.78(2.58)**	68.25(6.81)	61.63(16.13)	49.24(11.15)	61.41(15.97)	65.32(6.36)
8	70.91(1.61)	70.56(1.93)	66.86(4.43)	66.25(2.51)	66.55(4.63)	**71.31(0.83)**
9	50.07(13.69)	64.91(10.13)	61.36(15.64)	67.89(9.22)	**70.54(9.40)**	64.80(11.94)
10	32.71(0.26)	41.18(1.94)	59.25(12.16)	56.65(10.95)	**64.50(10.10)**	56.00(7.47)
11	35.90(0.06)	56.56(10.87)	45.52(17.42)	**73.44(5.23)**	35.90(0.06)	70.56(4.94)
12	10.03(2.36)	53.30(4.71)	41.59(10.19)	50.43(13.21)	50.05(4.35)	**70.93(7.62)**

Table 3 reports comparison with the FRR classifier, and Table 4 reports the same for MFPC.

Considering the accuracy, the FPC-(L)GA classifiers outperform the heuristic FRR and MFPC classifiers on most datasets with a clear higher average ranking. The Friedman test was used to compare pairs of classifiers [4]. The average rankings are shown above in Tables 3 and 4.

For comparison to FRR it was found $\chi_F^2 = 18.1026$ with 5 degrees of freedom and a p-value $= 0.00282$. In the comparison to MFPC it was found $\chi_F^2 = 24.284$ again with 5 degrees of freedom and a p-value $= 0.0001915$. Hence, the H_0 hypothesis can be dismissed in both cases, and there is significant difference between some of the classifiers. Further, the Nemenyi post-hoc test was conducted to find which classifiers contained the significant difference. Significant difference at $p < 0.05$ was found for the following classifiers:

Table 4. Mean accuracy and standard deviation. Best classifier is marked with bold.

#	$MFPC_{AIWA}$	$MFPC_{OWA}$	$FPC\text{-}GA_{AIWA}$	$FPC\text{-}GA_{OWA}$	$FPC\text{-}LGA_{AIWA}$	$FPC\text{-}LGA_{OWA}$
Avg rank	10.04167	8.95833	7.25000	5.75000	4.08333	3.33333
1	88.00(3.67)	92.96(2.35)	76.00(8.38)	84.74(8.14)	**93.33(2.32)**	92.30(3.72)
2	63.59(6.48)	67.90(4.84)	71.59(1.56)	71.67(2.37)	**72.15(1.03)**	71.48(1.21)
3	90.66(5.76)	91.20(6.42)	92.79(2.12)	93.67(1.78)	95.44(1.49)	**95.72(1.01)**
4	52.01(3.23)	54.72(3.93)	57.33(4.91)	56.75(2.82)	58.84(2.62)	**60.45(2.87)**
5	76.04(9.28)	77.98(9.44)	70.05(8.41)	71.28(7.96)	84.46(4.66)	**87.27(4.14)**
6	53.55(6.57)	57.07(6.87)	58.96(6.81)	57.64(7.61)	52.62(7.07)	**64.75(1.98)**
7	57.80(9.91)	54.50(12.98)	56.17(12.16)	54.55(15.41)	60.95(10.25)	**66.07(11.60)**
8	58.02(6.16)	60.83(5.76)	65.15(2.85)	65.07(3.37)	70.98(1.22)	**71.12(0.98)**
9	66.49(13.94)	**69.50(14.10)**	66.16(11.66)	69.46(6.16)	66.24(11.50)	67.46(11.75)
10	31.96(8.06)	31.80(8.03)	52.47(7.74)	**57.89(8.13)**	57.76(8.80)	55.41(7.28)
11	35.90(0.06)	35.90(0.06)	72.53(6.79)	73.82(3.72)	**74.01(3.26)**	69.83(2.76)
12	37.58(6.82)	40.94(8.15)	50.62(9.62)	54.53(7.45)	58.91(6.02)	**68.68(7.58)**

Table 5. Nemenyi statistic p-values and Friedman average ranks. Significant differences ($p \leq 0.05$) marked with bold.

	FPC-GA$_{PROD}$	FPC-GA$_{MEAN}$	FPC-GA$_{AIWA}$	FPC-GA$_{OWA}$	FPC-LGA$_{PROD}$	FPC-LGA$_{MEAN}$	FPC-LGA$_{AIWA}$	Avg. rank
FPC-GA$_{PROD}$	-	-	-	-	-	-	-	6.00000
FPC-GA$_{MEAN}$	0.99990	-	-	-	-	-	-	5.58333
FPC-GA$_{AIWA}$	0.99998	1.00000	-	-	-	-	-	5.66667
FPC-GA$_{OWA}$	0.84991	0.97458	0.96042	-	-	-	-	4.58333
FPC-LGA$_{PROD}$	0.76055	0.94135	0.91676	1.00000	-	-	-	4.41667
FPC-LGA$_{MEAN}$	0.19522	0.42572	0.37213	0.96042	0.98457	-	-	3.50000
FPC-LGA$_{AIWA}$	0.16187	0.37213	0.32179	0.94135	0.97458	1.00000	-	3.41667
FPC-LGA$_{OWA}$	**0.03318**	0.10798	0.08690	0.65411	0.76055	0.99781	0.99907	2.83333

- FRR$_{prod}$ vs. FPC-LGA$_{MEAN}$ – 0.00096
- MFPC$_{AIWA}$ vs. FPC-LGA$_{AIWA}$ – 0.00525
- MFPC$_{AIWA}$ vs. FPC-LGA$_{OWA}$ – 0.00096
- MFPC$_{OWA}$ vs. FPC-LGA$_{OWA}$ – 0.02305

However, at borderline significance, $p < 0.1$, several other classifiers were different as well, favoring FPC-GA and FPC-LGA over FRR and MFPC. Full details have been omitted for space constraints.

4.5   Difference Between FPC-GA and FPC-LGA

Comparing FPC-GA and FPC-LGA was also done by using Friedman and Nemenyi statistics. The results are shown in Table 5. At $p < 0.05$ the only difference is with different fuzzy aggregation operators, namely FPC-GA$_{PROD}$ and FPC-LGA$_{MEAN}$. However, considering borderline significance, also with different averaging there is difference between FPC-GA and FPC-LGA. However, in ranking, FPC-LGA is a clear winner over FPC-GA. Again, the differences between the fuzzy aggregation operators suggest that it is an important parameter to optimize by the GA as well.

4.6   Overfitting/underfitting

A potential problem with learning membership functions over the heuristic approach is that the classifier may adjust the membership functions too well to the training examples, and hence give poor accuracy on the unseen testing examples; this problem is known as overfitting. A common way to determine if a classifier is overfitting, is to consider the difference in accuracy between the training data and testing data. A larger difference means that the model is overfitting, while a small difference means no overfitting [1]. Table 6 shows the mean difference between test and training.

The FRR and MFPC classifiers have the lowest mean difference between the test and training data, indicating least overfitting, however their classification accuracy is consistently ranked below the FPC-(L)GA classifiers. The primary parameter for deciding how accurate to learn the μ-functions, is the number of

Table 6. Mean difference between test and training accuracy (all datasets)

Classifier	Training	Testing	Difference
FRR$_{prod}$	58.81(17.30)	47.94(19.53)	10.87(10.81)
FRR$_{mean}$	73.28(8.99)	64.47(13.54)	8.81(7.64)
FPC-GA$_{PROD}$	89.50(12.79)	63.65(13.26)	25.85(13.72)
FPC-GA$_{MEAN}$	94.88(5.05)	66.38(12.46)	28.50(11.50)
FPC-LGA$_{PROD}$	85.06(17.03)	65.92(15.02)	19.13(14.86)
FPC-LGA$_{MEAN}$	88.05(8.23)	72.07(12.06)	15.98(10.06)
MFPC$_{AIWA}$	68.61(19.22)	59.30(18.29)	9.31(9.13)
MFPC$_{OWA}$	73.75(20.65)	61.27(18.98)	12.48(10.20)
FPC-GA$_{AIWA}$	95.07(4.81)	65.82(11.36)	29.25(10.71)
FPC-GA$_{OWA}$	94.80(5.06)	67.59(11.92)	27.21(10.50)
FPC-LGA$_{AIWA}$	90.77(7.14)	70.48(13.59)	20.30(12.49)
FPC-LGA$_{OWA}$	89.09(8.04)	72.54(12.03)	16.55(10.51)

iterations in the GA. Intuition is that a large number of iterations may learn μ-functions too well, resulting in overfitting. To test this, an experiment with the FPC-GA and FPC-LGA classifiers was conducted, where number of iterations for the GA was $\{10, 50, 100, 500, 1000\}$. Figure 4 show the results for the classifiers FPC-GA$_{PROD}$ and FPC-LGA$_{PROD}$ on all datasets. Again the experiment was repeated 100 times and the mean accuracy used. Except for a single dataset (Ionosphere) the classifiers reach a stable accuracy around 50 to 100 iterations, and performance does not seem to decrease over number of iterations, indicating the number of iterations can be chosen freely. The Ionosphere dataset has the most features, which reflects in the number of genes required. This has effect on the number of iterations needed for the search to complete.

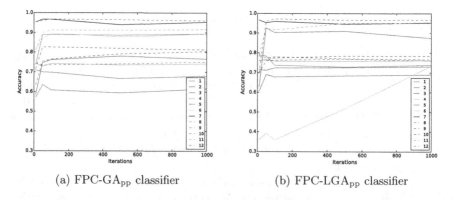

(a) FPC-GA$_{pp}$ classifier   (b) FPC-LGA$_{pp}$ classifier

Fig. 4. Accuracy for all datasets for two classifiers using a variable number of iterations.

Number of iterations is hard to control in a real setting. Other techniques such as incorporating cross-validation in the fitness function may be used to overcome overfitting. However, this has to be tested further.

5 Conclusions

In this paper new a new fuzzy pattern classifier with genetic learning of fuzzy membership functions and fuzzy aggregation operator was proposed, based on the existing recent work in fuzzy pattern classifiers.

The experimental results had both the Friedman and Nemenyi tests conducted to evaluate the significance of the proposed classifiers, and results were encouraging – the experiment present a potential usefulness of GA learned FPCs and may give direction for what next steps could be. The experiment addressed the overfitting problem where GA learned classifier slightly overfits the original classifier, however, with a better accuracy in all cases.

As indicated by the average ranking, the choice of fuzzy aggregation operator is important, but in the experiments does not allow FPC-(L)GA to search it, as given in our initial limitations. As the membership functions are only one part of tuning a FPC, this area is a future direction for research. Another interest is the overfitting problem, where mentioned alternatives to reducing number of iterations is to be done.

References

1. Gonçalves, I., Silva, S.: Experiments on controlling overfitting in genetic programming. In: Proceedings of the 15th Portuguese Conference on Artificial Intelligence, pp. 152–166 (2011)
2. Herrera, F.: Genetic fuzzy systems: taxonomy, current research trends and prospects. Evol. Intell. **1**(1), 27–46 (2008)
3. Hüllemeier, E.: Fuzzy sets in machine learning and data mining. Appl. Soft Comput. **11**(2), 1493–1505 (2011)
4. Japkowicz, N., Shah, M.: Evaluating Learning Algorithms: A Classification Perspective. Cambridge University Press, Cambridge (2011)
5. Klir, G.J., Yuan, B.: Fuzzy Sets and Fuzzy Logic: Theory and Applications. Prentice Hall, New York (1995)
6. Kuncheva, L.I.: Fuzzy Classifier Design. Stud Fuzz, vol. 49. Springer, Heidelberg (2000)
7. Meher, S.K.: A new fuzzy supervised classification method based on aggregation operator. In: Proceedings of the 3rd International IEEE Conference on Signal-Image Technologies and Internet-Based Systems, pp. 876–882 (2008)
8. Mönks, U., Larsen, H.L., Lohweg, V.: Aggregation operator based fuzzy pattern classifier design. In: Proceedings of the Machine Learning in Real-Time Applications (2009)
9. Rutkowski, L.: Flexible Neuro-Fuzzy Systems: Structures, Learning and Performance Evaluation. Kluwer, Boston (2004)
10. Whitley, D.: A genetic algorithm tutorial. Stat. Comput. **4**, 65–85 (1994)

IKLTSA: An Incremental Kernel LTSA Method

Chao Tan[1], Jihong Guan[2]([✉]), and Shuigeng Zhou[3]

[1] School of Computer Science and Technology, Nanjing Normal University,
Nanjing, China
chtan@njnu.edu.cn
[2] Department of Computer Science and Technology, Tongji University,
Shanghai, China
jhguan@tongji.edu.cn
[3] School of Computer Science, Fudan University, Shanghai, China
sgzhou@fudan.edu.cn

Abstract. Since 2000, manifold learning methods have been extensively studied, and demonstrated excellent performance in dimensionality reduction in some application scenarios. However, they still have
some drawbacks in approximating real nonlinear relationships during the
dimensionality reduction process, thus are unable to retain the original
data's structure well. In this paper, we propose an incremental version of
the manifold learning algorithm LTSA based on kernel method, which is
called *IKLSTA*, the abbreviation of *Incremental Kernel LTSA*. IKLTSA
exploits the advantages of kernel method and can detect the explicit
mapping from the high-dimensional data points to their low-dimensional
embedding coordinates. It is also able to reflect the intrinsic structure
of the original high dimensional data more exactly and deal with new
data points incrementally. Extensive experiments on both synthetic and
real-world data sets validate the effectiveness of the proposed method.

Keywords: Manifold learning · Dimensionality reduction · Kernel
method · Explicit mapping

1 Introduction

Dimensionality reduction is one of important tasks in machine learning. Manifold learning aims at constructing the nonlinear low-dimensional manifold
from sampled data points embedded in high-dimensional space, is a class of
effective nonlinear dimension reduction algorithms. Typical manifold learning
algorithms include Isomap [16] and LLE [14] etc. These algorithms are easy to
implement and can obtain satisfactory mapping results in some application scenarios. However, they also have some limitations. When projecting complicate
high-dimensional data to low-dimensional space, they can keep only some characteristics of the original high-dimensional data. For example, the local neighborhood construction in Isomap is quite different from that of the original dataset,
and the global distance between data points cannot be maintained well by the
LLE method.

© Springer International Publishing Switzerland 2015
P. Perner (Ed.): MLDM 2015, LNAI 9166, pp. 70–83, 2015.
DOI: 10.1007/978-3-319-21024-7_5

Kernel methods are a kind of approaches that use kernel functions to project the original data as inner product to high dimensional eigen-space where subsequent learning and analysis are done. Kernel functions are used to approximate real data's non-linear relations. From either global or local viewpoint [17], the low dimensional mapping can reflect the intrinsic structure of the original high dimensional data more accurately after dimension reduction. It is significant in many fields such as pattern recognition and so on.

Research works in recent years indicate the close inherent relationship between manifold learning and dimension reduction algorithms based on kernel method [2,8]. Ham [4] studied the relationships between several manifold learning algorithms including Isomap, LLE and Laplacian Eigenmap (LE) [1] etc. and kernel method. These algorithms all can be seen as the problem of calculating eigenvectors of a certain matrix, which satisfies the kernel matrix's conditions and reveals the characteristics to be preserved of the dataset. So all these methods can be regarded as kernel methods that keep the geometric characteristics of the original dataset after dimension reduction under the kernel framework.

For the out-of-sample learning problems, the key point lies in that most manifold learning algorithms find low-dimensional coordinates by optimizing a certain cost function, while the mappings of data points in high dimensional space to their low dimensional coordinates lack explicit formulation. When new samples are added into the dataset, it is unable to get its corresponding low dimensional coordinates directly by the obtained mappings, so the re-running of the algorithm is required. If some specific mapping function can be obtained, we can get newly-added data points' low dimensional coordinates through the mapping function directly. Then the processing efficiency can be significantly improved for the new-coming data.

So if we bring the mapping function from high-dimensional input data to the corresponding low-dimensional output coordinates into manifold learning methods under the kernel framework, given the explicit mapping relationships of kernel methods, we can combine the advantages of kernel methods and the explicit mapping function. As a result, the learning complexity can be reduced, and new data points' low dimensional coordinates can be calculated more efficiently, which makes the algorithms suitable for incremental manifold learning. In addition, dimensionality reduction algorithms served as a kind of feature extraction methods have attracted wide attention in applications. If we can apply the new method into these scenarios, human cost can be cut off greatly.

The rest of the paper is organized as follows. We review related work in Sect. 2. We present the proposed method in detail in Sect. 3. In Sect. 4, experiments on synthetic and real datasets are carried out to evaluate the proposed method. Finally, we conclude the paper in Sect. 5.

2   Related Work

As Liu and Yan pointed out [10], linear manifold learning algorithms that can extract local information, such as locality preserving projection (LPP) [5], neighborhood preserving embedding (NPE) [6], orthogonal neighborhood preserving projection (ONPP) [7] have sprung up in recent years. Recently, Zheng et al. used a kind of nonlinear implicit representation as a nonlinear polynomial mapping, and applied it to LLE and thus proposed the neighborhood preserving polynomial embedding algorithm (NPPE) [20]. This algorithm can keep the nonlinear characteristics of high dimensional data. Although these algorithms can detect the local characteristics very well [11], they cannot deal with noise effectively. Considering that real-world datasets inevitably contain outliers and noise, it is challenging for these manifold learning algorithms to process real-world data effectively.

Manifold learning algorithms based on kernel method are effective to solve the out-of-sample learning problem. Considering that data inseparable in the linear space may be separable in a nonlinear space, people tend to study manifold learning methods based on kernel functions (or kernel methods). Here, the selection of kernel function is the key step. Kernel function describes the dataset's characteristics. By doing eigen-decomposition on the kernel matrix, we can get the original dataset's characteristics, which improves the learning ability of the out-of-sample problem. Furthermore, the compute cost can be decreased consequently by doing eigen-decomposition on the kernel matrix instead of in high dimensional eigen-space.

2.1   LLE Under the Kernel Framework

Several works combining kernel functions and manifold learning methods appear in recent years. Take kernel LLE [4] for example, the last step of LLE is to search the low dimensional embedding coordinates Y that can maintain the weight matrix W optimally. The cost function is as follows:

$$\Phi(Y) = \sum_i ||Y_i - \sum_j w_{ij}Y_j|| = ||Y(I - W)||^2 = Trace(Y(I - W)(I - W)^T Y^T)$$

$$= Trace(YMY^T), where M = (I - W)(I - W)^T. \tag{1}$$

We try to find the coordinates Y minimizing the cost function, which are d-dimensional embedding coordinates, i.e., the d eigenvectors according to the d smallest eigenvalues of matrix M. Ham [4] supposed that LLE's kernel matrix can be represented as $K = (\lambda_{max}I - M)$, λ_{max} is the largest eigenvalue of M. Because M is positive definite, it can be proved that K is also positive definite, which satisfies the requirement of kernel matrix consequently. The original cost function can be rewritten as $\Phi(Y) = Trace(Y^T KY)$. Then, the first d largest eigenvectors of kernel matrix K that minimize the cost function are proved to be the low dimensional embedding coordinates.

The LTSA [18,19] algorithm is similar to LLE in essence, both come down to the eigen-decomposition of a matrix, which can be described as the kernel matrix form. We put LTSA under the framework of kernel method to pursue dimension reduction based on kernel technique.

2.2   Manifold Learning Based on Explicit Mapping

The mapping relationship F between a high-dimensional dataset and its low dimensional representation is usually nonlinear and cannot be represented in an exact form [13]. One commonly used method is to use a linear function [3] to approximate the real nonlinear mapping F in order to get the mapping between the high dimensional input dataset and its low dimensional coordinates. For example, in the locality preserving projection (LPP) [5] method, a linear function $Y = A^T X$ is used, where $X \in \mathbb{R}^D$, $Y \in \mathbb{R}^d$, $A \in \mathbb{R}^{D \times d}$, X and Y represent the input and output data respectively, and A represents the linear transformation matrix. This linear function is then substituted into the manifold learning method's optimization objective function and we can get $\arg\min(a^T X)L(a^T X)^T$. The optimal linear transformation matrix $A = [a_i]$ can be solved by minimizing the cost function $Trace(Y^T LY) = Trace(a^T XLX^T a)$. The mapping representation $y_i = A^T x_i$ gotten from the linear transformation matrix A reflects the nonlinear mapping relationship from X to Y. So we can get a newly-added data point's corresponding low dimensional coordinates according to the explicit mapping formula.

In the neighborhood preserving projections (NPP) [12] algorithm, based on the LLE method, the authors used a linear transformation matrix to build the linear connection $y_i = U^T x_i$ between the input dataset $X = [x_1, x_2, ..., x_N]$ and the corresponding output $Y = [y_1, y_2, ..., y_N]$ after dimension reduction by LLE. Then, the linear connection is put to a generic procedure of dimension reduction of manifold learning to calculate the optimal linear transformation matrix U, so as to minimize the cost function $\Phi(Y) = Trace(Y(I - W)(I - W)^T Y^T) = Trace(YMY^T)$. After that, we can get $Trace(YMY^T) = Trace(U^T XMX^T U)$. By doing eigen-decomposition on matrix XMX^T, we get the d smallest eigenvectors $u_1, u_2, ..., u_d$, which can be represented as a matrix $U = [u_1, u_2, ..., u_d]$. Finally, the low dimensional output coordinates Y are computed directly from the linear function $Y = U^T X$.

3   Incremental Kernel LTSA

3.1   LTSA Under Kernel Framework

For dimensionality reduction, both LTSA and LLE do eigen-decomposition on a certain cost function to determine the low dimensional output coordinates of a high-dimensional dataset. So we can use kernel in the third step of LTSA to align the global coordinates. The optimization objective of the original cost function $\min_U tr(YMY^T)$ with $M = WW^T$ can be formulated as $\min_U tr(YKY^T)$. Now we define the kernel matrix as $K = \lambda_{\max}I - M$, where $\lambda_{\max}$ is M's largest eigenvalue. K is positive definite because $M = WW^T$ is positive definite, which can be proved as follows.

The eigen-decomposition on matrix M is $YM = Y\lambda$, multiplying both sides by Y^T, we get $YMY^T = YY^T\lambda = \lambda$, substituting $M = \lambda_{\max}I - K$ into the equation above, we can get

$$Y(\lambda_{\max}I - K)Y^T = \lambda \Rightarrow Y\lambda_{\max}IY^T - YKY^T = \lambda \Rightarrow \lambda_{\max}I - \lambda = YKY^T \quad (2)$$

Considering that M is positive definite, we know M's eigenvalues λ are positive by $YM = Y\lambda$. While $\lambda_{\max}$ is M's largest eigenvalue, so it is positive and $\lambda_{\max}I - \lambda > 0$. So Eq. (2) is positive on both sides. That is, K is positive definite, which satisfies the condition of kernel matrix.

3.2   Putting Explicit Mapping Function into the Kernel Framework

On the other hand, LTSA assumes that the coordinates of the input data points' neighbors in the local tangent space can be used to reconstruct the global embedding coordinates' geometric structure. So there exist one-to-one correspondences between global coordinates and local coordinates. We put the explicit mapping function $Y = U^TX$ into the optimization function by kernel: $\min tr(YKY^T)$ where $K = \lambda_{\max}I - M$, $M = WW^T = \sum_{i=1}^{N} W_iW_i^T$, $W_i = (I - \frac{1}{k}ee^T)(I - \Theta_i^\dagger\Theta_i)$. The constraint on Y is $YY^T = I$. So the optimization function turns to $\min_U tr(U^TXK(U^TX)^T) = \min_U tr(U^T(XKX^T)U)$. Rewriting the constraint as $YY^T = U^TX(U^TX)^T = I$, i.e., $U^TXX^TU = I$. Then, we put the constraint into the optimization function and apply the Lagrangian multiplier to get the following equation:

$$L(U) = U^T(XKX^T)U + \lambda(I - U^TXX^TU)$$
$$\Rightarrow \frac{\partial L(U)}{\partial U} = 2(XKX^T)U - 2\lambda XX^TU = 0 \Rightarrow (XKX^T)U = \lambda XX^TU \quad (3)$$

The problem can be transformed to eigen-decomposition on matrix XKX^T. The solution of the optimization problem $\min_U tr(U^T(XKX^T)U)$ is $U = [u_1, u_2, ..., u_d]$, where $u_1, u_2, ..., u_d$ are corresponding eigenvectors of the d largest eigenvalues $\lambda_1, \lambda_2, ..., \lambda_d$. With the coefficient matrix U, we can work out the low dimensional output coordinates Y according to the function $Y = U^TX$ between the local coordinates and the low-dimensional global coordinates. With the explicit mapping function, a newly-added point x_{new}'s low-dimensional coordinates can be obtained by $y_{new} = U^Tx_{new}$. This is the key point of incremental manifold learning in this paper.

3.3   Procedure of the IKLTSA Algorithm

In this paper, the kernel method with an explicit mapping formulation $Y = U^TX$ has the ability of processing newly-added data points incrementally. The kernel matrix can be used to recover the original high-dimensional data's intrinsic structure effectively. We call the method Incremental Kernel LTSA — IKLTSA in short. The procedure of IKLSTA is outlined as follows.

Input: Dataset X, neighborhood parameter k
Output: Low-dimensional coordinates Y
Procedure:

- **Step 1 (Extract local information)**: Use the k-nearest neighbor method to evaluate each data point x_i's neighborhood X_i. Compute the eigenvectors v_i corresponding to the d largest eigenvalues of the covariance matrix $(X_i - \bar{x}_i e^T)^T (X_i - \bar{x}_i e^T)$ with respect to point x_i's neighborhood, and construct matrix V_i, then get each point's coordinates in the local tangent space $\Theta_i = V_i^T X_i (I - \frac{1}{k} ee^T)$.

- **Step 2 (Align local coordinates)**: Minimize the following local reconstruction error to align the global embedding coordinates Y_i corresponding to the coordinates in the local tangent space: $E_i = Y_i (I - \frac{1}{k} ee^T)(I - \Theta_i^\dagger \Theta_i)$. Let

$$W_i = (I - \frac{1}{k} ee^T)(I - \Theta_i^\dagger \Theta_i), \quad M = \sum_{i=1}^{N} W_i W_i^T.$$

- **Step 3 (Obtain the kernel function)**: Construct the kernel function by $K = \lambda_{\max} I - M$, $\lambda_{\max}$ is M's largest eigenvalue.

- **Step 4 (Use the explicit mapping function)**: Do eigen-decomposition on matrix XKX^T and get the eigenvectors $u_1, u_2, ..., u_d$ corresponding to the d largest eigenvalues by using the explicit mapping function $Y = U^T X$.

- **Step 5 (Compute the low dimensional output coordinates)**: After getting the coefficient matrix $U = [u_1, u_2, ..., u_d]$, calculate the low dimensional output coordinates Y in accordance with the function $Y = U^T X$. For a newly-added point x_{new}, the corresponding low dimensional coordinate are evaluated directly by $y_{new} = U^T x_{new}$.

4   Experimental Results

We conduct a series of experiments in this section to validate the proposed algorithm. Tested datasets include both simulated benchmarks and real world datasets of face images, handwritten digits etc.

4.1   Performance of Dimensionality Reduction

Here we show the dimensionality reduction effect of IKLTSA on datasets Swiss Roll and Twin Peaks [15]. We compare a series of manifold learning algorithms: LTSA [18], LE [1], LLE [14], ISOMAP [16], LPP [5], NPP [12], and the proposed method IKLTSA.

We first use these methods to reduce the dimensionality of 1000 points in the Swiss Roll dataset and map the points to two-dimensional space with k=14. The results are shown in Fig. 1. We can see clearly that our algorithm preserves best the mutual spatial relationships among data points after dimensionality reduction, which indicates that our method can reflect the intrinsic structure

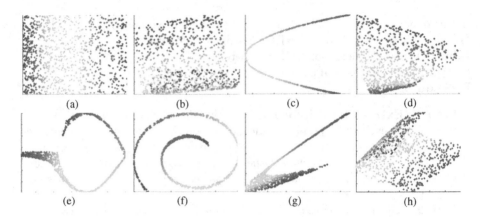

Fig. 1. Dimensionality reduction results of several manifold learning methods on Swiss Roll dataset. (a) The original dataset (1000 points), (b) Result of LTSA, (c) Result of LE, (d) Result of LLE, (e) Result of ISOMAP, (f) Result of LPP, (g) Result of NPP, (h) Result of IKLTSA.

of the original high-dimensional data accurately. We then test these methods on dataset Twin Peaks, and the two-dimensional results after dimensionality reduction are demonstrated in Fig. 2. We can see that the proposed algorithm is still superior to the other methods.

One major advantage of the proposed algorithm IKLTSA is that we can use it to evaluate the low dimensional coordinates of a newly-added point by the explicit formulation directly. To show this, we use three datasets: Swiss Roll, Punctured Sphere [15] and Twin Peaks [15], each of which contains 1000 points.

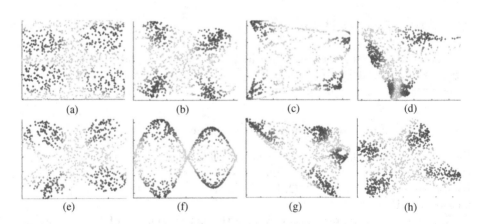

Fig. 2. Dimensionality reduction results of several manifold learning methods on twin peaks dataset. (a) The original dataset (1000 points), (b) Result of LTSA (c) Result of LE (d) Result of LLE (e) Result of ISOMAP (f) Result of LPP (g) Result of NPP (h) Result of IKLTSA.

Each dataset is divided into a testing subset (containing 20 points) and a training subset (containing 980 points). First, we reduce the dimensionality of points in the training subset, then compute the corresponding low dimensional coordinates of points in the testing subset by using the mapping function from the original high dimensional data to the low dimensional coordinates. Furthermore, we reduce the dimensionality of each dataset by using kernel function. All results are shown in Fig. 3.

From Fig. 3, we can see that the coordinates of testing points computed by the mapping function are close to the coordinates obtained by dimension reduction on the whole dataset. This means that our algorithm is able to process new data incrementally. For a new point, we can compute the corresponding low dimensional coordinates directly by the explicit mapping function. Our method has an obvious advantage in processing high dimensional data streams over the other methods.

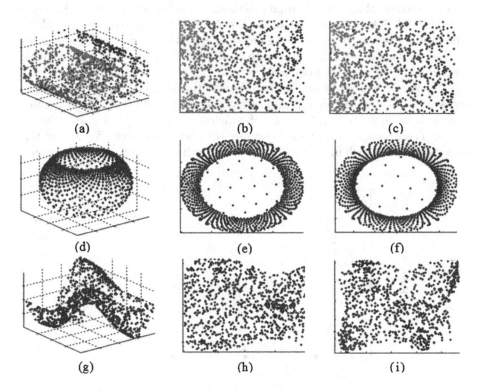

Fig. 3. Comparison between the low dimensional coordinates obtained by explicit mapping function and the coordinates obtained by direct dimension reduction on the whole dataset. (a, d, g) The original datasets; (b, e, h) Results of dimension reduction in training subset and the coordinates of points in the testing subset obtained by mapping function directly; (c, f, i) The low dimensional coordinates obtained by dimension reduction over the whole dataset.

4.2   Classification Performance

Dimension reduction is often used as a preprocessing step of classification task. So dimension reduction outputs play an important role in classification. In this section we first do dimension reduction on two real-world datasets using different algorithms, then classify the low dimensional outputs using kNN (k-nearest neighborhood) classifier. Finally, we compare the classification performance to validate the effectiveness of our algorithm.

First, we use a human face image dataset sampled from Olivetti Faces [21] as input. This dataset contains 8 individuals' face images of size 64 by 64 pixels; each individual has 10 face images of different expressions, from which 5 images are used for training and the others for testing. The results are shown in Fig. 4, where the horizontal axis represents the dimensionality of face images after dimension reduction. From Fig. 4, we can see that our method IKLTSA performs better than the other algorithms. Then we do the same experiments with major incremental manifold learning methods, including incremental LE, incremental LTSA, incremental Isomap, incremental LLE and incremental HLLE [9]. The results are shown in Fig. 5. Again we can see that our algorithm achieves the best performance. This indicates that our algorithm can detect the intrinsic structure hidden in the dataset well.

The second real-world dataset used here is the MNIST Handwritten Digits dataset [21], which contains images of hand-written digits 0–9. Here we select only 5 digits 0, 1, 3, 4 and 6, each of which has 980 images of size 28 by 28 pixels. These images are divided training set and testing set. We reduce the dimensionality of each image to 1–8 dimensions using different manifold learning

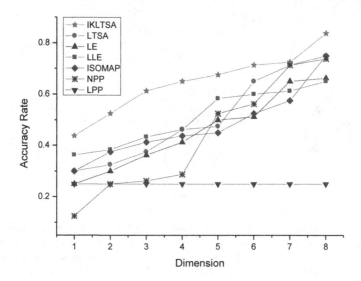

Fig. 4. Classification performance comparison on the Olivetti Faces dataset with major manifold learning algorithms.

algorithms, and then classify the low-dimensional images using kNN classifier. The classification results are shown in Fig. 6. We can see that our algorithm achieves a higher accuracy than the other algorithms. Similarly, we compare IKLTSA with major incremental methods on the same dataset, and get roughly similar results, which are shown in Fig. 7.

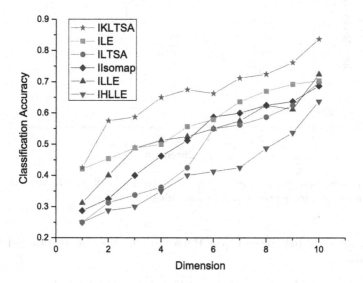

Fig. 5. Classification performance comparison on the Olivetti Faces dataset with major incremental manifold learning algorithms.

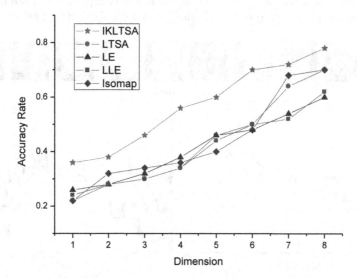

Fig. 6. Classification performance comparison on the MNIST Handwritten Digits dataset with major manifold learning algorithms.

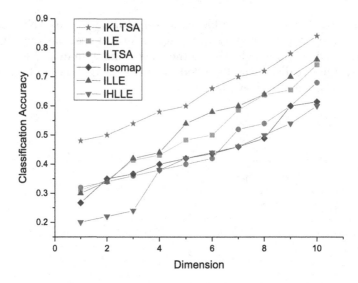

Fig. 7. Classification performance comparison on the MNIST Handwritten Digits dataset with major incremental manifold learning algorithms.

4.3 Dimensionality Reduction Performance on the Rendered Face Dataset

Here we check the dimensionality reduction performance on the rendered face dataset [16]. The results are shown in Fig. 8. The dataset contains 698 facial sculpture images of 64×64 pixels. These images have 2 groups of pose parameters: up to down and left to right. All images are transformed to 4096-dimension

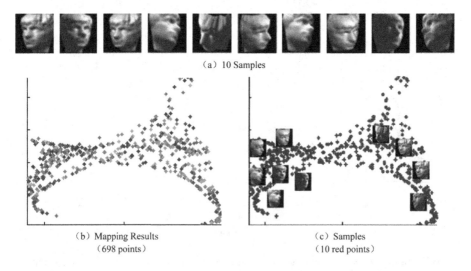

(a) 10 Samples

(b) Mapping Results
(698 points)

(c) Samples
(10 red points)

Fig. 8. Dimensionality reduction results on the Rendered face dataset.

vectors. We reduce the 698 high dimensional images to 2D using IKLTSA, and are shown in Fig. 8. Here, each point represents a facial image. 10 images are selected randomly and marked as red points in Fig. 8. We can see that the facial poses are from right to left along the horizontal axis, and from look-up to look-down along the vertical axis (posture is from up to down). So the low dimensional projections obtained by our algorithm keep the original data's intrinsic structure very well. The selected images are mapped to the low dimensional space

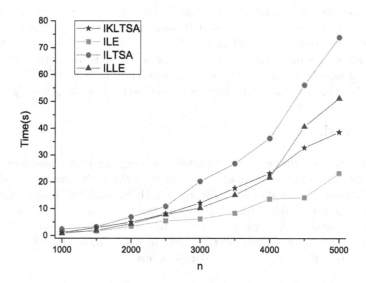

Fig. 9. Time cost comparison on the Swiss Roll dataset

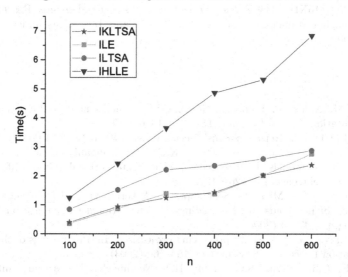

Fig. 10. Time cost comparison on the Rendered face dataset

accurately in accordance with their positions and poses in the original data. This shows that the proposed algorithm is effective in detecting the implicit structures of high dimensional images.

4.4 Time Cost

Here we compare time cost on dimension reduction of our IKLTSA method with three existing incremental manifold learning algorithms, including incremental LE, incremental LTSA and incremental HLLE, over the Swiss Roll and the Rendered face datasets. Figures 9 and 10 show the time cost as the number of data points or images processed increases. On the Swiss Roll dataset, although IKLTSA consumes more time than incremental LE, it is faster than the other incremental manifold learning algorithms to process the whole dataset. On the Rendered face datasets. our algorithm is always faster than the other algorithms.

5   Conclusion

In this paper, we propose a new manifold learning algorithm based on kernel that combines an explicit mapping function from the high dimensional data to its low dimensional coordinates. Compared with existing batch and incremental manifold learning algorithms, the new method can directly compute newly-added data points' low dimensional coordinates by the explicit mapping function. This enables the proposed method to process new data incrementally, which has extensive applications in processing data streams.

Acknowledgement. This work was supported by the Program of Shanghai Subject Chief Scientist (15XD1503600) and the Key Projects of Fundamental Research Program of Shanghai Municipal Commission of Science and Technology under grant No. 14JC1400300.

References

1. Belkin, M., Niyogi, P.: Laplacian eigenmaps for dimensionality reduction and data representation. Neural Comput. **15**, 1373–1396 (2003)
2. Choi, H., Choi, S.: Kernel isomap. Electron. Lett. **40**, 1612–1613 (2004)
3. Chen, M., Li, W., Zhang, W., Wang, X.G.: Dimensionality reduction with generalized linear models. In: Proceedings of the International Joint Conference on Artificial Intelligence, pp. 1267–1272 (2013)
4. Ham, J., Lee, D., Mika, S., Scholkopf, B.: A kernel view of the dimensionality reduction of manifolds. In: Proceedings of International Conference on Machine Learning, pp. 47–54 (2004)
5. He, X.F., Niyogi, P.: Locality preserving projections. In: Proceedings of the Neural Information Processing Systems, pp. 153–160 (2003)
6. He, X.F., Cai, D., Yan, S.C., Zhang, H.J.: Neighborhood preserving embedding. In: Proceedings of the 10th IEEE International Conference on Computer Vision, pp. 1208–1213 (2005)

7. Kokiopoulou, E., Saad, Y.: Orthogonal neighborhood preserving projections. In: Proceedings of the 5th IEEE International Conference on Data Mining, pp. 1–7 (2005)

8. Langone, R., Agudelo, O., Moor, B., Suykens, J.: Incremental kernel spectral clustering for online learning of non-stationary data. Neurocomputing **139**(2), 246–260 (2014)

9. Li, H., Jiang, H., et al.: Incremental manifold learning by spectral embedding methods. Pattern Recogn. Lett. **32**, 1447–1455 (2011)

10. Liu, S.L., Yan, D.Q.: A new global embedding algorithm. Acta AUTOMATICA Sinica **37**(7), 828–835 (2011)

11. Li, L., Zhang, Y.J.: Linear projection-based non-negative matrix factorization. Acta Automatica Sinica **36**(1), 23–39 (2010)

12. Pang, Y., Zhang, L., Liu, Z., Yu, N., Li, H.: Neighborhood Preserving Projections (NPP): a novel linear dimension reduction method. In: Huang, D.-S., Zhang, X.-P., Huang, G.-B. (eds.) ICIC 2005. LNCS, vol. 3644, pp. 117–125. Springer, Heidelberg (2005)

13. Qiao, H., Zhang, P., Wang, D., Zhang, B.: An explicit nonlinear mapping for manifold learning. IEEE Trans. Cybern. **43**(1), 51–63 (2013)

14. Roweis, S.T., Saul, L.K.: Nonlinear dimensionality reduction by locally linear embedding. Science **290**, 2323–2326 (2000)

15. Saul, L., Roweis, S.: Think globally, fit locally: Unsupervised learning of nonlinear manifolds. J. Mach. Learn. Res. **4**, 119–155 (2003)

16. Tenenbaum, J.B., de Silva, V., Langford, J.C.: A global geometric framework for nonlinear dimensionality reduction. Science **290**, 2319–2323 (2000)

17. Tan, C., Chen, C., Guan, J.: A nonlinear dimension reduction method with both distance and neighborhood preservation. In: Wang, M. (ed.) KSEM 2013. LNCS, vol. 8041, pp. 48–63. Springer, Heidelberg (2013)

18. Zhang, Z.Y., Zha, H.Y.: Principal manifolds and nonlinear dimensionality reduction via tangent space alignment. SIAM J. Sci. Comput. **26**(1), 313–338 (2005)

19. Zhang, Z.Y., Wang, J., Zha, H.Y.: Adaptive manifold learning. IEEE Trans. Pattern Anal. Mach. Intell. **34**(2), 253–265 (2012)

20. Zheng, S.W., Qiao, H., Zhang, B., Zhang, P.: The application of intrinsic variable preserving manifold learning method to tracking multiple people with occlusion reasoning. In: Proceedings of the IEEE/RSJ International Conference on Intelligent Robots and Systems, pp. 2993–2998 (2009)

21. Supporting webpage. http://www.cs.nyu.edu/~roweis/data.html

22. Supporting webpage. http://archive.ics.uci.edu/ml/

Sentiment Analysis

SentiSAIL: Sentiment Analysis in English, German and Russian

Gayane Shalunts[✉] and Gerhard Backfried

SAIL LABS Technology GmbH, Vienna, Austria
{Gayane.Shalunts,Gerhard.Backfried}@sail-labs.com

Abstract. Sentiment analysis has been well in the focus of researchers in recent years. Nevertheless despite the considerable amount of literature in the field the majority of publications target the domains of movie and product reviews in English. The current paper presents a novel sentiment analysis method, which extends the state-of-the-art by trilingual sentiment classification in the domains of general news and particularly the coverage of natural disasters in general news. The languages examined are English, German and Russian. The approach targets both traditional and social media content. The extensive experiments demonstrate that the performance of the proposed approach outperforms human annotators, as well as the original method, on which it is built and extended.

Keyword: Multilingual sentiment analysis

1 Introduction

Sentiment analysis methods may be divided into two types: machine-learning-based and lexicon-based [9]. Machine learning methods are implemented as supervised binary (i.e. positive or negative) classification approaches. Here labeled data is employed to train classifiers [9,14]. Learning-based methods are advantageous, as they can produce trained models for specific purposes and contexts, whereas their drawback is the demand of labeled data and thus the low applicability of the method on new data. Data labeling is usually costly or even impracticable in some cases [9]. On the other hand, lexicon-based methods use a predefined list of words as features, also referred to as *sentiment dictionary* or *lexicon*, where each word is associated with a specific sentiment [9]. The lexical methods vary according to the context in which they were created [9]. Regarding the advantages and disadvantages of lexical methods, it is a plus that no costly labeled data is required, whereas the challenge lies in obtaining a sentiment dictionary applicable in various contexts. Thus lexicon-based methods are tuned to cover specific target domains and media types, since traditional media conveys formal language and social media - colloquial language, slang.

The authors in [9] compare 8 popular sentiment analysis methods in terms of coverage (i.e. the fraction of messages whose sentiment is identified) and agreement (i.e. the fraction of identified sentiments that agree with ground truth).

© Springer International Publishing Switzerland 2015
P. Perner (Ed.): MLDM 2015, LNAI 9166, pp. 87–97, 2015.
DOI: 10.1007/978-3-319-21024-7_6

The experiments are carried on two English datasets of Online Social Networks (OSNs) messages. The methods compared are SentiWordNet [8], SASA [21], PANAS-t [10], Emoticons, SentiStrength [20], LIWC [19], SenticNet [4] and Happiness Index [7]. The conclusions drawn from the comparison are the following:

1. Existing sentiment analysis methods have varying degrees of coverage, ranging between 4 % and 95 % when applied to real events.

2. None of the methods observed had high coverage and high agreement meanwhile.

3. When it comes to the predicted polarity, the methods varied widely in their agreement, ranging from 33 % to 80 %, indicating that the same social media text could be interpreted very differently depending on the method choice.

4. The methods examined varied widely in their sentiment prediction of notable social events. E.g. for the case of an airplane crash, half of the methods predicted dominance of positive affect in the relevant tweets. Similarly on the tweet dataset related to a disease outbreak, only 2 out of 8 methods predicted prevailing negative sentiment.

Whereas considerable research has been devoted to sentiment analysis, the majority of works in the field target the domain of movie and product reviews in English. The authors in [15] present a German language sentiment analysis method called SentimentWortschatz or SentiWS. The approach targets the domain of financial newspaper articles and respective blog posts on a German stock index [15]. The sentiment lexicon is developed from General Inquirer (GI) lexicon [17] by semiautomatic translation into German using Google Translate[1] and is manually revised afterwards. The lexicon post-translation revision included the removal of inappropriate words and addition of words from the finance domain [15]. The usage of the GI lexicon as a base is justified by the fact that it is widely accepted in the sentiment analysis community and has a broad coverage. Another method in German, introduced by [12], targets the domain concerning German celebrities. The approach utilizes the SentiStrength tool [20] and also permits the classification of mixed sentiments. Here also the English opinion dictionary was automatically translated to German and manually revised afterwards by two German native speakers. The publications [5,6] present the recently emerged sentiment analysis research in Russian. Authors in [5] propose an approach for domain specific sentiment lexicon extraction in the meta-domain of products and services. Reference [6] describes and evaluates the state-of-the-art sentiment analysis systems in Russian.

Our method is called *SentiSAIL* and is integrated into the Sail Labs Media Mining System (MMS) for Open Source Intelligence [1]. MMS is a state-of-the-art Open-Source-Intelligence system, incorporating speech and text-processing technologies. Sentiment analysis makes a part of MMS automatic multifaceted processing of unstructured textual and speech data. The objective of the development of SentiSAIL is to create a multilingual sentiment analysis tool addressing the domain of general news and particularly the coverage of natural disasters in general news. News articles compared to product reviews represent a

---

[1] http://translate.google.com.

much broader domain, requiring a larger sentiment lexicon. SentiSAIL employs the algorithm of one of the state-of-the-art sentiment analysis methods - SentiStrength. SentiStrength, likewise [18], is a lexicon-based approach, using dictionaries of words annotated with semantic orientation (polarity and strength) and incorporating intensification and negation. The innovative contribution of SentiSAIL lies in expanding SentiStrength algorithm into new domains, multiple languages and granularity level of full articles. Whereas SentiStrength is optimized for and evaluated on social media content, SentiSAIL targets both social and traditional media data. In the current work SentiSAIL is evaluated on a trilingual traditional media corpus, as well as on an English OSN dataset for performance comparison with other state-of-the-art methods. The performance evaluation of SentiSAIL on a German social media corpus related to natural disasters is reported in [16].

Reference [8] categorized the literature related to sentiment detection into 3 classes, identifying : (1) Text SO-Polarity (Subjective-Objective), (2) Text PN-Polarity (Positive-Negative), (3) The strength of text PN-Polarity. SentiSAIL falls into both category (2) and (3). The strength of the text PN-Polarity is reflected in a pair of numbers, thresholding of which determines the sentiment class.

The overwhelming majority of sentiment analysis methods solve a binary classification task by classifying a query text into either positive or negative class. [22] presents a two stage lexicon-based approach of automatic distinguishing of prior and contextual polarity. At the first stage phrases, containing a clue with prior polarity, are classified as neutral or polar. At the second stage the polar phrases pass to further classification of contextual polarity into one of the classes positive, negative, neutral or both. SentiSAIL, likewise [22], instead of a binary classification problem presents a dual classification framework, comprising positive, negative, neutral and mixed classes. The neutral class indicates the absence of both positive and negative sentiments in the query text, whereas the mixed class - the presence of both. The dual classification scheme choice is justified, as psychological research affirms human ability to experience positive and negative emotions simultaneously [13]. Our approach determines the polarity of the sentiment of the input text, not aiming to perform opinion mining or distinguishing between objective (facts) and subjective (opinions) aspects of the observed data.

The authors in [11] performed granularity-based (word, sentence/passage, document) categorization of the related literature. Whereas our approach supports all three levels of granularity, the current paper focuses on sentiment analysis of whole news articles on the document level. The authors in [2] also target the domain of news limited only to English. They examine the performance of different sentiment dictionaries, as well as attempt to distinguish between positive or negative opinion and good or bad news.

The remainder of the paper is organized as follows: Sect. 2 clarifies the methodology behind SentiSAIL. Section 3 presents the experimental setup, performance evaluation and results. And finally Sect. 4 draws conclusions from the work presented.

2   SentiSAIL

SentiSAIL employs the methodology of SentiStrength [20] and extends it to cover the domains of general news and particularly natural disasters (floods, earthquakes, volcanoes, avalanches, tsunamis, typhoons, tornadoes, wildfires, etc.) related news in multiple languages (English, German, Russian). A brief description of the algorithm behind SentiStrength is outlined below, followed by the description of the novel contribution of SentiSAIL.

SentiStrength is a lexical-based method, for which the discriminative features are predefined lists of words. The main list is a dictionary of sentiment expressing words - the sentiment lexicon, with a score associated to each word. The positive words are weighed in the range [1; 5], the negative ones - [-5; -1] in a step 1, e.g. "charming" 4, "cruel" -4. Thus sentiment detection on the granularity level of word [11] is achieved by the sentiment lexicon. Further lists attempt to model the grammatical dependencies of the given language. Sentiment words may be intensified or weakened by words called *boosters*, which are represented in the booster list. E.g. "less charming" will weigh 3 and "very cruel" -5. *Negation* words are the next list. Here it is assumed that negating a positive word inverts the sentiment to negative and weakens it twice, whereas negating a negative word neutralizes the sentiment. E.g. "not charming" scores -2, whereas "isn't cruel" equals 0. The boosters and negations affect up to 2 following words. Though in Russian and German the position of negations is flexible, e.g. negations may also follow the word they negate, as well as negate the meaning of the complete sentence. Negations in such cases are hard to model, thus remain unidentified. To overcome the issue of formation of diverse words from the same stem (inflection and declension), stemming is performed on the sentiment lexicon, e.g. "sympath*" will match all the words starting with "sympath", e.g. "sympathize", "sympathizes", "sympathized", "sympathy", "sympathetic", etc.

Phrases and idioms form the next lists of features carrying sentiment. In the current context we define a *phrase* as a set of words, expressing sentiment only while combined. An *idiom* is also a combination of words, but unlike a phrase it expresses figurative and not literal meaning. Idioms and phrases score as a whole, overriding the scores of their component words. Whereas the idiom list is available in SentiStrength, the phrase list is newly incorporated by our SentiSAIL approach. The sentiment lexicon comprises the polarity of individual words (*prior polarity*) [11]. The polarity of a word in a sentence (*contextual polarity*) may be different from its prior polarity [11] and is determined in the context of negations, phrases and idioms.

Further feature lists address only the processing of social media data. In the list of emotion icons or *emoticons* each emoticon is associated with a sentiment score, e.g. :-) 1. The list of the *slang words* and *abbreviations* is formed likewise, e.g. lol (laughing out loud) 2. Single or multiple exclamation marks (!) boost the immediately preceding sentiment word. Repeated letters (e.g. niiiiiiiice), capitalizing of the whole word (e.g. HAPPY) add emphasis to the sentiment words and are processed similarly to boosters.

Though SentiStrength comprises the mentioned feature lists in 14 languages, the lists in languages other than English are limited and, most importantly, the sentiment lexicons lack stemming. Since there is a cost to automation and the best path to long-term improvement is achieved by development of language-specific resources [3], we choose to expand the multilingual sentiment lexicons manually. The development of our SentiSAIL tool, aiming to achieve multilingual (English, German, Russian) sentiment analysis in the domain of general news and natural disasters, is started by taking as a base and improving the initial SentiStrength sentiment lexicons in German and Russian.

At first a human expert, fluent in German and Russian, went through both lexicons and performed stemming. Unlike in English both in German and Russian adjective declension takes place depending on the linguistic gender, case, number of the characterized noun. In German also the article type (definite, indefinite, null) is added to the factors mentioned. The stemming of adjectives accomplishes the matching of all declension forms. Stemming is also needed and implemented on German and Russian booster and negation word lists. Meanwhile lexicon words, that were false hits or may convey sentiments of different strength and polarity depending on the context, were removed.

The next step is the extension of the trilingual sentiment lexicons to cover the domains of general news and natural disasters. A database of articles from the observed domains and in the target languages was collected from the web with that purpose. A training dataset per language was separated from the whole database. Afterwards additional domain-specific sentiment terms were manually extracted from each training dataset and added to the lexicon of the corresponding language together with their associated scores. To obtain richer lexicons articles covering diverse topics were included in the training datasets. As a result the SentiSAIL sentiment lexicon in English grew from 2546 original terms to 2850 terms, in German - from 1983 to 2176 and in Russian from 1684 to 2024.

A pair of positive and negative sentiment scores is calculated for each text line. Herewith we perform sentiment detection on granularity level of sentence/passage [11], where a single line corresponds to a passage. We employ and compare (see Sect. 3) three algorithms to calculate the line scores:

(1) *Maximization.* The scores of the most positive and the most negative terms of the line are assigned to its positive and negative scores respectively. SentiStrength uses this algorithm of phrase scoring.

(2) *Averaging.* Positive and negative scores of each line are calculated respectively as the average of its all positive and negative word scores.

(3) *Aggregation.* Positive and negative scores of each line are obtained from respective aggregation of the scores of all positive and negative words of the line, bounded by 5 (positive score) and −5 (negative score) values.

The positive and negative sentiment scores of the full text (sentiment detection on granularity level document [11]) are obtained by averaging the respective scores of all lines. And finally the sentiment class of the text is predicted by double (positive and negative) thresholding: the classification of the "Positive" and "Negative" classes is straightforward. The texts passing both thresholds are classified into the "Mixed" class, whereas those exceeding neither one of the thresholds belong to the "Neutral" class. We define the positive threshold equal to 1 and the negative one equal to −1, since those are the minimum scores expressing sentiment in our framework. The thresholds are parametrized and may be adjusted to achieve weaker or stronger sentiment detection, i.e. raising the thresholds will lead to detection of stronger affects.

3   Experimental Setup, Evaluation and Results

The evaluation of SentiSAIL in the current context has a number of specificities:

(1) Since human opinions on a particular event or phenomenon may vary depending on cultural background, political views, emotional state, etc., obtaining the ground truth for the evaluation of a sentiment analysis approach is not a trivial task. Because of the subjective nature of the ground truth, annotations by multiple experts is a common practice in the field. The performance of a method is considered satisfactory, if its average agreement rate with the annotators is close to or even outperforms the average inter-annotator agreement rate.

(2) As expected the distribution of the observed 4 classes is not equal, as natural disasters are strongly negative events and in general news the negative sentiment dominates.

SentiSAIL is implemented in Perl. SentiSAIL performance speed is proportional to $log_2 N$, where N is the number of sentiment lexicon terms in the language observed. Logarithmic performance speed is the result of running binary search on sentiment lexicons. While processing traditional media social media features (emoticons, slang words, abbreviations, exclamation marks, repeated letters and capitalizing) can be disabled via a parameter to gain speed. The language of the data to be classified is provided via another parameter, so that the feature lists of the corresponding language can be selected.

Since the current work is the first to tackle the problem of sentiment analysis on the traditional media coverage of general and natural disasters related news trilingually and on the granularity level of full articles, we had to collect and label our own trilingual text corpus. Performance evaluation of SentiSAIL is carried out for English, German and Russian training datasets separately. Further sentiment analysis is performed on the testing datasets in each language. The training dataset includes 32 news articles in English, 32 - in German and 48 - in Russian. The testing dataset comprises 50 news articles in each language.

The English, German and Russian training datasets were labeled by 3 annotators. Afterwards SentiSAIL was run on each annotated dataset employing aggregation, averaging and maximization algorithms (see Sect. 2) on granularity level line. The average inter-annotator agreement on English training dataset yielded 80 % (Table 1), whereas the average SentiSAIL-annotator agreement rate scored 89.59 %, 91.15 % and 90.11 % for aggregation, averaging and maximization algorithms respectively (Table 1). Thus in case of all three algorithms SentiSAIL-annotator agreement rate outperforms inter-annotator agreement rate. This dominance holds true also on the German training dataset. On the Russian

Table 1. SentiSAIL evaluation on the training datasets.

	Annotator 1	Annotator 2	Annotator 3	Average
English training set				
Annotator 1	-	84.38 %	76.56 %	**80 %**
Annotator 2	-	-	79.69 %	
Aggregation	92.19 %	90.63 %	85.94 %	89.59 %
Averaging	96.88 %	89.06 %	87.5 %	91.15 %
Maximization	93.75 %	92.19 %	84.38 %	90.11 %
Maximization, Titles	98.44 %	87.5 %	85.94 %	90.63 %
Maximization, SentiStrength features	84.38 %	82.81 %	78.13 %	81.77 %
German training set				
Annotator 1	-	78.13 %	79.69 %	**79.17 %**
Annotator 2	-	-	79.69 %	
Aggregation	96.88 %	78.13 %	79.69 %	84.9 %
Averaging	92.19 %	73.44 %	78.13 %	81.25 %
Maximization	92.19 %	73.44 %	78.13 %	81.25 %
Maximization, Titles	98.44 %	78.13 %	76.56 %	84.38 %
Maximization, SentiStrength features	51.56 %	42.19 %	40.63 %	44.79 %
Russian training set				
Annotator 1	-	84.38 %	79.17 %	**81.95 %**
Annotator 2	-	-	82.29 %	
Aggregation	83.33 %	83.33 %	79.17 %	81.94 %
Averaging	82.29 %	81.25 %	76.04 %	79.86 %
Maximization	86.46 %	84.38 %	78.13 %	82.99 %
Maximization, Titles	85.42 %	77.08 %	87.5 %	83.33 %
Maximization, SentiStrength features	39.58 %	44.79 %	43.75 %	42.71 %

training dataset SentiSAIL performance in the case of the aggregation algorithm (81.94 %) is almost equal to the inter-annotator performance, whereas in case of the averaging algorithm falls behind (76.86 %). The experiments employing aggregation, averaging and maximization algorithms had the objective to reveal the best performing algorithm. But, as reported in Table 1, none of the methods achieved considerable superiority over the others and for each language one of the algorithms scored maximum agreement rate. Since only the maximization algorithm achieved slight excellence over inter-annotator agreement rate on Russian training dataset, all further experiments were conducted employing it.

Another experiment is carried out to observe how well the article titles reflect the affect of their content. The experiment outcome is surprisingly good,

Table 2. SentiSAIL evaluation on the testing datasets.

	Annotator 1	Annotator 2	Annotator 3	Average
English testing set				
Annotator 1	-	83 %	76 %	**76.67 %**
Annotator 2	-	-	71 %	
Maximization	77 %	76 %	73 %	75.33 %
Maximization, Titles	75 %	78 %	75 %	76 %
Maximization, SentiStrength features	73 %	72 %	69 %	71.33 %
German testing set				
Annotator 1	-	85 %	76 %	**76.67 %**
Annotator 2	-	-	69 %	
Maximization	81 %	80 %	77 %	79.33 %
Maximization, Titles	84 %	89 %	72 %	81.67 %
Maximization, SentiStrength features	43 %	34 %	63 %	46.67 %
Russian testing set				
Annotator 1	-	93 %	93 %	**92.66 %**
Annotator 2	-	-	92 %	
Maximization	92 %	88 %	90 %	90 %
Maximization, Titles	86 %	85 %	85 %	85.33 %
Maximization, SentiStrength features	43 %	46 %	48 %	45.67 %

indicating comparable and even higher SentiSAIL-annotator average agreement rate opposed with the processing of the full articles: in English 90.63 % vs 90.11 %, in German 84.38 % vs 81.25 %, in Russian 83.33 % vs 82.99 % (Table 1).

In order to evaluate the improvement of SentiSAIL features (sentiment lexicon, boosters, phrases, etc.) over SentiStrength [20] original features, SentiSAIL is run also with the original SentiStrength features using the line maximization algorithm. The experiment outcome in each language observed is presented in Table 1, indicating minor improvement of the average system-annotators agreement rate in English (from 81.77 % to 90.63 %) and major improvement in German (from 44.79 % to 84.38 %) and Russian (from 42.71 % to 83.33 %). This result was to be expected, since besides lexicon extension we also performed stemming in German and Russian.

To evaluate SentiSAIL performance on unfamiliar data from the domains the same sets of experiments, employing only the maximization algorithm on granularity level of line, are conducted on ground truth labeled testing datasets per language. The experiment results are satisfactory and are presented in Table 2. Firstly, SentiSAIL-annotator average agreement rate surpasses the average

Table 3. Average accuracy comparison of SentiSAIL and 8 other methods.

Method	Average accuracy
PANAS-t	0.677
Emoticons	0.817
SASA	0.649
SenticNet	0.590
SentiWordNet	0.643
Happiness Index	0.639
SentiStrength	0.815
LIWC	0.675
SentiSAIL	**0.736**

inter-annotator agreement rate on the German testing dataset (79.33 % vs 76.67 %) and falls slightly behind it on English (75.33 % vs 76.67 %) and Russian (90 % vs 92.66 %) testing datasets. Secondly, whereas sentiment analysis only on the titles of the testing datasets articles insignificantly outperforms the results of the processing of the full articles in English (76 % vs 75.33 %) and German (81.67 % vs 79.33 %), it falls behind in Russian (85.33 % vs 90 %) (Table 2). Thirdly, likewise the identical experiment on the training datasets, SentiSAIL feature improvement over the original SentiStrength features in English is slight (75.33 % vs 71.33 %), whereas it is considerable in German (79.33 % vs 46.67 %) and Russian (90 % vs 45.67 %) (Table 2). As an overall conclusion the empiric results on the testing datasets are comparable and in tune with the results obtained over the training datasets, showing that SentiSAIL performance is acceptable on unfamiliar data.

Whereas the current work focuses on sentiment analysis on full articles from traditional media, we take a further step by comparing SentiSAIL performance to other state-of-the-art sentiment analysis methods on an OSN ground-truth annotated dataset. The dataset was introduced by the authors of SentiStrength and comprises 11814 short social media texts from MySpace, Twitter, Digg, BBC forum, Runners World forum and YouTube. The comparison metric is the agreement rate with the ground truth (i.e. average accuracy). Please note that we compute the agreement rate according to our 4-class classification scheme, but not in a binary classification scheme, as in the other approaches. Table 3 presents the average accuracy obtained by 8 sentiment analysis methods on the above mentioned OSN dataset, as reported in [9]. The average accuracy, achieved by SentiSAIL is equal to 0.736 and is shown in the last row of Table 3. It outperforms all the methods, except Emoticons and SentiStrength. Comparison against SentiStrength is not fair, since SentiStrength was trained on this dataset. Though Emoticons' score is the highest, it lacks language processing, is based only on analysis of emoticons and thus has limited application.

4   Conclusion

The paper presented the SentiSAIL sentiment analysis tool, addressing the domains of general and natural disasters related news in English, German and Russian on the granularity level of full articles. SentiSAIL employed the algorithm of another state-of-the-art sentiment analysis tool, called SentiStrength, and extended it to the target domains trilingually. The extension was performed in two steps: firstly, stemming was performed on the original SentiStrength lexicons in German and Russian, secondly, the domains-specific lexicon was extracted from trilingual training datasets. Extensive experiments were conducted to evaluate SentiSAIL performance in each language: (1) aggregation, averaging and maximization algorithms were employed on granularity level line, none of them demonstrating significant advantage over another, (2) contrasting sentiment analysis on article titles with full contents led to similar outcome, (3) improvement estimation over the original SentiStrength features proved the refinement in English to be slight, whereas in German and Russian - considerable, (4) the results of identical experiments on training and testing datasets were comparable, demonstrating satisfactory performance on unknown data from the target domains. The experimental setup was completed by SentiSAIL performance comparison to eight other state-of-the-art sentiment analysis approaches on an OSN dataset, exhibiting SentiSAIL excellence over six out of eight methods observed. The performance evaluation criterion in all experiments was the comparability of average SentiSAIL-annotators agreement rate to the average inter-annotator agreement rate.

The future work will be in the direction of lengthening the list of the processed languages.

References

1. Backfried, G., Schmidt, C., Pfeiffer, M., Quirchmayr, G., Glanzer, M., Rainer, K.: Open source intelligence in disaster management. In: Proceedings of the European Intelligence and Security Informatics Conference (EISIC), pp. 254–258. IEEE Computer Society, Odense, Denmark (2012)
2. Balahur, A., Steinberger, R., Kabadjov, M., Zavarella, V., van der Goot, E., Halkia, M., Pouliquen, B., Belyaeva, J.: Sentiment analysis in the news. In: Proceedings of the 7th International Conference on Language Resources and Evaluation (LREC 2010). European Language Resources Association (ELRA), Valletta, Malta (2010)
3. Brooke, J., Tofiloski, M., Taboada, M.: Cross-linguistic sentiment analysis: from english to spanish. In: Proceedings of Recent Advances in Natural Language Processing (RANLP), pp. 50–54. RANLP 2009 Organising Committee / ACL, Borovets, Bulgaria (2009)
4. Cambria, E., Speer, R., Havasi, C., Hussain, A.: Senticnet: a publicly available semantic resource for opinion mining. In: AAAI Fall Symposium: Commonsense Knowledge, pp. 14–18 (2010)
5. Chetviorkin, I., Loukachevitch, N.: Extraction of russian sentiment lexicon for product meta-domain. In: Proceedings of the 24th International Conference on Computational Linguistics (COLING), Bombay, India, pp. 593–610 (2012)

6. Chetviorkin, I., Loukachevitch, N.: Evaluating sentiment analysis systems in russian. In: Proceedings of the 4th Biennial International Workshop on Balto-Slavic Natural Language Processing, Sofia, Bulgaria, pp. 12–17 (2013)

7. Dodds, P.S., Danforth, C.M.: Measuring the happiness of large-scale written expression: songs, blogs, and presidents. J. Happiness Stud. **11**(4), 441–456 (2009)

8. Esuli, A., Sebastiani, F.: Sentiwordnet: a publicly available lexical resource for opinion mining. In: Proceedings of the 5th Conference on Language Resources and Evaluation (LREC06), pp. 417–422 (2006)

9. Gonçalves, P., Araújo, M., Benevenuto, F., Cha, M.: Comparing and combining sentiment analysis methods. In: Proceedings of the 1st ACM Conference on Online Social Networks (COSN 2013), pp. 27–38. ACM, Boston, USA (2013)

10. Gonçalves, P., Benevenuto, F., Cha, M.: Panas-t: A pychometric scale for measuring sentiments on twitter. CoRR abs/1308.1857 (2013)

11. Missen, M.M.S., Boughanem, M., Cabanac, G.: Opinion mining: reviewed from word to document level. Soc. Netw. Anal. Min. **3**(1), 107–125 (2013)

12. Momtazi, S.: Fine-grained german sentiment analysis on social media. In: Proceedings of the 8th International Conference on Language Resources and Evaluation (LREC 2012), pp. 1215–1220. European Language Resources Association (ELRA), Istanbul, Turkey (2012)

13. Norman, G.J., Norris, C.J., Gollan, J., Ito, T.A., Hawkley, L.C., Larsen, J.T., Cacioppo, J.T., Berntson, G.G.: Current emotion research in psychophysiology: the neurobiology of evaluative bivalence. Emot. Rev. **3**(3), 349–359 (2011)

14. Pang, B., Lee, L., Vaithyanathan, S.: Thumbs up?: sentiment classification using machine learning techniques. In: Proceedings of the ACL Conference on Empirical Methods in Natural Language Processing (EMNLP 2002), Philadelphia, PA, USA, pp. 79–86 (2002)

15. Remus, R., Quasthoff, U., Heyer, G.: Sentiws - a german-language resource for sentiment analysis. In: Proceedings of the 7th conference on International Language Resources and Evaluation (LREC), Valletta, Malta, pp. 1168–1171 (2010)

16. Shalunts, G., Backfried, G., Prinz, K.: Sentiment analysis of German social media data for natural disasters. In: Proceedings of the 11th International Conference on Information Systems for Crisis Response and Management (ISCRAM), pp. 752–756. University Park, Pennsylvania, USA (2014)

17. Stone, P.J., Dunphy, D.C., Smith, M.S., Ogilvie, D.M.: The General Inquirer: A Computer Approach to Content Analysis. MIT Press, Cambridge (1966)

18. Taboada, M., Brooke, J., Tofiloski, M., Voll, K., Stede, M.: Lexicon-based methods for sentiment analysis. Comput. Linguist. **37**(2), 267–307 (2011)

19. Tausczik, Y.R., Pennebaker, J.W.: The psychological meaning of words: liwc and computerized text analysis methods. J. Lang. Soc. Psychol. **29**(1), 25–54 (2010)

20. Thelwall, M., Buckley, K., Paltoglou, G., Cai, D., Kappas, A.: Sentiment strength detection in short informal text. J. Am. Soc. Inf. Sci. Technol. **61**(12), 2544–2558 (2010)

21. Wang, H., Can, D., Kazemzadeh, A., Bar, F., Narayanan, S.: A system for real-time twitter sentiment analysis of 2012 U.S. presidential election cycle. In: ACL (System Demonstrations), pp. 115–120 (2012)

22. Wilson, T., Wiebe, J., Hoffmann, P.: Recognizing contextual polarity: an exploration of features for phrase-level sentiment analysis. Comput. Linguist. **35**(3), 399–433 (2009)

Sentiment Analysis for Government: An Optimized Approach

Angelo Corallo[1], Laura Fortunato[1(✉)], Marco Matera[1],
Marco Alessi[2], Alessio Camillò[3], Valentina Chetta[3],
Enza Giangreco[3], and Davide Storelli[3]

[1] Dipartimento di Ingegneria Dell'Innovazione,
University of Salento, Lecce, Italy
{angelo.corallo,laura.fortunato}@unisalento.it,
marco.matera@studenti.unisalento.it
[2] R&D Department, Engineering Ingegneria Informatica SPA, Palermo, Italy
marco.alessi@eng.it
[3] R&D Department, Engineering Ingegneria Informatica SPA, Lecce, Italy
{alessio.camillo,valentina.chetta,enza.giangreco,
davide.storelli}@eng.it

Abstract. This paper describes a Sentiment Analysis (SA) method to analyze tweets polarity and to enable government to describe quantitatively the opinion of active users on social networks with respect to the topics of interest to the Public Administration.

We propose an optimized approach employing a document-level and a dataset-level supervised machine learning classifier to provide accurate results in both individual and aggregated sentiment classification.

The aim of this work is also to identify the types of features that allow to obtain the most accurate sentiment classification for a dataset of Italian tweets in the context of a Public Administration event, also taking into account the size of the training set. This work uses a dataset of 1,700 Italian tweets relating to the public event of "Lecce 2019 – European Capital of Culture".

Keywords: Sentiment analysis · Machine learning · Public administration

1 Introduction

Recently, Twitter, one of the most popular micro-blogging tools, has gained significant popularity among the social network services. Micro-blogging is an innovative form of communication in which users can express in short posts their feelings or opinions about a variety of subjects or describe their current status.

The interest of many studies aimed at SA of Twitter messages, tweets, is remarkable. The brevity of these texts (the tweets cannot be longer than 140 characters) and the informal nature of social media, have involved the use of slang, abbreviations, new words, URLs, etc. These factors together with frequent misspellings and improper punctuation make it more complex to extract the opinions and sentiments of the people.

© Springer International Publishing Switzerland 2015
P. Perner (Ed.): MLDM 2015, LNAI 9166, pp. 98–112, 2015.
DOI: 10.1007/978-3-319-21024-7_7

Despite this, recently, numerous studies have focused on natural language processing of Twitter messages [1–4], leading to useful information in various fields, such as brand evaluation [1], public health [5], natural disasters management [6], social behaviors [7], movie market [8], political sphere [9].

This work falls within the context of the public administration, with the aim to provide reliable estimates and analysis of what citizens think about the institutions, the efficiency of services and infrastructures, the degree of satisfaction about special events. The paper proposes an optimized sentiment classification approach as a political and social decision support tool for governments.

The dataset used in the experiments contains Italian tweets relating to public event "Lecce 2019-European Capital of Culture" which were collected using the Twitter search API between 2 September 2014 and 17 November 2014.

We offer an optimized approach employing a document-level and a dataset-level supervised machine learning classifier to provide accurate results in both individual and aggregated classification. In addition, we detect the particular kind of features that allow obtaining the most accurate sentiment classification for a dataset of the Italian tweets in the context of a Public Administration event, considering also the size of the training set and the way this affect results.

The paper is organized as follows: the research background is described in Sect. 2, followed by a discussion of the public event "Lecce 2019: European Capital of Culture" in Sect. 3, a description of the dataset in Sect. 4 and a description of the machine learning algorithms and optimization employed in Sect. 5. In Sect. 6, the results are presented. Section 7 concludes and provides indications towards future work.

2 Research Background

The related work can be divided into two groups, general SA research and research which is devoted specifically to the government domain.

2.1 Sentiment Analysis Approach

SA and Opinion mining studies, analyzes, classifies documents about the opinions expressed by the people about a product, a service, an event, an organization or a person. The objective of this area is the development of linguistic analysis methods that allows identifying the polarity of the opinions.

In the last decade, SA had a strong development, thanks to the large presence on the World Wide Web, to an increasing number of documents generated by users and thanks to the diffusion of social network.

Every day, millions of users share opinions about their lives, providing an inexhaustible source of data on which it is possible to apply the techniques for opinions analysis.

In 2001 the paper of Das [10] and Tong [11] began to use the term "sentiment" in reference to the automatic analysis of evaluative text and tracking of the predictive judgments. Since that time, the awareness of the research problems and the opportunity related to SA and Opinion Mining has been growing. The growing interest in SA and Opinion Mining is partly due to the different application areas: in commercial field to the analysis of the review of products [12], in political field to identify the electorate mood and therefore the trend in the voting (or abstaining) [13], etc. In social environments, the SA can be used as a survey tool that allows to understand the existing points of view: for example, to understand the opinion that some people have about a subject, to predict the impact of a future event or to analyze the influence of a past occurrence [14]. The big data technologies, the observation methods and the analysis of the behavior on the web, make SA an important decision making tool for the analysis of social network, able to develop relation, culture and sociological debate.

2.2 Sentiment Analysis for Government

In the public administrations, SA is a technique capable to facilitate the creation of relationship between public body and citizens. SA is able to discover the criticalities of this relationship in order to focus on taking right actions.

The SA carried out in social networks allows public administrations to identify and meet user's needs and enables citizens to affect the service delivery and to participate in the creation of a new service, or even to identify innovative uses of existing services [15].

The basic principle of this work is the Open Services Innovation [16], where service innovation is generated by the contribution of citizen judgments and opinions. The center of this system moves from the services to the citizens with their emotions and feelings.

Capturing the citizen opinions using social media can be a mind-safe approach less expensive and more reliable than surveys, where there is the risk of false statements of opinion. Moreover, the social networks reveal a more extensive and widespread involvement of users or citizens; it allows an automatic interception of topics or key events. Analyses of the sentiment of citizen opinions are crucial for the analysis of issues relating to the services provided by public administration and the development of new processes. The objective is to support the public decision maker in the decision process in order to facilitate the growth and innovation aiming at improving the daily life of the community. In this way, the public institution will have the opportunity to acquire, identify, organize, and analyze the "noise" about a public sector and will highlight the quantitative and qualitative aspects that determine the positive or negative sentiment image for the qualification/requalification of its activities.

The information floating within social networks assist public administration to better understand the current situation and to make predictions about the future with respect to many social phenomena.

3 Lecce 2019: European Capital of Culture

Counting 90,000 citizens, Lecce is a mid-sized city, which represents the most important province of Salento located in the "heel" of the Italian "boot". Even though Lecce is known for its enormous cultural artistic and naturalistic heritage, it can also be considered as a typical example of southern Italian city from a socio-economic point of view: poor in infrastructure, with high and increasing unemployment rates. However, despite this disadvantageous context, remarkable investments in research university education and tourism sector have been taking place during the last few years, making Lecce an area of attraction at the international scale. In fact, the only possibility of improvement for the territory is to bet on a radical change.

The change is aiming at obtaining a deep innovation of Lecce and Salento, starting from a concrete enhancement of their resources, where citizens are extremely important.

Three opportunities can be highlighted in order to reconsider the territory in a more innovative and respectful way for the citizenship: the participation to the National Smart City Observatory where Lecce is one of the pilot cities together with Beneven-to, Pordenone and Trento, an urban planning co-created with the citizens, the candidacy as European Capital of Culture 2019.

The European Capital of Culture is a city designated by the European Union for a period of one calendar year during which it organizes a series of cultural events with a strong European dimension [17].

As stated previously, the public administration of Lecce is trying to change approach, creating a shared path towards a social model in which a direct participation and collaboration of the citizens is included. In the guidelines of the candidacy as European Capital of Culture, one of the main criteria of the bid book evaluation is "the city and citizens" [18], referring to concrete initiatives that must be launched to attract local neighboring and foreign citizens' interest and participation. Moreover, these initiatives should be long-term and should be integrated in cultural and social inno-vation strategies.

The challenge of the candidacy is making Lecce a Smart City, which means moving towards an innovative ecosystem that improves citizens' quality of life through an efficient use of resources and technologies. A fundamental aspect is citizens' par-ticipation aimed at collecting their needs as beneficiaries and main characters in the open innovation process.

The urgent need to express a strong break with the past is well summarized in the slogan for Lecce as European Capital of Culture: "Reinventing Eutopia", that means reinterpreting the European dream from a political, social, cultural and economic perspective. That concept is composed by eight utopias for change, the main of which is DEMOCRAtopia until 2019. As described in the bid book (Lecce 2019, 2013), DEMOCRAtopia refers to the creation of a climate of trust, awareness, collaboration, responsibility, and ownership, with a special emphasis on the collective knowledge and on a development perspective oriented to citizens' dreams and needs [19].

The title of European Capital of Culture may provide the government and the communities with the opportunity to thrive together, in order to achieve a medium or long-term goal. This requires a constant dialogue, which is an essential component in this process activation. The candidacy is the beginning of a journey, a playground for the future, a dream, a laboratory of ongoing and future experiments, and an opportunity to become what we are, if we really want to [18].

The title of European Capital of Culture is a part of Open Service Innovation Process because have as objective the creation of new processes, services and events in the city of Lecce.

The Open Service Innovation Process was realized by allowing citizens to contribute to the creation and design of new services, infrastructure and events.

The realization of the idea Management System and of events for promotion allowed the involvement of citizen.

This paper is an instrument of analysis of an Open Service Innovation Process that allows observing and evaluating the sentiment of citizen about Lecce 2019 event. The opinion classification can be used to understand the strengths and weaknesses of the process.

4   Dataset

We collected a corpus of tweets using the Twitter search API between 2 September 2014 and 17 November 2014, period in which there were more Twitter messages about the event. We extracted tweets using query-based search to collect the tweets relating to "#Lecce 2019" and "#noisiamolecce2019", hashtag most used for this topic. The resulting dataset contains 5,000 tweets. Duplicates and retweets are automatically removed leaving a set of 1,700 tweets with a class distribution as shown in Table 1.

Table 1. Class distribution

Positive	391	23 %
Neutral	1241	73 %
Negative	68	4 %
Total	1,700	

In order to achieve a training set for creating a language model useful to the sentiment classification, a step of manual annotation was performed. This process involved three annotators and a supervisor. The annotators were asked to identify the sentiment associated with the topic of the tweet and the supervisor has developed a classification scheme and created a handbook to train annotators on how to classify text documents.

The manual coding was performed using the following 3 labels:

- Positive: tweets that carry positive sentiment towards the topic Lecce 2019;
- Negative: tweets that carry negative sentiment towards the topic Lecce 2019;
- Neutral: tweets which do not carry any sentiment towards the topic Lecce 2019 or tweets which do not have any mention or relation to the topic.

Each annotator has evaluated all 1,700 tweets. For the construction and the dimension of the training set, see next section.

Analysis of the inter-coder reliability metrics demonstrates that annotators were agreed for more than 80 % of the documents (the agreement is average of 0.82) with good inter-coder reliability coefficients (Multi-Pi (Siegel and Castellan), Multi-Kappa (Davis and Fleiss), Alpha (Krippendorf)) [20].

These measures are very important: a low inter-coder reliability means to generate results that could not be considered "realistic" for sentiment analysis, whereas "realistic" means conceptually shared with the common thought. However, these metrics do not affect the accuracy values of classification algorithms. It can therefore happen that, despite the accuracy of the classification algorithm is equal to 100 %, the analysis results are not conceptually consistent with the reality.

5 Method

5.1 Features Optimization

Before performing the classification of sentiment, the text has been processed by a preprocessing component, using the following features:

- *Identification of expressions with more than one word (n-grams).* Often opinions are transmitted as a combination of two or more consecutive words. This makes incomplete the analysis performed on tokens that consists of only single words. Generally, single terms (unigrams) are considered as features, but in several researches, albeit with contrasting results, also n-grams of words were considered. From some experimental results, it seems that the features formed by the more frequent n-grams in the collection of documents, added to those consisting unigrams, increase the classifier performance because they are grasped as a whole, parts of the expression that can't be separate. However, sequences with length n > 3 are not helpful and could lead to a reduction in performance. For this reason, the choice fell on the use of unigrams, bigram and their combination.
- *Conversion of all letters in lowercase.* The preservation of lowercase and uppercase letters for all words can create a useless feature chipping with an increment of processing time and a reduction of system performance. One of the most successful approaches in the texts preprocessing for SA is to edit all letters in lowercase; even if a term in all caps may want to communicate an intensification of an opinion, this feature degrades performance and does not justify its use.
- *Hashtags inclusion with the removal of the "#" character.* Often hashtags are used to characterize the sentiment polarity of an opinion expressed, as well as to direct the text towards a specific subject; consider, for example, the hashtag "#happy" and "#bad" that communicate a direct sentiment. An idea may be, therefore, to include the text of the hashtag among valid elements to characterize the language model.
- *Twitter user names removal.*
- *Emoticon removal or emoticon replacement with their relevant category.* Emoticons are extremely common elements in social media and, in some cases, are carriers of reliable sentiment. The replacement of emoticons with a specific tag according to their sentiment polarity (positive or negative) can be a winning approach in some contexts. However, we must be careful: these representations firstly express emotions and not sentiment intended as being for or against a particular topic. For example, being disappointed for a particular event, but still

favorable, creates contrast between the emoticons seen as negative (⊗) and the sentiment favorable for that topic.

- *URL replacement with the string "URL"*;
- *Removal of words that do not begin with a letter of the alphabet.*
- *Removal of numbers and punctuation.* Another element that can characterize text polarity is punctuation, in particular exclamation points, question marks and ellipsis. The inclusion of these elements in the classification process, can lead to a more accurate definition of sentiment, also taking into account the repetitions that intensify the opinion expressed. However, the inclusion of punctuation slows the classifier speed.
- *Stopwords removal.* Some words categories are very frequent in the texts and are generally not significant for sentiment analysis. This set of textual elements includes articles, conjunctions and prepositions. The common use, in the field of sentiment analysis, is to remove these entities from the text before the analysis.
- *Removal of token with less than 2 characters*;
- *Shortening of repeated characters.* Sometime words are lengthened with the repetition of characters. This feature can be a reliable indicator of intensified emotion. In the Italian language there are no sequences of three or more identical characters in a word, so we can consider that such occurrences are an extension of the base word. Since the number of repeated characters is not predictable and because the probability that small differences are significant is low, it is possible to map sequences of three or more repeated characters in sequences of only two characters.
- *Stemming execution.* Stemming is the process of reducing words to their word stem. This can help to reduce the vocabulary size and thus to increase the classifier performance, especially for small datasets. However stemming can be a double-edged sword: the reduction of all forms of a word can eliminate the sentiment nuances that, in some cases, make the difference, or it can lead to the unification of words that have opposite polarities. It seems that the benefits of stemmer application are more evident when the documents of training are few, although the differences are generally imperceptible.
- *Part-of-Speech Tagging.* There are many cases in which there is an interpretation conflict among words with same representation but different role in the sentence. This suggests that it may be useful to run a PoS tagger on the data and to use the pair word-tag as a feature. In the literature, there is often a slight increase in accuracy with the use of a PoS tagger at the expense of processing speed, which slows the preprocessing phase.

As specified in the following paragraph, 8 classification stages were performed, each with different approaches of text preprocessing and features selection, using training set of different sizes. For each cycle a 10-fold cross-validation was performed, dividing the manually annotated dataset into a training set and a fixed test set of 700 tweets. Every 10-fold validation was repeated 10 times. This was necessary to obtain reliable results in terms of accuracy and suffering from a minimum amount of error [29].

All stages contain the following preprocessing steps: all letters are converted to lowercase; user names, URLs, numbers, punctuation, hashtags, words that do not begin with a letter of the Latin alphabet and those composed of less than two characters, are removed. In addition to these, the sets of features that characterize the 8 classification stages are the following:

- set 1: uni-grams;
- set 2: bi-grams;
- set 3: uni-grams + bi-grams;
- set 4: set 1 + stopwords removal + repeated letters shortening;
- set 5: set 4 + stemming;
- set 6: set 5 + emoticon inclusion with replacement;
- set 7: set 6 + hashtags inclusion with character "#" removal;
- set 8: set 7 + PoS tag.

5.2   Training Set Dimension

To classify the documents in the test, the supervised machine learning methods, as described below, uses relationships between features in the training set. The training set then is a random sample of documents, representative of the corpus to be analyzed [21].

In order to achieve acceptable results with supervised machine learning approaches, the training set must be constituted by a sufficient amount of documents. Hopkins and King argue that five hundred documents are a general threshold in order to have a satisfactory classification. Generally, this amount is effective in most cases, but we must always consider the specific application of interest. For example, to the increase of the number of categories in a classification scheme, necessarily increases the number of information crucial to learn the relationships between the words and each category, and, consequently, this requires the increase of the documents number in the training set.

The creation phase of the training set is the most costly in terms of human labor. In order to identify the minimum amount of documents to be annotated manually maintaining a good accuracy of the results, it was performed an analysis that compares different evaluation metrics of the various classifiers for each set of features identified, varying the size of the training set from time to time.

10 classification runs were performed, bringing the documents of the training set from 100 to 1,000 in steps of 100. For each cycle, a 10-fold cross-validation, repeated 10 times, was performed to minimize the error estimation accuracy.

5.3   Sentiment Classification

Today there are many algorithms of SA and this amount of solutions evidences how hard it is to determine the right solution for every context. The algorithm that provides 100 % of efficiency does not exist; each approach tries to reach an improvement in a particular aspect of the analysis, in a specific context and in a particular type of data to process [22].

The main classes of SA algorithm, lexicon-based and machine learning-based, have different fields of application and are based on different requirements [22].

The creation of a dictionary of terms previously classified has low accuracy of results, considering the continuous changing of the language and the change in the meaning of words in different contexts.

Moreover, it is arduous to find, in this type of dictionary, the words used in the slang of the social networks and so, it's very difficult to identify the polarity of the words. Surely, the strength of this type of approach is the speed with which the results can be obtained, allowing to obtain real-time analysis. Moreover, the absence of a manually annotated training set makes the use of these methods more attractive.

The supervised machine learning approach allows obtaining an improved accuracy but requires a training phase.

Using specific training set in these methods it is possible to adapt the tool to the context of interest. If a supervised machine learning algorithm is trained on a language model, specific for a given context and for a particular type of document, it will provide less accurate results by moving the analysis to another context of interest [22].

A tool for SA in the context of public administration must be able to provide reliable estimates and analysis on everything that citizens think about the institutions, the efficiency of services and infrastructure and the level of satisfaction about events. We believe, therefore, that the most suitable approach to the analysis of the sentiment in this context, involves the use of supervised machine learning algorithms.

This work implements the classification of sentiment as follows:

(1) Single text document-level, using algorithms Naive Bayes Multinomial (NB) in binary version [23] and Support Vector Machine (SVM) in the "one-vs-rest" variant [24].
(2) Globally, at the entire dataset-level, using ReadMe [25].

Document-level Classification. Generally, the classification of sentiment is performed on each distinct document. This is very useful when you want to characterize the performance of sentiment over time or when you want to identify the motivations that drive people to publish their opinions. The more accurate supervised algorithms are Naive Bayes (NB) in binary Multinomial version and Support Vector Machine (SVM) of the type "one-vs-rest". This choice derives from the different analyses in the literature showing the duality of these algorithms: Naive Bayes is characterized by a rapid classification and the need for a training set of lower size than required by SVM, which, however, shows results that are more accurate when there are many training documents [26].

Dataset-level classification. It is also helpful to have accurate estimations of global sentiment of the entire dataset of document. While the state of the art algorithms of supervised machine learning achieve high accuracy in the individual classification of documents, on the other hand tend to provide results affected by greater error when individual results are aggregated to obtain an overall global measure. This requirement has led to select, as an additional classification algorithm, the approach developed by Hopkins and King, ReadMe; this technique allows obtaining an estimate of the proportions of global sentiment of a dataset with a low margin of error, exceeding the predictive performance of the other algorithms [25].

To evaluate the algorithms that classify individual text documents, the following performance metrics were measured: Accuracy (A), Precision (P), Recall (R), F1-measure [27].

In order to compare and evaluate classification algorithms that predict the overall proportion of the categories of sentiment, the following statistics metrics are used: Mean Absolute Error (MAE) and Root Mean Square Error (RMSE) [25].

6   Results

The aim of this analysis is to identify which combination of features and training set size produces an optimal sentiment classification of short text messages related to the event "Lecce 2019".

The Root Mean Square Error value has been calculated for the three algorithms (svm, nb and readme) and for the first three sets of features (1, 2, 3), by varying the size of the training set from 100 to 1,000 in steps of 100.

Table 2. Root mean square error value for three different sets of features and varying the size of the training set

		svm01	nb01	readme01	svm02	nb02	readme02	svm03	nb03	readme03
	100	0,115	0,140	0,115	0,151	0,153	0,150	0,152	0,147	0,116
	200	0,136	0,142	0,094	0,144	0,145	0,112	0,142	0,142	0,098
	300	0,132	0,139	0,089	0,137	0,139	0,093	0,134	0,136	0,086
	400	0,127	0,134	0,081	0,130	0,134	0,083	0,128	0,132	0,079
Traing set dimension	500	0,123	0,130	0,075	0,125	0,129	0,077	0,123	0,128	0,073
	600	0,119	0,125	0,071	0,121	0,125	0,073	0,119	0,123	0,071
	700	0,115	0,121	0,069	0,116	0,120	0,070	0,114	0,118	0,069
	800	0,111	0,116	0,067	0,112	0,116	0,068	0,110	0,114	0,067
	900	0,108	0,113	0,066	0,108	0,112	0,067	0,107	0,111	0,066
	1000	0,105	0,110	0,065	0,105	0,109	0,065	0,104	0,108	0,065

As seen from Table 2, the Root Mean Square Error value for the three classification algorithms shows that, varying the size of the training set, the first three set of features performs almost the same way. The use of bi-grams (set 2) generates slightly worse results than the other two features sets, while the joint use of uni-grams and bi-grams (set 3) produce a greater number of features that slows down a little the classification step. For this reasons, the subsequent sets were constructed starting from the set 1 and adding other methods of text preprocessing and features selection.

In all cases, ReadMe appears to be the best algorithm for the aggregate sentiment (dataset-level) classification, even with a training set of small dimension. This result validate our choice.

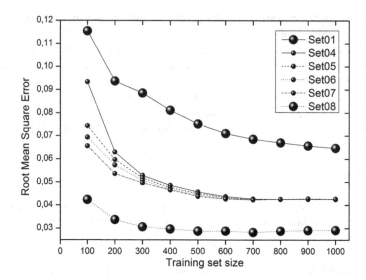

Fig. 1. Root mean square error values for ReadMe algorithm using different sets of features and varying the size of the training set

Figure 1 shows how Root Mean Square Error of ReadMe classification varies by changing the set of features.

The Root Mean Square Error decreases adding to unigrams other feature selection and extraction methods, that is decreasing the number of words taken into account or characterizing them in different ways. This reduction is particularly evident from set 4 onwards, where stopwords removal and repeated characters shortening allow to obtain a Root Mean Square Error value of 4.2 %. However, the application of further pre-processing methods (sets 5, 6, 7) is effective only with a small training set. It can be also noted that for sets 4, 5, 6 and 7, the increase of the training tweets number over 600 has no effect on the trend of RMSE.

It is rather remarkable the reduction of Root Mean Square Error with the application of set 8. For this set, we reach the least error (2.8 %) with a training set of 700 documents. However, having a training set greater than 400 tweets is not very useful in terms of error reduction, since the Root Mean Square Error reduction is about 0.1 %.

There is not much difference in terms of computational complexity to create training sets with sets of features 4, 5, 6, 7. All these sets are quite similar, but the set 7 give the best results. The use of PoS tagging (set 8) instead, introduces a slowdown in the preprocessing stage, but reaching the best results among all feature sets.

The measures of Accuracy (Fig. 2), carried out for SVM and NB algorithms using the set 7, shows almost the same trend for the two algorithms. This result, which see the weakly NB next to the best state-of-art classification algorithm SVM, is probably due to the kind of documents analyzed, namely short text messages (tweets), and is in agreement with the literature that also point out how SVM perform better with longer texts [28].

As shown in Fig. 3 and as already pointed out previously, the analysis of the accuracy for the other sets of features doesn't lead to significant increases. For all sets,

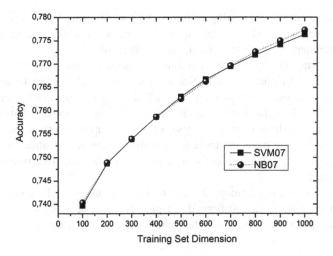

Fig. 2. Accuracy of SVM and NB using features set 7 varying the training set size

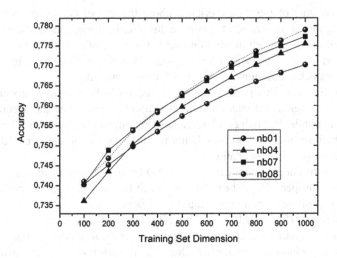

Fig. 3. Accuracy values of NB with different sets of features varying the training set size

the accuracy increases as the size of the training set, up to a value of 78 % for the NB algorithm with the use of PoS tagging (set 8).

The same trends are obtained in the measures Precision (P), Recall (R), F1-measure.

In summary, for the dataset-level sentiment analysis of the tweets, the choice of unigrams features with the stopwords and repeated characters shortening, stemming, emoticon replacement, hashtags inclusion with "#" character removal and PoS tagging (set 8), proved to be the most successful. We believe that the best number of tweets to be included in the training set to get good results with a good compromise between error and human labor is 300, using features set 8, or 500 using features sets 4, 5, 6 or 7. However, even with 200–300 training tweets you can achieve good results.

For a document-level classification, an accuracy of 78 % is achieved with NB algorithm by using the 1,000 tweets training set with feature set 8; however, for a document-level classification it is sufficient to use set 6 or 7, since these generates almost the same accuracy values than set 8, eliminating slowness of the PoS tagging step. Here, unlike the dataset-level classification, there seems no be a flattening of the accuracy increase as the size of training set grows. So, up to a certain limit, the more training tweets you will have, the more accurate the sentiment classification. However, we believe that a training set consisting of about 300–400 tweets can generate acceptable results.

In both types of classification, the use of PoS tagging must be carefully chosen according to the type of application; if the goal is real-time sentiment analysis, this preprocessing approach must be avoided otherwise it can be used.

7   Conclusion

Following the state of the art experience about the use of algorithms for sentiment classification, this paper intends to propose an optimized approach for the analysis of tweets related to a public administration event. The possibility to extract opinions from social networks and classify sentiment using different machine learning algorithms, make this a valuable decision support tool for Government.

To meet this need, this paper proposes an approach that considers document-level and dataset-level sentiment classification algorithms to maximize the accuracy of the results in both single document and aggregated sentiment classification. The work also point out which features sets produce better results compared to the size of the training set and to the level of classification.

We have introduced a new dataset of 1,700 tweets relating to the public event of "Lecce 2019: European Capital of Culture". Each tweet in this set has been manually annotated for positive, negative or neutral sentiment.

An accuracy of 78 % is achieved using NB document-level sentiment classification algorithm and unigrams features with stopwords removal, repeated characters shortening, stemming, emoticon replacement, hashtags inclusion with "#" character removal and PoS tagging with a training set of 1,000 tweets. A training set of 300–400 tweets can be a reasonable lower limit to achieve acceptable results.

Our best overall result for a dataset-level classification is obtained with the ReadMe approach using a feature set that included also the PoS tagging and a training set of 700 tweets. Using this optimal set of features, a dataset-level sentiment classification reports a low Root Mean Square Error value, equal to 2.8 %. However, with a training set of 400 tweets can be obtained almost the same results.

In a context such as public administration, the emotional aspect of the opinions can be crucial. Future work involves carrying out algorithms that allow extracting and detecting the type of emotions or moods of citizens in order to support the decisions for the public administration.

References

1. Jansen, B., Zhang, M., Sobel, K., Chowdury, A.: Twitter power: tweets as electronic word of mouth. J. Am. Soc. Inf. Sci. Technol. **60**(11), 2169–2188 (2009)
2. O'Connor, B., Balasubramanyan, R., Routledge, B., Smith, N.: From tweets to polls: linking text sentiment to public opinion time series. In: Proceedings of the Fourth International Conference on Weblogs and Social Media, ICWSM 2010, Washington, DC, USA (2010)
3. Tumasjan, A., Sprenger, T., Sandner, P., Welpe, I.: Predicting elections with Twitter: what 140 characters reveal about political sentiment. In: Proceedings of the Fourth International Conference on Weblogs and Social Media, ICWSM 2010, Washington, DC, USA (2010)
4. Kouloumpis, E., Wilson, T., Moore, J.: Twitter sentiment analysis: the good the bad and the OMG! In: Proceedings of the Fifth International Conference on Weblogs and Social Media, ICWSM 2011, Barcelona, Catalonia, Spain (2011)
5. Salathe, M., Khandelwal, S.: Assessing vaccination sentiments with online social media: implications for infectious disease dynamics and control. PLoS Comput. Biol. **7**(10), 1002199 (2011)
6. Mandel, B., Culotta, A., Boulahanis, J., Stark, D., Lewis, B., Rodrigue J.: A Demographic analysis of online sentiment during hurricane irene. In: Proceedings of the Second Workshop on Language in Social Media, LSM 2012, Stroudsburg (2012)
7. Xu, J.-M., Jun, K.-S., Zhu, X., Bellmore, A.: Learning from bullying traces in social media. In: HLT-NAACL, pp. 656–666 (2012)
8. Asur, S., Huberman, B.A.: Predicting the future with social media. In: Proceedings of the 2010 International Conference on 132 Web Intelligence and Intelligent Agent Technology, WI-IAT 2010, vol. 01, pp. 492–499. IEEE Computer Society, Washington, D.C., USA (2010)
9. Bakliwal, A., Foster, J., van der Puil, J., O'Brien, R., Tounsi, L., Hughes, M.: Sentiment analysis of political tweets: towards an accurate classifier. In: Proceedings of the Workshop on Language in Social Media (LASM 2013), pp. 49–58. Atlanta, Georgia (2013)
10. Sanjiv Das, M.C.: Yahoo! for Amazon: extracting market sentiment from stock message boards. In: Proceedings of the Asia Pacific Finance Association Annual Conference (APFA) (2001)
11. Tong, R.M.: An operational system for detecting and tracking opinions in on-line discussion. In: Proceedings of the SIGIR Workshop on Operational Text Classification (OTC) (2001)
12. Dave, K., Lawrence, S., Pennock, D.M.: Mining the peanut gallery: opinion extraction and semantic classification of product reviews. In: Proceedings of WWW, pp. 519–528 (2003)
13. Neri, F., Aliprandi, C., Camillo, F.: Mining the web to monitor the political consensus. In: Wiil, U.K. (ed.) Counterterrorism and Open Source Intelligence. LNSN, pp. 391–412. Springer, Vienna (2011)
14. Kale, A., Karandikar, A., Kolari, P., Java, A., Finin, T., Joshi, A.: Modeling trust and influence in the blogosphere using link polarity. In: Proceedings of the International Conference on Weblogs and Social Media (ICWSM) (2007)
15. Dolicanin, C., Kajan, E., Randjelovic, D.: Handbook of Research on Democratic Strategies and Citizen-Centered E-Government Services, pp. 231–249. IGI Global, Hersey (2014)
16. Chesbrough, H.: Open Services Innovation. Wiley, New York (2011)
17. http://ec.europa.eu/programmes/creative-europe/actions/capitals-culture_en.htm
18. http://www.capitalicultura.beniculturali.it/index.php?it/108/suggerimenti-per-redigere-una-proposta-progettuale-di-successo
19. http://www.lecce2019.it/2019/utopie.php

20. Koch, G.G., Landis, J.R.: The measurement of observer agreement for categorical data. Biometrics **33**, 159–174 (1977)
21. Hopkins, D., King, G.: Extracting systematic social science meaning from text. Unpublished manuscript, Harvard University (2007). http://gking.harvard.edu/files/abs/words-abs.shtml
22. Liu, B.: Sentiment analysis and opinion mining. Synthesis Lectures on Human Language Technologies. Morgan & Claypool Publishers (2012)
23. Narayanan, V., Arora, I., Bhatia, A.: Fast and accurate sentiment classification using an enhanced naive Bayes model. In: Yin, H., Tang, K., Gao, Y., Klawonn, F., Lee, M., Weise, T., Li, B., Yao, X. (eds.) IDEAL 2013. LNCS, vol. 8206, pp. 194–201. Springer, Heidelberg (2013)
24. Yang, Y., Xu, C., Ren, G.: Sentiment Analysis of Text Using SVM. In: Wang, X., Wang, F., Zhong, S. (eds.) EIEM 2011. LNEE, vol. 138, pp. 1133–1139. Springer, London (2011)
25. King, G., Hopkins, D.: A method of automated nonparametric content. Am. J. Polit. Sci. **54** (1), 229–247 (2010)
26. Hassan, A., Korashy, H., Medhat, W.: Sentiment analysis algorithms and applications: a survey. Ain Shams Eng. J. **5**, 1093–1113 (2011)
27. Huang, J.: Performance Measures of Machine Learning. University of Western Ontario, Ontario (2006)
28. Wang, S., Manning, C.D.: Baselines and bigrams: simple, good sentiment and topic classification. In: Proceedings of the 50th Annual Meeting of the Association for Computational Linguistics: Short Papers, vol. 2. Association for Computational Linguistics (2012)
29. Refaeilzadeh, P., Tang, L., Liu, H.: Cross-validation. In: Liu, L., Özsu, M.T. (eds.) Encyclopedia of Database Systems, pp. 532–538. Springer, New York (2009)

Data Preparation and Missing Values

A Novel Algorithm for the Integration
of the Imputation of Missing Values
and Clustering

Roni Ben Ishay[1,2(✉)] and Maya Herman[1]

[1] The Open University of Israel, Ra'anana, Israel
roni@ash-college.ac.il, maya@openu.ac.il
[2] The Ashkelon Academic College, Ashkelon, Israel
roni@ash-college.ac.il

Abstract. In this article we present a new and efficient algorithm to handle missing values in databases applied in data mining (DM). Missing values may harm the calculation of the clustering algorithm, and might lead to distorted results. Therefore missing values must be treated before the DM. Commonly, methods to handle missing values are implemented as a separate process from the DM. This may cause a long runtime and may lead to redundant I/O accesses. As a result, the entire DM process may be inefficient. We present a new algorithm (*km-Impute*) which integrates clustering and imputation of missing values in a unified process. The algorithm was tested on real Red wine quality measures (from the UCI Machine Learning Repository). *km-Impute* succeeded in imputing missing values and in building clusters as a unified integrated process. The structure and quality of clusters which were produced by *km-Impute* were similar to clusters of *k-means*. In addition, the clusters were analyzed by a wine expert. The clusters represented different types of Red wine quality. The success and the accuracy of the imputation were validated using another two datasets: White wine and Page blocks (from the UCI). The results were consistent with the tests which were applied on Red wine: The ratio of success of imputation in all three datasets was similar. Although the complexity of *km-Impute* was the same as *k-means*, in practice it was more efficient when applying on middle sized databases: The runtime was significantly shorter than *k-means* and fewer iterations were required until convergence. *km-Impute* also performed much less I/O accesses in comparison to *k-means*.

Keywords: Clustering · Data mining · Pre-processing · Missing values · Imputation · k-means · Integration

1 Introduction

Databases which are applied in Data Mining (DM) are usually incomplete, and may contain type-value inconsistencies, errors, outliers, and missing values. Running DM on an incomplete database (such as missing values) may cause miscalculations, and bias the results. For example, the calculation of centroids of clusters involves the entire cluster's data points. If data points contain missing values, the algorithm may produce

© Springer International Publishing Switzerland 2015
P. Perner (Ed.): MLDM 2015, LNAI 9166, pp. 115–129, 2015.
DOI: 10.1007/978-3-319-21024-7_8

wrong or inaccurate centroids. Hence, the structure of clusters may be inaccurate which will lead to wrong conclusions and insights. To avoid this, the database (DB) must be cleaned and prepared (pre-processed), prior to the DM process. The pre-processing involves actions such as: data type conversion, outliers handling, value normalizing, and handling missing values. One of the challenging issues in the pre-processing stage is handling missing values in attributes. An attribute may be at a "missing value" status due to a human error, or a value unavailability (for example, unknown address). In the literature, there are various methods to handle missing values [2, 3]. These methods may be classified into three types:

1. Ignoring/discarding incomplete records [2].
2. Imputing arbitrary values (such as average value, median, etc.) [2].
3. Estimating the value using a prediction model [2].

Ignoring incomplete records (type 1) is a fast and simple method. But it may lead to data loss. This will have an adverse effect on the quality of the DM model. Imputing arbitrary values (type 2) is easy to apply. Yet, the DM model may be biased because possible associations between attributes may be ignored. Using a prediction model (type 3) is relatively accurate, but requires more I/O transactions compared to other methods. Therefore, choosing the right method to handle missing values is a very significant step [1, 4]. Usually, missing values are handled as a standalone process, which is applied during the pre-processing. The process requires a series of I/O accesses to the DB. Later, when building the DM model, more I/O accesses are applied, regardless of the previous process. This may lead to multiplication of I/O accesses. The result is an increased runtime, hence - an inefficient DM process.

Several imputation techniques are mentioned in the literature, and they can be classified into three main groups:

1. Simple imputation [2].
2. Estimation by model [2].
3. Clustering or k-nearest-neighbor algorithms [1, 8, 9].

Simple imputation technique (group 1) involve arbitrary imputation (such as the average value) or manual imputation. Although these are simple methods, they are inaccurate, and may cause a loss of vital information. Estimation by model (group 2) applies statistics methods such as linear regression, or prediction to estimate a suitable value to impute [2]. Linear regression is relatively accurate, but suitable when only one attribute has a missing value [2]. Clustering-based, and k-nearest-neighbor-based imputation algorithms (group 3) are based on the similarity between data points. A missing value of data point may be imputed by sampling the nearest cluster, or from one of the k-nearest-neighbors [1, 8, 9]. Since these algorithms are related to our study, we present several versions [3, 4, 6–9]. K-Nearest Neighbors Imputation (KNNI) [1, 7, 9] is composed of 3 main steps: (1) Randomly selecting K complete data objects as centroids. (2) Reducing the sum of distances of each object from the centroid. (3) Imputing incomplete attributes, based on the clusters' information. However, KNNI is suitable usually when only one attribute has a missing value. Therefore, some heuristics are required: imputing the average value of the respective attribute of nearest neighbors [9]. An improvement to KNNI is Imputation Fuzzy K-means Clustering

(FKMI) algorithm [1, 9]. FKMI provides a membership function which determines the degree of a data point to belong to a specific cluster [9]. Note however that these clusters cannot be used for a DM model, since data points may appear in several clusters. Clustering Missing Value Imputation (CMI) Algorithm [6] implements a kernel-based method to impute values. First, the DB is divided into clusters, including data points with missing values. Next, the impute value is generated based on within clusters' value. The drawback of CMI is possible inaccurate imputations. This is because missing value attributes are included in generation process, and may bias the value. A Dynamic Clustering-Based Imputation (DCI) technique is described in [7]. Missing values are imputed based on the mean value of the corresponding attribute in other instances within the same cluster. However, unlike typical clustering algorithms, DCI allows instances to become members of many clusters. The drawback of DCI is that the clusters may be significantly different from clusters which are produced by classic clustering algorithm. Therefore, DCI can be applied for pre-processing only.

These imputation methods represent the common approach with regard to preparing DB for DM. They are applied during the pre-processing phase, as a standalone process. This approach has two main disadvantages:

1. The DB is parsed at least twice. First, during the imputation, second during running the DM algorithm. It causes a multiplication of I/O accesses, and impacts the overall runtime.
2. When applying clustering for imputation, the produced clusters are not used later to build the DM model. Re-building the clusters from scratch is inefficient and time consuming.

We argue that these issues can be addressed by integrating the imputation process and the clustering process. This article is organized as follows: Sect. 2 presents our new *km-Impute* algorithm which integrates the imputation and clustering. Section 3 is dedicated to experiments on *km-Impute* algorithm using Red wine quality measurements database [5, 10], analysis of the results and conclusions. Section 4 concludes this article with a description of suggestions for further research.

2   The *km-Impute* Algorithm

2.1   The Rationale

The main idea is to implement both clustering and the imputation of missing values as an integrated process. Suppose that a data point (*dp*) has one (or more) attribute(s) with missing value(s). Also, assume that *dp* is very close to a specific cluster. Therefore, it is likely that within this cluster one (or more) data point(s) which is/are relatively similar to *dp* will be found. These data points are considered as sampling candidates for imputation, and their respective values will be imputed into *dp*. Once the imputation is complete, *dp* will be joined to the cluster. This integration approach is more efficient than the common approaches because the algorithm produces fewer I/O accesses and the overall runtime is shorter.

2.2   The Pseudo Code

Notations

- orgDB - A database which may contain incomplete data points;
- completeDB – A complete version of orgDB which was built during the process of imputation of missing values;
- sampleDB – A random sample of complete data points, where sampleDB $\subseteq$ orgDB
- and {sampleDB} $\ll$ {orgDB}
- N - the size of orgDB
- D - the dimensions of orgDB
- K - number of clusters
- c_j - a cluster where $0 < j <= K$
- C - a group of K clusters {c_j}, where $0 < j <= K$
- dp_i - a data point, where $dp_i \in$ orgDB and $0 < i <= N$
- $centroid_j$ - the centroid of c_j, where $0 < j <= K$
- $radius_j$ - the radius of c_j, where $0 < j <= K$
- $randomDp_j$ - a random sampled data point, where random $Dp_j \in c_j$ $0 < j <= K$, and $0 < i <= |c_j|$
- maxSamples - the maximum data points to sample for imputation

Pseudo code

```
Program kmImpute()
1.  input: orgDB
2.  output: C,completeDB
3.  begin
4.    C=kmeans(sampDb); // create initial clusters

5.     for (each dpᵢ ∈ orgDB)
6.        find the nearest cluster cⱼ to dpᵢ ;
7.        if (dpᵢ is complete) then
8.           addDataPoint(dpᵢ, cⱼ); // add dpᵢ to cⱼ,
9.        else // dpᵢ has (a)missing value(s)
10.         repeat
11.            if (euclidianDistance(dpᵢ,centroidⱼ)< radiusⱼ)then
12.               imputeDp(dpᵢ ,centroidⱼ);
13.               addDataPoint(dpᵢ ,cⱼ); // add dpᵢ to cⱼ.
14.            else // randomly pick a complete data point
15.               If(euclidianDistance(dpᵢ,randomDpⱼ)< radiusⱼ)
16.                  then
17.                  imputeDp(dpᵢ,randomDpⱼ)
18.                  addDataPoint(dpᵢ ,cⱼ); // add dpᵢ to cᵢ
19.         until((impute is done)or(reached maxSamples))
20.         if (dpᵢ completed) then
21.            update centroidⱼ;
22.            update radiusⱼ;
23.         else
24.            discard dpᵢ from orgDB;
25.   end for
26.   C=kmeans(completeDB); // re-converge the clusters
27. end program
```

Pseudo code description

Line 4: Start with an initial K seeds (which represents the clusters' centroids).

Lines 5–25: Grow cluster(s) as follows:

Lines 6–8: For a given data point, if the data point is complete find nearest cluster and add it to the cluster.

Lines 10–20: If the data point has missing attribute values (incomplete), test if it lies inside a cluster's calculated radius, and impute the cluster's centroid value.

Else, pick a random data point which is a member of the nearest cluster. If the distance between the random data point and the incomplete data point lies inside the cluster's radius - impute values and add the completed data point to the cluster. Otherwise, pick another random data point, and try again.

Lines 21, 22: Update the centroid and radius of the cluster.

Line 26: Run *k-means* on the completed DB to re-converge the clusters, (data points may swap between clusters, until convergence is achieved and *km-Impute* terminates).

2.3   Complexity Analysis

The complexity of *km-Impute* is similar to *k-means* complexity: $O(N^{Dk+1}\log N)$ where N is the DB size, K is the number of clusters and D is the number of attributes.

Nevertheless, *km-Impute* runtime is significantly shorter than *k-means*. Since the clusters in *km-Impute* are built iteratively, fewer swaps of data points between clusters are applied during the buildup, in comparison to *k-means*. Therefore, *km-Impute* requires much less I/O accesses than *k-means*, which gains better runtime than *k-means*. The above issues will be demonstrated in Sect. 3.

3   Experiments and Results

3.1   General

The goals of the experiments were to test *km-Impute* using a real database, and compare the results to *k-means*. The tests were conducted under the following assumptions:

1. The database contains numeric type values only.
2. All attributes have equal weight with regard to the clustering process.

The tests were done using Red wine quality measure database (from the UCI repository) [5, 10]. The Red wine contains 12 wine quality (numeric) attributes. It contains 1,599 complete records. 20 % of Red wine records were randomly picked and the values of random attributes were removed. Three versions of Red wine dataset were prepared as follows:

1. Red-Wine-1: One attribute with missing value per record.
2. Red-Wine-2: Two attributes with missing value per record
3. Red-Wine-3: Three attributes with missing value per record

Note that the original values were stored for later comparison.

The experiments involved the following tests:

1. Imputation test
2. Correctness of algorithm test
3. Clustering and imputation test
4. Performance test

3.2 Imputation Test

Objective. To assess the imputation success and accuracy of *km-Impute*, using an incomplete database.

Results and Discussion. *km-Impute* ran three times using Red-Wine-1, Red-Wine-2 and Red-Wine-3 datasets. The algorithm succeeded in imputing value into attributes with missing values. The success and accuracy of imputation varied according to the database which was used. For example, with Red-Wine-1 (1 attribute with missing value), *km-Impute* succeeded in imputing 93 % of the missing values (Fig. 1). 214 values (49 + 71 + 94 = 214) were imputed with an accuracy above 80 % (Fig. 2). When testing with Red-Wine-2 and Red-Wine-3, the success and accuracy of imputation decreased accordingly. For example, with Red-Wine-3 the success of imputation was 42 %. Fewer values (122) were imputed with an accuracy above 80 % compared to the imputation with Red-Wine-1.

Most of the records which contained 2 or 3 attributes with missing values, were imputed very accurately (Fig. 2). For example, 87 records which contained 2 attributes with missing values, have been imputed with accuracy of above 90 %. Fewer records (25 and 57) were imputed with an accuracy of 80 % and 90 % respectively.

For additional validation of the success and accuracy of the *km-Impute* algorithm, we ran the algorithm on a white wine quality measures dataset, and block page dataset

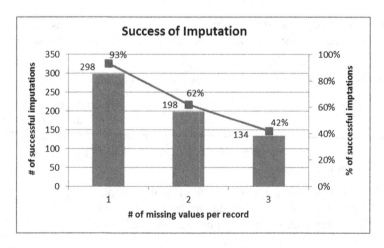

Fig. 1. The success of imputations. Red wine database with 1, 2 and 3 attributes with missing values.

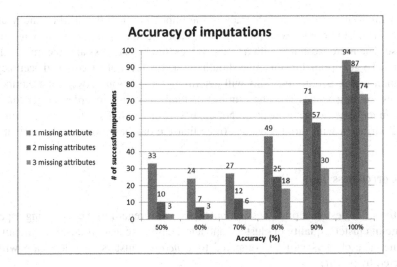

Fig. 2. The accuracy of imputation, with red wine dataset which contains 1, 2 and 3 attributes with missing values.

[5]. The White wine dataset contains 4,898 records, with 12 continuous attributes. The Block page dataset contains all the block of page layout of a document (text and graphic areas). This dataset contains 5,474 records, with 11 integer and continuous attributes. Similar trends were found when comparing the ratios of success of imputation in all three datasets (Tables 1 and 2). In all three datasets, the success of imputation decreases as more attributes contain missing values.

For example, in Red wine, the algorithm has imputed 43.94 % of the missing values with ~100 % accuracy. In Page blocks dataset (which is larger), the percentage of ~100 % accurate imputation was 87.60 %.

Table 1. Comparison of the imputation ratio, using 1, 2 and 3 attributes with missing value, using red wine, white wine, and page blocks datasets.

Name of dataset/# of missing values	1	2	3
Red wine	93 %	62 %	42 %
White wine	98 %	81 %	52 %
Page block	87 %	70 %	51 %

Table 2. The distribution of accuracy of imputation, on datasets were 2 attributes contain missing values.

Name of dataset/distribution of imputation accuracy	Size	50 %	60 %	70 %	80 %	90 %	100 %
Red wine	1999	5.05 %	3.54 %	6.06 %	12.63 %	28.79 %	43.94 %
White wine	4898	3.03%	3.17%	6.60%	14.12%	24.93%	48.15%
Page block	5474	2.23 %	1.25 %	0.97 %	1.67 %	6.27 %	87.60 %

Conclusion. The success and accuracy of the imputation is influenced by the relation of the number of attributes with missing values and the total number of attributes in the database. The more attributes with missing values, the less success and accuracy will be achieved. To get optimal results, one should define the level of success and accuracy of imputation which is required. This will determine the relation between the number of attributes missing values and the number of total attributes in order to get the best results. In the specific experiment we got the best result with a database which contains up to 2 attributes with missing values (note that the database contains 12 attributes).

3.3   Correctness Test

Objectives. To test the correctness of *km-Impute* with regard to the following aspects: structure of clusters, quality of clusters, and a test case to reveal groups of the quality of the wine. The clusters will be compared to *k-means'* clusters. The test case will be conducted by a wine expert.

Results and Discussion. We ran the *km-Impute* on a complete Red wine database. The clusters were compared with *k-means'* clusters. The quality of clusters was assessed according to the following parameters:

1. Density - the radius of the clusters.
2. Distance between clusters - based on Euclidian distance [1] between *centroids*.
km-Impute produced similar clusters to clusters which were produced by *k-means*. The *centroids* of clusters which were produced by *k-means* and *km-Impute* were almost identical (Table 3). For example the *centroid* of cluster-1 was identical in both algorithms: (0.65, 2.24, 3.37). In addition, the sizes of clusters which were produced by

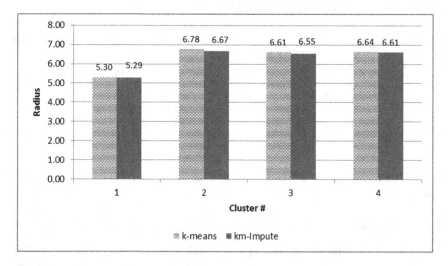

Fig. 3. Similar radius for each of the clusters produced by both *k-means* and *km-Impute* using a complete database.

km-Impute were very similar to *k-means'* clusters: the size of cluster-2 in *k-means* was 381 and 379 in *km-Impute*.

The radius of each of the clusters in *km-Impute* was similar to the radius of clusters in *k-means* (Fig. 3). For example, the radius of cluster 2 in *k-means* is 6.83, and in *km-Impute* it is 6.78.

The distance between clusters in *km-Impute* was similar to the distances which were measured in *k-means* (Table 4). For example, the distance between Cluster-2 and Cluster-3 is identical in both algorithms. The distance between Cluster-1 and Cluster-4 is 5.11 in *k-means* and 5.12 in *km-Impute*.

The clusters were also analyzed by a wine expert. According to the analysis, each cluster represents a distinguished quality of wine. In Cluster-1 the quality of wine is the worst because of the high level of volatile acidity (0.65). The Wine of Cluster-4 has better quality than Cluster-1, since it contains a lower level (yet too high) of volatile acidity (0.41). Cluster-2 is ranked as the second-best with regard to the wine quality, with a better level of volatile acidity and residual sugar (0.41, 2.72 respectively). Since the pH level is relatively low (3.18), it may lead to over acidity. The best wine was detected in Cluster-3: the pH level (3.39) indicates that the acidity of the wine is balanced. In addition, the levels of the volatile acidity (0.46) and the residual sugar (2.25) are acceptable.

Table 3. Similar *centeroids* and cluster size in *k-means* and *km-Impute*, using a complete red wine database.

Cluster	Algorithm	Cluster size	Volatile acidity	Residual sugar	pH
1	k-means	549	0.65	2.24	3.37
1	km-Impute	552	0.65	2.24	3.37
2	k-means	381	0.41	2.71	3.18
2	km-Impute	379	0.41	2.72	3.18
3	k-means	338	0.46	2.25	3.39
3	km-Impute	330	0.46	2.25	3.40
4	k-means	332	0.53	3.13	3.28
4	km-Impute	339	0.53	3.11	3.28

Conclusion. The results show that *km-Impute* successfully produced 4 clusters. Both the structure and the quality of clusters were similar to *k-means* clusters. Each cluster revealed a different level of wine quality. Therefore, *km-Impute* can be implemented as a regular clustering algorithm.

Table 4. Similar distance between clusters in both *k-means* and *km-Impute* using complete Red wine database.

k-means	Cluster-2	Cluster-3	Cluster-4
Cluster-1	4.36	5.15	5.11
Cluster-2		3.43	4.65
Cluster-3			3.94

km-Impute	Cluster-2	Cluster-3	Cluster-4
Cluster-1	5.16	4.34	5.12
Cluster-2		3.43	3.94
Cluster-3			4.68

3.4   Clustering and Imputation Test

Objective. To test *km-Impute* in performing both imputation and clustering as an integrated process.

Results and Discussion. The algorithm was tested on Red-Wine-3 (which contains 3 attributes with missing values). The results were compared to clusters which were produced by *k-means*. Note that *k-means* was running on a complete Red-Wine version, since it cannot handle incomplete databases. The quality of clusters was assessed according to the following parameters:

1. Density – the radius of the cluster.
2. Distance between clusters – based on Euclidian distance [1] between *centroids*.

km-Impute produced similar clusters to *k-means* (Table 5). For example, the *centroids* of Cluster-2 in *k-means* was (0.41, 2.72, 3.18) and (0.44, 2.32, 3.38) in *km-Impute*. Also the sizes of *km-Imputes* clusters were similar to *k-means* clusters. For example, the size of Cluster-1 was 549 in *k-means* and 542 in *km-Impute*. Note however that the similarity between *k-means* and *km-Impute* was less than that achieved in the previous test using a complete database (see Sect. 3.3).

The qualities of the clusters were measured by the cluster's internal radii and by the distance between clusters (Fig. 4; Table 6). The internal radiuses of *km-Imputes'* clusters were similar to clusters which were produced by *k-means*. For example Cluster-2 is 6.78 in *k-means* and 6.66 in *km-Impute*.

Table 5. Similar *centeroids* produced by *k-means* (using a complete red wine database), and *km-Impute*, (using red-wine-3 which contains 3 attributes with a missing value).

Cluster	Algorithm	Cluster size	Volatile acidity	Residual sugar	pH
1	k-means	549	0.65	2.24	3.37
1	km-Impute	542	0.64	2.24	3.37
2	k-means	379	0.41	2.72	3.18
2	km-Impute	300	0.44	2.32	3.38
3	k-means	339	0.53	3.11	3.28
3	km-Impute	344	0.41	2.76	3.18
4	k-means	333	0.47	2.24	3.40
4	km-Impute	291	0.53	30.5	3.29

Table 6. Similar distance between clusters in *k-means* (using a complete red wine database), and *km-Impute* (using red-wine-3 which contains 3 attributes with a missing value).

k-means	Cluster-2	Cluster-3	Cluster-4	km-Impute	Cluster-2	Cluster-3	Cluster-4
Cluster-1	3.93	4.67	5.12	Cluster-1	3.42	5.14	4.01
Cluster-2		3.43	5.17	Cluster-2		4.27	4.65
Cluster-3			4.34	Cluster-3			4.86

The distances between clusters in *km-Impute* was also similar to those in *k-means* (Table 6). For example, the distance between Cluster-1 and Cluster-3 in *k-means* was 3.93, while the distance between and in *km-Impute* was 3.42.

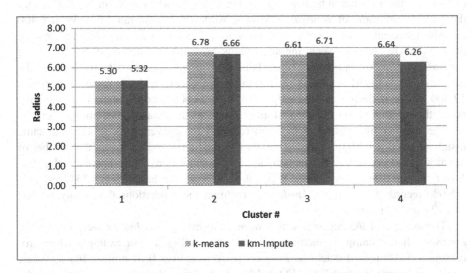

Fig. 4. Similar radius of clusters have been produced by both *k-means* (using a complete Red wine database) and *km-Impute* (using Red-Wine-3 which contains 3 attributes with a missing value).

Conclusion. *km-Impute* successfully accomplished its main task: integrating the clustering and the imputation. The structure and the quality of the clusters are similar to clusters which were produced by *k-means*. Note however, that the similarity between the clusters of *k-means* and *km-Impute* is less than that achieved in the previous test (3.3). Also note that *k-means* ran on a complete database while *km-Impute* on an incomplete one (Red-Wine-3). The relatively high number of attributes with missing values may influence the accuracy of imputation. This may lead to a certain degree of difference between the algorithms' clusters.

3.5 Performance

Objective. To compare the scalability of *km-Impute* and *k-means* in terms of runtime, number of iterations and the number of I/O accesses which were required for convergence.

Results and Discussion. The test was applied on 5 synthetic databases. Their values were randomly generated based on values of Red wine dataset. The size of the databases ranged from 1,600 to 25,600 records each. In order to simulate an incomplete database, 20 % of the records were picked randomly and the value of one randomly selected attribute was removed. The tests ran on an HP Compaq Elite 8300 MT PC: Intel Core™ i7 3770 CPU 3.47 MHz, 16 GB RAM, 1 TB 7200 RPM hard drive, Windows 7 Professional OS 64 bit.

The runtime of *km-Impute* grew according to the size of the database. The runtime of *k-means* has grew much higher (Fig. 5). For example, when applying a 3,200 record database, the runtime of *k-means* was 64 s, while *km-Impute* ran 1.5 s. With 6,400 records, the runtime of *k-means* has grew to 1,455 s and *km-Impute* was finished after 5 s. The gap between the algorithms' runtime is illustrated by the ratio: *k-meansRunTime/km-ImputeRunTime*. This ratio grows according to the size of the database: 26, 43, 291, and 638. A turning point occurs with databases larger than 12,800 where it decreases gradually (427 with 25,600 record database). From that point, the runtime of *km-Impute* will grow faster, as the database becomes larger.

The number of iterations which are required for convergence, reflects the same insights gained regarding runtime (Fig. 6). On medium sized databases the number of iterations until convergence was similar on both algorithms. Note however, that *km-Impute* required fewer iterations. When the database was larger than 12,800 records (25,600 records for example) *km-Impute* required more iterations than *k-means*: 169 and 123 respectively.

The number of I/O accesses which were performed by *km-Impute* were fewer than *k-means* when running on medium sized databases (Fig. 7) For example when processing a 3,200 record database, *k-means* performed over 0.70 million I/O accesses, while *km-Impute* accessed the I/O 0.44 million times. Similar ratios between the algorithms' I/O accesses were measured with 6,400 and 12,800 record databases. With a database which contained 25,600 records, the difference between the numbers of I/O accesses which were performed by the algorithms was smaller than with medium sized databases.

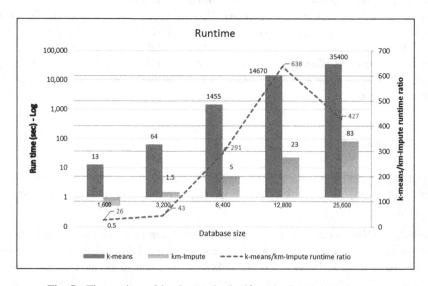

Fig. 5. The runtime of *km-Impute* is significantly shorter than *k-means*

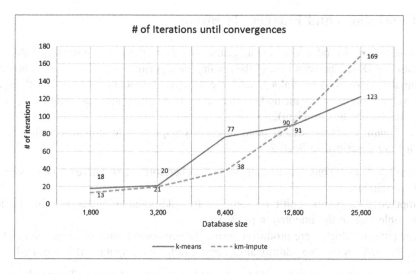

Fig. 6. Both *k-means* and *km-Impute* ran a similar number of iterations until convergences

Conclusion. The *km-Impute* algorithm is significantly more efficient than *k-means* when running on medium sized databases. As the database grows, the efficiency of *km-Impute* is decreased. On very large DB (and large clusters), the algorithm requires more iterations until locating a suitable data point. As a result, the number of iterations may exceed the number of iterations of *k-means*.

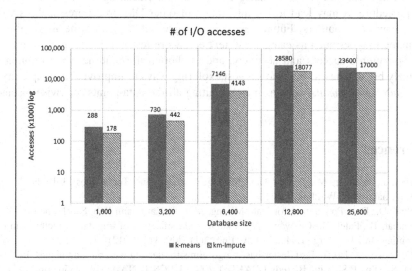

Fig. 7. *km-Impute* performed fewer I/O accesses compared to *k-means*.

4 Conclusions and Further Research

In this study, we introduced a new algorithm (*km-Impute*) to handle missing values in database for DM. The main idea of the *km-Impute* algorithm is to integrate the process of imputation of missing values, with the clustering process. This approach is an improvement on common methods which apply imputation and clustering as separate processes (hence less efficient).

km-Impute contains several advantages in comparison to imputation algorithms which were described in Sect. 2:

1. Data points may belong to one cluster only. Therefore the structure of the clusters is more reliable.
2. Imputed data points won't be sampled for additional imputations. This avoids possible bias in the imputation process.
3. The clusters which were produced during the imputation process, are considered to be the DM results. No additional clustering (after the imputation) is required.

The algorithm was tested on Red wine quality measures database. The results show that *km-Impute* successfully produced clusters using both complete and incomplete databases. The structure of clusters was similar to clusters which were produced by *k-means* algorithms. When running on an incomplete DB, *km-Impute* successfully imputed missing values during clustering. The algorithm was validated using two additional datasets: White wine and Page blocks. The results were consistent with the previous experiments (with Red wine). Although the complexity of *km-Impute* is similar to *k-means*, its runtime was shorter. In particular, the algorithm was significantly faster on medium sized databases and performed fewer I/O accesses.

This new approach of integrating imputation and clustering, improves the overall DM procedure. It may lead to the integration of other DM procedures such as value conversions, and outliers. Future research is suggested to handle the imputation of non-numeric types such as categorical, text etc. More research is required to improve the imputation's success and accuracy, and to shorten the runtime when running on relatively big databases. We also suggest exploring ways to improve the complexity of *km-Impute* and *k-means*, such as pre-computing all the data-points' pairwise distance.

References

1. Han, J., Kamber, M., Pei, J.: Data Mining: Concepts and Techniques, 3rd edn. Morgan Kaufmann Pub, Waltham (2012)
2. Pyle, D.: Data Preparation for Data Mining. Morgan Kaufmann Pub, San Francisco (1999)
3. Suthar, B., Patel, H., Goswami, A.: A survey: classification of imputation methods in data mining. Int. J. Emerg. Technol. Adv. Eng. 2(1), 309–312 (2012)
4. Fujikawa, Y., Ho, T.-B.: Cluster-based algorithms for dealing with missing values. In: Chen, M.-S., Yu, P.S., Liu, B. (eds.) PAKDD 2002. LNCS (LNAI), vol. 2336, pp. 549–554. Springer, Heidelberg (2002)
5. Bache, K., Lichman, M.: UCI Machine Learning Repository (2013). http://archive.ics.uci.edu/ml/

6. Zhang, S., Zhang, J., Zhu, X., Qin, Y., Zhang, C.: Missing value imputation based on data clustering. In: Gavrilova, M.L., Kenneth Tan, C.J. (eds.) Transactions on Computational Science I. LNCS, vol. 4750, pp. 128–138. Springer, Heidelberg (2008)

7. Ayuyev, V.V., Jupin, J., Harris, P.W., Obradovic, Z.: Dynamic Clustering-Based Estimation of Missing Values in Mixed Type Data. In: Pedersen, T.B., Mohania, M.K., Tjoa, A.M. (eds.) DaWaK 2009. LNCS, vol. 5691, pp. 366–377. Springer, Heidelberg (2009)

8. Miller, L.D., Stender, N., Soh, L.K., Samal, A., Kupzyk, K.: Hierarchical clustering algorithm with dynamic tree cut for data imputation (2011). http://ponca.unl.edu/facdb/csefacdb/TechReportArchive/TR-UNL-CSE-2011-0003.pdf

9. Luengo, J., Garcia, S., Herrera, F.: Imputation of missing values : methods' description. University of Granada, Granada, Spain (2011). http://sci2s.ugr.es/MVDM/pdf/MV-methods-description-Complementary-material.pdf

10. Cortez, P., Cerdeira, A., Almeida, F., Matos, T., Reis, J.: Modeling wine preferences by data mining from physicochemical properties. Decis. Support Syst. **47**(4), 547–553 (2009)

Improving the Algorithm for Mapping of OWL to Relational Database Schema

Chien D.C. Ta$^{(\boxtimes)}$ and Tuoi Phan Thi

Faculty of Computer Science and Engineering, Ho Chi Minh City University of
Technology, Ho Chi Minh City, Vietnam
{chientdc, tuoi}@cse.hcmut.edu.vn

Abstract. Ontologies are applied to many applications in recent years, especially on the semantic web, information retrieval, information extraction, and
question answering. The purpose of domain-specific ontology is to get rid of
conceptual and terminological confusion. It accomplishes this by specifying a
set of generic concepts that characterizes the domain as well as their definitions
and interrelationships. There are some languages in order to represent ontologies, such as RDF, OWL. However, these languages are only suitable with
ontologies having small data. For representing ontologies having big data,
database is usually used. Many techniques and tools have been proposed over
the last years. In this paper, we introduce an improved approach for mapping
RDF or OWL to relational database based on the algorithm proposed by Kaunas
University of Technology. This approach can be applied to ontologies having
many classes and big data.

Keywords: Ontology · OWL/RDF · Transform to RDB

1 Introduction

Representing ontologies by relational database has been became an active field of
research in recent years. Depending on different applications and the size of data in
ontologies, converting either relational database to RDF/OWL or RDF/OWL to relational database is applied. There are a lot of research and tools relevant to this field. In
September, 2011 Ricardo [1] developed a prototype and define a mapping algorithm
for converting between the database and OWL. Antonie [2] presented X2R, a system
for integrating heterogeneous data sources in an ontological knowledge base. Their
system created a unified view of information stored in a relational database, XML and
LDAP data sources within an organization, expressed in RDF using a common
ontology and valid according to a prescribed set of integrity constraints. Barzdins and
Kirikova [3] presented a RDB2OWL mapping specification language that is aimed at
presenting RDB-to-RDF/OWL mappings possibly involving advanced correspondences between the database and ontology in a human comprehensible way. Michel
et al. [4] introduced some approaches and tools for converting RDF/OWL to relational
database and from a relational database to RDF/OWL. In this paper, he classified the
different approaches into four major axes: (i) motivation of the approach; (ii) mapping
description language and expressiveness; (iii) mapping implementation; and (iv) data

© Springer International Publishing Switzerland 2015
P. Perner (Ed.): MLDM 2015, LNAI 9166, pp. 130–139, 2015.
DOI: 10.1007/978-3-319-21024-7_9

retrieval method. Ramathilagam and Valarmathi [5] proposed an RDB to ontology mapping system framework which can generate an ontology based on the proposed mapping rules for a banking domain. The mapping rules are generated both for, (1) direct mapping, such as, tables, attributes (2) integrity constraints mapping, such as, primary key, foreign key. Vysniauskas and Nemuraite [6] proposed an algorithm, which fully automatically convert ontologies, represented in OWL language to a relational database schema. Some concepts, e.g. ontology classes and properties are mapped to relational tables, relations and attributes. Other constraints are stored like metadata in special tables. In general, there is much research relevant to transforming OWL/RDF to relational database (RDB) or from RDB to RDF/OWL. The complexity of these tasks depends on the size of data and the structure of the ontology. In this paper's context, we introduce an improved algorithm for mapping OWL language to relational database based on the algorithm proposed by the Kaunas University of Technology. We have improved some steps in this algorithm in order to make it suitable to our ontology and increase the performance of data processing.

Our key contributions are as follows: (i) we improve mapping ontology classes into RDB tables; (ii) we also improve mapping OWL object properties into foreign keys or tables; (iii) we propose an approach for mapping computing domain ontology, which covers 170 distinct classes relevant to computing domain; (iv) we create indexes on primary keys having numeric type instead of class ID.

The rest of this paper is organized as follows: Sect. 2 examines related work and overviews a sample of approaches; Sect. 3 introduces the proposed methodology; Sect. 4 illustrates the experimental results; Sect. 5 discusses the conclusions and future works.

2 Related Work

As outline from Vysniauskas and Nemuraite [6], for ontology development, the semantic web languages and technologies are dedicated: resource description framework RDF and schema RDFS; web ontology language (OWL) that consists of three sub-languages – OWL lite, OWL description logic (DL) and OWL full. However, the representation of ontology based on these semantic web languages is insufficient to address semantic interoperability problems that arise in various concrete applications. Besides, these languages are not enough strong in order to process ontologies having big data. They proposed a framework for transforming ontology representation from OWL to RDB. In his other research [7], he proposed the principles of mapping OWL concepts to RDB schema, along with an algorithm for the transformation of OWL ontology descriptions to the RDB. He also mentioned that there are some proposals for transforming ontology to RDB; however, these approaches are mainly straightforward and still incomplete, or obtained relational structures are not applicable to the real information system. He proposed an algorithm which fully automatically transforms ontologies, represented in OWL, to RDB schema. Concepts and properties are mapped to relations tables, relations and attributes are stored like metadata in special tables. However, his algorithm is not suitable in case ontologies having complex structure and many classes, such as our computing domain ontology (CDO), which is a domain ontology focusing on computing domain. To solve this problem, we propose the improved algorithm based on his algorithm. In additional, we use Oracle database for demo purpose only.

3   Algorithm for Transforming OWL Language to RDB Schema

3.1   Introduce Computing Domain Ontology

Ontology is a formal and explicit specification of a shared conceptualization of a domain of interest. Their classes, relationships, constraints and axioms define a common vocabulary to share knowledge. Conceptualization refers to an abstract model of some phenomenon in the world. Explicit, means that the type of concepts used and the limitations of their use are explicitly defined. Formal, refers to the fact that the ontology should be machine-readable. Shared, reflects the notion that ontology captures consensual knowledge, that is, it is not private to some individual but accepted by a group.

Formally, an ontology can be defined as the tuple [8]:

$$O = (C, I, S, N, H, Y, B, R)$$

where,

C, is set to consist of classes, i.e., concepts represent categories of computing domain (for example, "artificial Intelligent, hardware devices, NLP" $\in$ C)

I is set of instances belong to categories. Set I consist of computing vocabulary (for example, "robotic, random access memory" $\in$ I)

$S = N^S \cup H^H \cup Y^H$ is the set of synonyms, hyponyms and hypernyms of instances of set I.

$N = N^S$ is set of synonyms of instances of set I.

$H = H^H$ is set of hyponyms of instances of set I.

$Y = Y^H$ is set of hypernyms of instances of set I. (e.g., "ADT", "data structure", "ADT is a kind of data structure that is defined by programmer" are synonym, hyponym and hypernym of "abstract data type")

$B = \{belong\_to\ (i_1, c_1) \mid i_1 \in I, c_1 \in C\}$ is set of semantic relationships between concepts of set C and instances of set I and are denoted by $\{belong\_to\ (i1, c1) \mid i1 \in I, c1 \in C\}$ mean that i1 belong to category c1. (e.g., belong_to ("robotic", "artificial intelligent"))

$R = \{rel\ (s_1, i_1) \mid s_1 \in S, i_1 \in I\}$ is the set of relationships between terms of set S and instances of set I and are denoted by hierarchy and are denoted by $\{rel\ (s1, i1) \mid s1 \in S, i1 \in I\}$ mean that s1 is a relationship with i1. The relationships can be synonym, hyponym or hypernym. (e.g., synonym ("ADT", "abstract data type"), hyponym ("data structure", "abstract data type"), hypernym ("ADT is a kind of data structure that is defined by programmer", "abstract data type"). According to the previous ontology definition, from the ontology in Fig. 1, the following sets can be identified:

C = {Programming language, devices, data structure, database management}
I = {Abstract data type, Random access memory, Read only memory}
N = {ADT, RAM, ROM}
H = {Data structure, database, memory}
Y = {EPROM, EEPROM, DDRAM, DDRAM2, DDRAM3}

In addition, all concepts, instances of this ontology focus on computing domain; therefore this ontology is known as computing domain ontology (CDO).

We separate CDO into four layers.

The first layer is known as the topic layer. In order to build it, we extract terms from ACM categories [9]. We obtain over 170 different categories from this site and rearrange them in this layer.

Next layer is known as the ingredient layer. In this layer, there are many different instances, which are defined as nouns or compound nouns from vocabulary about computing domain, e.g., "robot", "support vector machine", "local area network", "wireless", "UML", etc. In order to setup this layer, we use Wikipedia with focus on English language and computing domain.

The third layer of CDO is known as the synset layer. To set up this layer, we use the WordNet ontology. Similar to Wikipedia, we only focus on computing domain. This layer encloses a set of synset. A synset includes synonyms, hyponyms, and hypernyms of instances of the ingredient layer. As we combine two ontologies, which are Wikipedia and WordNet, we have over 900,000 instances that belong to many different categories of computing domain. That is an advantage of this ontology.

The last layer of CDO is known as the sentence layer. Instances of this layer are sentences that represent syntactic relations extracted from preprocessing stage. Hence, these sentences are linked to one or many terms of the ingredient layer. This layer also includes sentences that represent semantic relations between terms of Ingredient layer, such as, IS-A, PART-OF, MADE-OF, RESULT-OF, etc. The overall hierarchy of CDO is shown as Fig. 1.

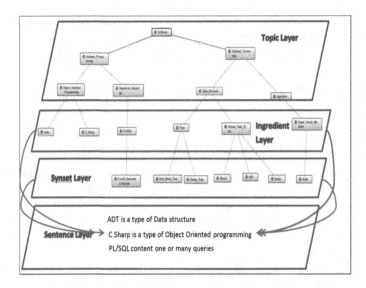

Fig. 1. CDO hierarchy is presented by Protégé

3.2 Algorithm for Transforming OWL Language to RDB Schema

OWL Class and RDB Table. OWL classes provide an abstraction mechanism for grouping resources with similar characteristics [7]. When Vysniauskas converts the OWL ontology description to RDB schema, he maps one ontology class to one database table. It is not suitable for ontologies that have many classes since it make the database systems becoming slow when processing on many tables. For instance, there are 170 ontology classes in our ontology which are corresponding to the 170 different topics belonging to topic layer of CDO. So instead of mapping to 170 database tables, we map these ontology classes into only one database table as follows. This task will improve performance of database accessing, as shown in Fig. 2.

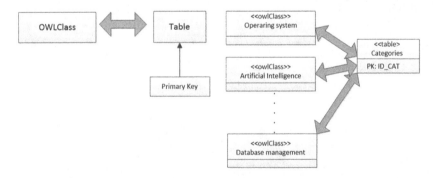

Fig. 2. Mapping ontology class into RDB table

In this Categories table, the data type of ID_CAT column is autonumber, which will automatically increase by 1 as records are added to the table. Furthermore, we create indexes on this primary key in order to increase search speed. In addition, the constructor < rdfs:subClassOf > in OWL is mapped to the same table and we create in RDB schema a new column "belong to" that has the integer data type and points to this constructor, as shown in Fig. 3.

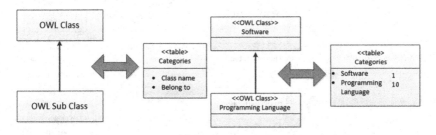

Fig. 3. Mapping OWL subclass to the table.

OWL Properties, RDB Column and Constrains. There are the kinds of properties that are defined by OWL as follows:

- Object properties, which relate individuals to other individuals
- Data type properties, which relate individuals to data values
- Annotation properties, which allow to annotate various constructs in ontology
- Ontology properties, which allow saying things about the ontology itself.
  The Fig. 4 is shown these properties [6].

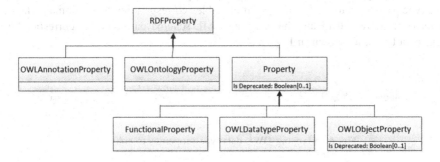

Fig. 4. Kinds of OWL properties

A data type property in OWL ontology relates an individual to a data value [6, 10]. A data type property is defined as the instance of the build-in OWL class owl: data type property. An object property relates to individual to other individuals. An object property is defined as the instance of the build-in OWL class owl: object property.

According to Vysniauskas, object property in OWL ontology relates an individual to other individuals, and the foreign key (relationships between tables) associates columns from one table with columns of another table, so for converting the OWL description to the RDB schema, he maps the object property to the foreign key. However, we cannot do that since there are a lot of different object properties in our ontology, so we map the object properties, which have the same semantic into one table and the foreign key is the primary key of the individual which are related to this object, as shown in Fig. 5.

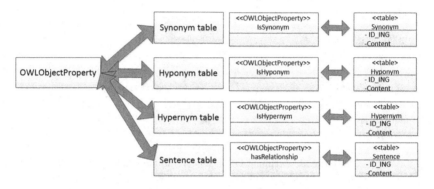

Fig. 5. Mapping OWL object properties to foreign keys

Where:

- ID_ING is a foreign key. It is also the primary key of a term belonging to the ingredient table (ingredient layer).
- Content is data value of a term, e.g., SVM, ADT, etc.
- Has relationship represents a semantic relations between terms of ingredient layer.

A data type property in OWL ontology relates an individual to a data value. We map data type properties to RDB columns of the tables corresponding to the domain classes of these properties. It means that we map XML schema data types to corresponding SQL data types, as shown in Fig. 6.

Fig. 6. Mapping OWL data type properties to columns

In the relational model, a key can be an attribute or a set of attributes that defines uniqueness among the instances of the class. The relational model extends UniqueKey class to UniqueConstraint [10].

Algorithm for Transforming OWL Language to RDB. We propose the algorithm for transforming OWL language to the RDB schema as follows

```
In:    O: Set of OWL classes
Out:   D: Data rows of RDB schema
Procedure OWLtoRDB(O,D)
  R ← OWLclass ∈ O
  While (R is not null)
    If R is Root
    Foreach C ∈ O
     Queue ←  all_sub_class of C
     Forech instance ∈ Queue
       Ins ← instance
       Case Ins ∈
       Case 1: Topic layer
               Category table ∈ D ← Ins
       Case 2: Ingredient layer
               Ingredient table ∈ D ← Ins
       Case 3: Synset layer
               Synset table ∈ D ← Ins
       Case 4: others
               Sentence table ∈ D ← Ins
       End Case
      End For
     End For
    End if
   R ← OWLclass ∈ O
 End While
End Procedure
```

4 Experiments

We have the entity relationship diagram as shown in Fig. 7 after applying the above algorithm. Additionally, we use Oracle Database in this research.

As shown in Fig. 7, there are only six tables for representation CDO. Moreover, we also index all primary keys of all tables, e.g., ID_CAT, ID, ID_SYN, etc. As the result, the performance of our ontology is improved.

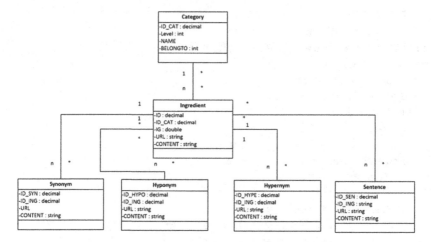

Fig. 7. Entity relationship diagram of CDO

5 Conclusions

In this paper, we introduce an improved algorithm for transforming OWL language to RDB schema based on the algorithm proposed by Kaunas University of Technology. Efforts were also invested in order to reduce the overall processing time of ontology. We reduce the number of tables while mapping OWL classes into tables of RDB. Furthermore, we improve the algorithm while mapping OWL subclass into tables and mapping OWL Object properties into foreign keys. The improved algorithm makes the database system performing better when it processes data of tables. The enriching of CDO, hence, becomes easier. Our approach can be applied to ontologies having big data and many classes.

There is no single best or preferred approach for transforming OWL language to RDB schema. The choice of a suitable approach depends on the size of data and classes of the ontology. In the future work, we will focus particularly on building Information Extraction system based on this ontology to support users in computing domain.

References

1. Ricardo, F.: Relational databases conceptual preservation. In: Fourth Workshop on Very Large Digital Libraries (VLDL2011), in Conjunction with the 1st International Conference on Theory and Practice of Digital Libraries (TPDL2011), Berlin (2011)
2. Antonie, M.: Integration of heterogeneous data sources in an ontological knowledge base. Comput. Inform. **31**, 189–223 (2012)
3. Barzdins, J., Kirikova, M.: RDB2OWL: a RDB-to-RDF/OWL mapping specification language. In: The 9th International Baltic Conference on Databases and Information Systems (Baltic DB & IS 2010), Riga, pp. 3–18, (2010)

4. Michel, F. et al. Semantic web journal, Dec 2013. http://www.semantic-web-journal.net/content/survey-rdb-rdf-translation-approaches-and-tools
5. Ramathilagam, C., Valarmathi, M.L.: A framework for OWL DL based ontology construction from relational database using mapping. Int. J. Comput. Appl. **76**(17), 31–37 (2013)
6. Vysniauskas, E., Nemuraite, L.: Transforming ontology representation from OWL to relational database. Inf. Technol. Control **35**(3A), 333–343 (2006)
7. Vysniauskas, E., Nemuraite, L.: Informactjos Sistermy Katedra (2009). http://isd.ktu.lt/it2009/material/Proceedings/OCM.pdf
8. Zhang, L.: Ontology based partial building information model extraction. J. Comput. Civ. Eng. **27**, 1–44 (2012)
9. Association for computing machinery. http://www.acm.org/about/class/ccs98-html
10. Vysniauskas, E., Nemuraite, L.: Reversible lossless transformation from OWL 2 ontologies into relational databases. Inf. Technol. Control **40**(4), 293–306 (2011)
11. Chieze, E., Farzindar, A., Lapalme, G.: An automatic system for summarization and information extraction of legal information. In: The 23rd Canadian Conference on Artificial Intelligence, Ottawa (2010)

Robust Principal Component Analysis of Data with Missing Values

Tommi Kärkkäinen and Mirka Saarela[✉]

Department of Mathematical Information Technology, University of Jyväskylä,
40014 Jyväskylä, Finland
mirka.saarela@jyu.fi

Abstract. Principal component analysis is one of the most popular machine learning and data mining techniques. Having its origins in statistics, principal component analysis is used in numerous applications. However, there seems to be not much systematic testing and assessment of principal component analysis for cases with erroneous and incomplete data. The purpose of this article is to propose multiple robust approaches for carrying out principal component analysis and, especially, to estimate the relative importances of the principal components to explain the data variability. Computational experiments are first focused on carefully designed simulated tests where the ground truth is known and can be used to assess the accuracy of the results of the different methods. In addition, a practical application and evaluation of the methods for an educational data set is given.

Keywords: PCA · Missing data · Robust statistics

1 Introduction

Principal component analysis (PCA) is one of the most popular methods in machine learning (ML) and data mining (DM) of statistical origin [12]. It is typically introduced in all textbooks of ML and DM areas (e.g., [1,10]) and is used in numerous applications [15]. It seems that the versatile line of utilization has also partly redefined the original terminology from statistics: in ML&DM, the computation of principal components and their explained variability of data, many times together with dimension reduction, is referred to as PCA, even if the term *analysis*, especially historically, refers to statistical hypothesis testing [12]. However, nowadays the use of the term PCA points to the actual computational procedure. Certainly one of the appealing facets of PCA is its algorithmic simplicity with a supporting linear algebra library: (a) create covariance matrix, (b) compute eigenvalues and eigenvectors, (c) compute data variability using eigenvalues, and, if needed, transform data to the new coordinate system determined by the eigenvectors. This is also the algorithmic skeleton underlying this work.

Even if much researched, the use of PCA for sparse data with missing values (not to be mixed with sparse PCA referring to the sparsity of the linear model [6])

© Springer International Publishing Switzerland 2015
P. Perner (Ed.): MLDM 2015, LNAI 9166, pp. 140–154, 2015.
DOI: 10.1007/978-3-319-21024-7_10

seems not to be a widely addressed topic, although [27] provides a comparison of a set of second-order (classical) methods. We assume here that there is no further information on the sparsity pattern so that the non-existing subset of data is *missing completely at random* (MCAR) [18]. As argued in [24,25], a missing value can, in principle, represent any value from the possible range of an individual variable so that it becomes difficult to justify assumptions on data or error normality, which underlie the classical PCA that is based on second-order statistics. Hence, we also consider the so-called nonparametric, robust statistical techniques [11,13], which allow deviations from normality assumptions while still producing reliable and well-defined estimators.

The two simplest robust estimates of location are median and spatial median. The median, a middle value of the ordered univariate sample (unique only for odd number of points, see [16]), is inherently one-dimensional, and with missing data uses only the available values of an individual variable from the marginal distribution (similarly to the mean). The spatial median, on the other hand, is truly a multidimensional location estimate and utilizes the available data pattern as a whole. These estimates and their intrinsic properties are illustrated and more thoroughly discussed in [16]. The spatial median has many attractive statistical properties; particularly that its breakdown point is 0.5, that is, it can handle up to 50% of the contaminated data, which makes it very appealing for high-dimensional data with severe degradations and outliers, possibly in the form of missing values. In statistics, robust estimation of data scattering (i.e., covariability) has been advanced in many papers [7,19,28], but, as far as we know, sparse data have not been treated in them.

The content of this work is as follows: First, we briefly derive and define basic and robust PCA and unify their use to coincide with the geometrical interpretation. Then, we propose two modifications of the basic robust PCA for sparse data. All the proposed methods are then compared using a sequence of carefully designed test data sets. Finally, we provide one application of the most potential procedures, i.e., dimension reduction and identifying the main variables, for an educational data set, whose national subset was analyzed in [24].

2   Methods

Assume that a set of observations $\{\mathbf{x}_i\}_{i=1}^N$, where $\mathbf{x}_i \in \mathbf{R}^n$, is given, so that N denotes the number of observations and n the number of variables, respectively. To avoid the low-rank matrices by the form of the data, we assume that $n < N$. In the usual way, define the data matrix $\mathbf{X} \in \mathbf{R}^{N \times n}$ as $\mathbf{X} = \left(\mathbf{x}_i^T\right), i = 1, \ldots, N$.

2.1   Derivation and Interpretation of the Classical PCA

We first provide a compact derivation underlying classical principal component analysis along the lines of [4]. For the linear algebra, see, for example, [8]. In general, the purpose of PCA is to derive a linear transformation to reduce the dimension of a given set of vectors while still retaining their information content (in practice, their variability). Hence, the original set of vectors $\{\mathbf{x}_i\}$ is to be transferred to a set of new vectors $\{\mathbf{y}_i\}$ with $\mathbf{y}_i \in \mathbf{R}^m$, such that $m < n$ but also $\mathbf{x}_i \sim \mathbf{y}_i$ in a suitable sense. Note that every vector $\mathbf{x} \in \mathbf{R}^n$ can be represented using a set of orthonormal basis vectors $[\mathbf{u}_1 \ldots \mathbf{u}_n]$ as $\mathbf{x} = \sum_{k=1}^{n} z_k \mathbf{u}_k$, where $z_k = \mathbf{u}_k^T \mathbf{x}$. Geometrically, this rotates the original coordinate system.

Let us consider a new vector $\tilde{\mathbf{x}} = \sum_{k=1}^{m} z_k \mathbf{u}_k + \sum_{k=m+1}^{n} b_k \mathbf{u}_k$, where the last term represents the residual error $\mathbf{x} - \tilde{\mathbf{x}} = \sum_{k=m+1}^{n} (z_k - b_k) \mathbf{u}_k$. In case of the classical PCA, consider the minimization of the least-squares-error:

$$J = \frac{1}{2} \sum_{i=1}^{N} \|\mathbf{x}_i - \tilde{\mathbf{x}}_i\|^2 = \frac{1}{2} \sum_{i=1}^{N} (\mathbf{x}_i - \tilde{\mathbf{x}}_i)^T (\mathbf{x}_i - \tilde{\mathbf{x}}_i) = \frac{1}{2} \sum_{i=1}^{N} \sum_{k=m+1}^{n} (z_{i,k} - b_k)^2. \quad (1)$$

By direct calculation, one obtains $b_k = \mathbf{u}_k^T \bar{\mathbf{x}}$, where $\bar{\mathbf{x}} = \frac{1}{N} \sum_{i=1}^{N} \mathbf{x}_i$ is the sample mean. Then (1) can be rewritten as $((\mathbf{u}^T \mathbf{v})^2 = \mathbf{u}^T (\mathbf{v}\,\mathbf{v}^T) \mathbf{u}$ for vectors $\mathbf{u}, \mathbf{v})$ so that

$$J = \frac{1}{2} \sum_{k=m+1}^{n} \sum_{i=1}^{N} \left(\mathbf{u}_k^T (\mathbf{x}_i - \bar{\mathbf{x}}) \right)^2 = \frac{1}{2} \sum_{k=m+1}^{n} \mathbf{u}_k^T \Sigma \mathbf{u}_k, \quad (2)$$

where Σ is the sample covariance matrix

$$\Sigma = \sum_{i=1}^{N} (\mathbf{x}_i - \bar{\mathbf{x}}) (\mathbf{x}_i - \bar{\mathbf{x}})^T. \quad (3)$$

Note that the standard technique (e.g., in Matlab) for sparse data is to compute (3) only for those data pairs where both values $(\mathbf{x}_i)_j$ and $(\mathbf{x}_i)_k$ exist. By setting $\mathbf{v}_i = \mathbf{x}_i - \bar{\mathbf{x}}$, we have for the quadratic form, with an arbitrary vector $\mathbf{x} \neq 0$:

$$\mathbf{x}^T \Sigma \mathbf{x} = \mathbf{x}^T \left[\mathbf{v}_1 \mathbf{v}_1^T + \ldots + \mathbf{v}_N \mathbf{v}_N^T \right] \mathbf{x} = (\mathbf{x}^T \mathbf{v}_1)^2 + \ldots + (\mathbf{x}^T \mathbf{v}_N)^2 \geq 0. \quad (4)$$

This shows that any matrix of the form of (3) is always at least positive semidefinite, with positive eigenvalues if $\mathbf{v}_i$'s span $\mathbf{R}^n$, that is, if $\text{rank}[\mathbf{v}_1 \ldots \mathbf{v}_N] \geq n$. The existence of missing values clearly increases the possibility of semidefiniteness.

Now, let $\{\lambda_k, \mathbf{u}_k\}$ be the kth eigenvalue and eigenvector of Σ satisfying

$$\Sigma \mathbf{u}_k = \lambda_k \mathbf{u}_k, \quad k = 1, \ldots, n. \quad (5)$$

This identity can be written in the matrix form as $\Sigma \mathbf{U} = \mathbf{U}\mathbf{D}$, where $\mathbf{D} = \text{Diag}\{\lambda_1, \ldots, \lambda_n\}$ (vector $\boldsymbol{\lambda}$ as the diagonal matrix) and $\mathbf{U} = [\mathbf{u}_1 \mathbf{u}_2 \ldots \mathbf{u}_n]$. Using (5) shows that (2) reduces to $J = \frac{1}{2} \sum_{k=m+1}^{n} \lambda_k$. This means that the reduced representation consists of those m eigenvectors that correspond to the m largest eigenvalues of matrix Σ. For the unbiased estimate of the sample covariance matrix $\Sigma \simeq \frac{1}{N-1} \Sigma$, one can use scaling such as in (3) because it does not affect eigenvectors or the relative sizes of the eigenvalues. Finally, for any $\mathbf{x} \in \mathbf{R}^n$ and $\mathbf{y} = \mathbf{U}^T \mathbf{x}$, we have

$$\mathbf{x}^T \Sigma \mathbf{x} = \mathbf{y}^T \mathbf{D} \mathbf{y} = \sum_{k=1}^{n} \lambda_k \mathbf{y}_k^2 = \sum_{k=1}^{n} \frac{\mathbf{y}_k^2}{\left(\lambda_k^{-\frac{1}{2}} \right)^2}. \quad (6)$$

Geometrically, this means that in the transformed coordinate system $\mathbf{U}^T\mathbf{e}_k$ ($\mathbf{e}_k$s are the base vectors for the original coordinates), the data define an n-dimensional hyperellipsoid for which the lengths of the principal semi-axis are proportional to $\sqrt{\lambda_k}$.

To this end, we redefine the well-known principle (see, e.g., [15]) for choosing a certain number of principal components in dimension reduction. Namely, the derivations above show that eigenvalues of the sample covariance matrix Σ represent *the variance* along the new coordinate system, $\lambda_k = \sigma_k^2$, whereas the geometric interpretation related to (6) proposes to use the standard deviation $\sigma_k = \sqrt{\lambda_k}$ to assess the variability of data.

Proposition 1. *The relative importance RI_k (in percentages) of a new variable y_k for the principal component transformation based on the sample covariance matrix is defined as $RI_k = 100\frac{\sqrt{\lambda_k}}{\sum_{i=1}^n \sqrt{\lambda_i}}$, where λ_k satisfy (5). We refer to $\sqrt{\lambda_i}$ as the estimated variability of the ith (new) variable.*

2.2 Derivation of Robust PCA for Sparse Data

Formally, a straightforward derivation of the classical PCA as given above is obtained from the optimality condition for the least-squares problem (1). Namely, assume that instead of the reduced representation, the problem $\min_{\mathbf{x}} \mathcal{J}(\mathbf{x})$ as in (1) is used to estimate the location of the given data $\{\mathbf{x}_i\}$. In second-order statistics, this provides the sample mean $\bar{\mathbf{x}} = \frac{1}{N}\sum_{i=1}^N \mathbf{x}_i$, whose explicit formula can be obtained from the optimality condition (see [16]):

$$\frac{d\mathcal{J}(\bar{\mathbf{x}})}{d\mathbf{x}} = \frac{d}{d\mathbf{x}}\frac{1}{2}\sum_{i=1}^N \|\mathbf{x}_i - \mathbf{x}\|^2 = \sum_{i=1}^N (\mathbf{x}_i - \bar{\mathbf{x}}) = \mathbf{0}. \tag{7}$$

The covariate form of this optimality condition $\sum_{i=1}^N (\mathbf{x}_i - \bar{\mathbf{x}})(\mathbf{x}_i - \bar{\mathbf{x}})^T$ readily provides us the sample covariance matrix up to the constant $\frac{1}{N-1}$.

Next we assume that there are missing values in the given data. To define their pattern, let us introduce the projection vectors $\mathbf{p}_i$, with $i = 1\ldots,N$ (see [2,17,24,25]), which capture the availability of the components:

$$(\mathbf{p}_i)_j = \begin{cases} 1, \text{if } (\mathbf{x}_i)_j \text{ exists}, \\ 0, \text{otherwise}. \end{cases} \tag{8}$$

We also define the corresponding matrix $\mathbf{P} \in \mathbf{R}^{N\times n}$ that contains these projections in the rows, being of compatible size with the data matrix $\mathbf{X}$.

The spatial median $\mathbf{s}$ with the so-called available data strategy can be obtained as the solution of the projected Weber problem

$$\min_{\mathbf{v}\in\mathbf{R}^n} \mathcal{J}(\mathbf{v}), \quad \text{where } \mathcal{J}(\mathbf{v}) = \sum_{i=1}^{n_j} \|\text{Diag}\{\mathbf{p}_i\}(\mathbf{x}_i - \mathbf{v})\|. \tag{9}$$

As described in [16], this optimization problem is nonsmooth, that is, it is not classically differentiable at zero. Instead, the so-called subgradient of $\mathcal{J}(\mathbf{v})$ always exists and is characterized by the condition

$$\partial \mathcal{J}(\mathbf{v}) = \sum_{i=1}^{N} \boldsymbol{\xi}_i \text{ for } \begin{cases} (\boldsymbol{\xi}_i)_j = \dfrac{\mathrm{Diag}\{\mathbf{p}_i\}(\mathbf{v} - \mathbf{x}_i)_j}{\|\mathrm{Diag}\{\mathbf{p}_i\}(\mathbf{v} - \mathbf{x}_i)\|}, \text{if } \|\mathrm{Diag}\{\mathbf{p}_i\}(\mathbf{u} - \mathbf{x}_i)\| \neq 0, \\ \|\boldsymbol{\xi}_i\| \leq 1, \text{ when } \|\mathrm{Diag}\{\mathbf{p}_i\}(\mathbf{u} - \mathbf{x}_i)\| = 0. \end{cases}$$

(10)

Then, the minimizer $\mathbf{s}$ of (9) satisfies $\mathbf{0} \in \partial \mathcal{J}(\mathbf{s})$. In [20] it is shown, for the complete data case, that if the sample $\{\mathbf{x}_i\}$ belongs to a Euclidean space and is not concentrated on a line, the spatial median $\mathbf{s}$ is unique. In practice (see [2]), one can obtain an accurate approximation for the solution of the nonsmooth problem by solving the following equation corresponding to the regularized form

$$\sum_{i=1}^{N} \frac{\mathrm{Diag}\{\mathbf{p}_i\}(\mathbf{s} - \mathbf{x}_i)}{\max\{\|\mathrm{Diag}\{\mathbf{p}_i\}(\mathbf{s} - \mathbf{x}_i)\|, \varepsilon\}} = \mathbf{0} \quad \text{for } \varepsilon > 0.$$

(11)

This can be solved using the SOR (Sequential Overrelaxation) algorithm [2] with the overrelaxation parameter $\omega = 1.5$. For simplicity, define $\|\mathbf{v}\|_\varepsilon = \max\{\|\mathbf{v}\|, \varepsilon\}$.

To this end, the comparison of (7) and (11) allows us to define the *robust covariance matrix* corresponding to the spatial median s:

$$\Sigma_R = \frac{1}{N-1} \sum_{i=1}^{N} \left(\frac{\mathrm{Diag}\{\mathbf{p}_i\}(\mathbf{s} - \mathbf{x}_i)}{\|\mathrm{Diag}\{\mathbf{p}_i\}(\mathbf{s} - \mathbf{x}_i)\|_\varepsilon} \right) \left(\frac{\mathrm{Diag}\{\mathbf{p}_i\}(\mathbf{s} - \mathbf{x}_i)}{\|\mathrm{Diag}\{\mathbf{p}_i\}(\mathbf{s} - \mathbf{x}_i)\|_\varepsilon} \right)^T.$$

(12)

This form can be referred to as the *multivariate sign covariance matrix* [5,7,28]. By construction, the nonzero covariate vectors have a unit length, so that they only accumulate the deviations of angles and not the sizes of the available variables. Such an observation is related to one perspective on statistical robustness that can be formalized using the so-called influence function [9]. Using Σ_R as the sample covariance matrix, one can, by again solving the corresponding eigenvalue problem (5), recover a new basis $\{\mathbf{u}_k\}$ for which the corresponding eigenvalues $\{\lambda_k\}$, again, explain the amount of variability along the new coordinates. Because Σ_R is based on the first-order approximation, the nonnegative eigenvalues readily correspond to the geometric variability represented by the standard deviation in the second-order statistics, and, then, we do not need to take any square roots when computing the relative importances of the robust procedure as in Proposition 1. Hence, the two PCA approaches are comparable to each other.

2.3   Projection Using PCA-Based Transformation

In the matrix form, the existence of a new basis in the columns of the given unitary matrix $\mathbf{U}$, and given a complete location estimate for the sparse data $\mathbf{s} \in \mathbf{R}^n$ (i.e., the spatial median), for which we define the corresponding matrix $\mathbf{S} \in \mathbf{R}^{N \times n}$ by replication of $\mathbf{s}^T$ in N rows, yields the transformed data matrix

$$\mathbf{Y} = (\mathbf{P} \circ (\mathbf{X} - \mathbf{S}))\, \mathbf{U},$$

(13)

where ∘ denotes the Hadamard product. When $\mathbf{U}$ is ordered based on RI_k's, the dimension reduction is obtained by selecting only m of the n coordinates (columns) in $\mathbf{Y}$. Hence, we see that even if there are missing values in the original data, the resulting new data vectors become complete. We also know from the basic linear algebra that, for complete data, both the length of the original vectors and the angle between any two vectors are preserved in (13) because $\mathbf{U}$ is unitary. However, in the case of missing data, some of the coordinate values of the original vectors are not present, and then, presumably, the transformed vectors in $\mathbf{Y}$ are of smaller length, i.e., closer to the origin in the transformed space. Moreover, the angles might also become degraded. These simple observations readily raise some doubts concerning the available data strategy in the form of incomplete data vectors as proposed in (12).

2.4   Two Modifications of the Robust PCA Procedure

Let us define two modifications of the robust PCA procedure that are based on the similar form of the covariance matrix as defined in (12). As discussed above, both the amount of variability of data and/or the main directions of variability might be underestimated due to sparse data vectors, that is, missing coordinate values. Our suggested modifications are both based on a simple idea: use only the "almost complete" data in estimation (cf. the cascadic initializations of robust clustering in [24,25]). Note that this is one step further than the typical way of using only the complete pairs or complete observations in the computation of a covariance matrix.

The first suggested modification, for the computation of the relative importances of the principal components, is related to using the actual projections along the new coordinate axis for this purpose. Similar to the alpha-trimmed mean [3], which presumably neglects outlying observations, we use (see the tests in [26]) the 10% and 90% percentiles, denoted as $\text{prc}_{10}(\cdot)$ and $\text{prc}_{90}(\cdot)$, related to the transformed data matrix $\mathbf{Y}$ in (13). Namely, for the each new variable $\{y_k\}$, its estimated variability is computed as

$$RI_k = 100(\text{prc}_{90}(\{y_k\}) - \text{prc}_{10}(\{y_k\})). \tag{14}$$

Moreover, because it is precisely the sparsity that diminishes the lengths and angles of the transformed data vectors, we restrict the computation of (14) to that subset of the original data, where at most one variable is missing from an observation $\mathbf{x}_i$. This subset satisfies $\sum_{j=1}^{n}(\mathbf{p}_i)_j \geq n - 1$.

Our second suggested modification uses a similar approach, but already directly for the robust covariance matrix (12), by taking into account only those observations of which at most one variable is missing. Hence, we define the following subsets of the original set of indices $\mathcal{N} = \{1, 2, \ldots, N\}$:

$$I_c = \{i \in \mathcal{N} \mid \mathbf{x}_i \text{ is complete}\},$$
$$I_j = \{i \in \mathcal{N} \mid \text{variable } j \text{ is missing from } \mathbf{x}_i\}.$$

We propose computing a reduced, robust covariance matrix $\widetilde{\Sigma}_R$ as

$$\widetilde{\Sigma}_R = \frac{1}{\widetilde{N}-1}\left(\sum_{i\in I_c}\mathbf{v}_i\mathbf{v}_i^T + \sum_{j=1}^{n}\sum_{i\in I_j}\mathbf{v}_i\mathbf{v}_i^T\right), \quad \mathbf{v}_i = \frac{\text{Diag}\{\mathbf{p}_i\}(\mathbf{s}-\mathbf{x}_i)}{\|\text{Diag}\{\mathbf{p}_i\}(\mathbf{s}-\mathbf{x}_i)\|_\varepsilon},$$

with $\widetilde{N} = |I_c| + \sum_j |I_j|$. Hence, only that part of the first-order covariability that corresponds to the almost complete observations is used.

3   Computational Results

Computational experiments in the form of simulated test cases, when knowing the target result, are given first. The parametrized test is introduced in Sect. 3.1, and the computational results for the different procedures are provided in Sect. 3.2. Finally, we apply the best methods to analyze the educational data of PISA in Sect. 3.3. As a reference method related to the classical, second-order statistics as derived in Sect. 2.1, with sparse data, we use the Matlab's PCA routine with the 'pairwise' option.

3.1   The Simulated Test Cases

For simplicity, we fix the number of observations as $N = 1000$. For the fixed size of an observation n, let us define a vector of predetermined standard deviations as $\sigma = [\sigma_1\ \sigma_2 \ldots \sigma_n]$. Moreover, let $\mathbf{R}_{a,b}(\theta) \in \mathbf{R}^{n\times n}$ be an orthonormal (clockwise) rotation matrix of the form

$$\mathbf{R}_{ab}(\theta) = \{\mathbf{M} = \mathbf{I}_n \wedge \mathbf{M}_{aa} = \mathbf{M}_{bb} = \cos(\theta),\ \mathbf{M}_{ab} = -\mathbf{M}_{ba} = -\sin(\theta)\},$$

where $\mathbf{I}_n$ denotes the $n \times n$ identity matrix. Then, the simulated data $\{\mathbf{d}_i\}_{i=1}^{N}$ is generated as

$$\begin{aligned}
\mathbf{d}_i^T \sim &\frac{\sigma}{2} + \left[\mathcal{N}(0,\sigma_1)\,\mathcal{N}(0,\sigma_2) \ldots \mathcal{N}(0,\sigma_n)\right] \\
&+ \eta_i\left[\mathbf{R}_n\left[\mathcal{U}([-\sigma_1,\sigma_1])\,\mathcal{U}([-\sigma_2,\sigma_2]) \ldots \mathcal{U}([-\sigma_n,\sigma_n])\right]^T\right]^T,
\end{aligned} \tag{15}$$

where $\mathcal{N}(0,\sigma)$ denotes the zero-mean normal distribution with standard deviation σ and $\mathcal{U}([-c,c])$ the uniform distribution on the interval $[-c,c]$, respectively. $\mathbf{R}_n$ defines the n-dimensional rotation that we use to orientate the latter noise term in (15) along the diagonal of the hypercube, that is, we always choose $\theta = \frac{\pi}{4}$ and take, for the actual tests in $2D, 3D, 4D$, and $6D$,

$$\begin{aligned}
\mathbf{R}_2 &= \mathbf{R}_{12}(\theta), \quad \mathbf{R}_3 = \mathbf{R}_{23}(\theta)\mathbf{R}_{12}(\theta), \quad \mathbf{R}_4 = \mathbf{R}_{14}(\theta)\mathbf{R}_{23}(\theta)\mathbf{R}_{34}(\theta)\mathbf{R}_{12}(\theta), \\
\mathbf{R}_6 &= \mathbf{R}_{36}(\theta)\mathbf{R}_{45}(\theta)\mathbf{R}_{56}(\theta)\mathbf{R}_{14}(\theta)\mathbf{R}_{23}(\theta)\mathbf{R}_{34}(\theta)\mathbf{R}_{12}(\theta).
\end{aligned}$$

Finally, a random sparsity pattern of a given percentage of missing values represented by the matrix $\mathbf{P}$ as defined in (8) is attached to data.

To conclude, the simulated data are parametrized by the vector $\boldsymbol{\sigma}$, which defines the true data variability. Moreover, the target directions of the principal components are just the original unit vectors $\mathbf{e}_k$, $k = 1, \ldots, n$. Their estimation is disturbed by the noise, which comes from the uniform distribution whose width coordinatewise coincides with the clean data. Because the noise is rotated towards the diagonal of the hypercube, its maximal effect is characterized by $\frac{\max_k \sigma_k}{\min_k \sigma_k}$. By choosing σ_k's as the powers of two and three for $n = 2, 3, 4, 6$, we are then gradually increasing the effect of the error when the dimension of the data is increasing. Finally, we fix the amount of noise to 10% so that $\eta_i = 1$ with a probability of 0.1 in (15). In this way, testing up to 40% of missing values randomly attached to $\{\mathbf{d}_i\}$ will always contain less than 50% of the degradations (missing values and/or noise) as a whole.

3.2   Results for the Simulated Tests

The test data generation was repeated 10 times, and the means and standard deviations (in parentheses) over these are reported. As the error measure for the directions of $\{\mathbf{u}_k\}$, we use their deviation from being parallel to the target unit vectors. Hence, we take $\text{DirE} = \max_k\{1 - |\mathbf{u}_k^T \mathbf{e}_k|\}$, $k = 1, \ldots, n$, such that $\text{DirE} \in [0, 1]$. In the result tables below, we report the relative importances of RI_k in the order of their importance. 'Clas' refers to the classical PCA, 'Rob' to the original robust formulation, 'RobP' to the modification using percentiles for the importances, and 'RobR' to the use of the reduced covariance matrix $\tilde{\Sigma}_R$.

Table 1. Results for $\boldsymbol{\sigma} = [3 \; 1]$

Missing	PC	True(Std)	Clas(Std)	Rob(Std)	RobP(Std)	RobR(Std)
0%	1	75.0(0.00)	73.7(0.8)	73.0(1.1)	73.0(1.3)	73.0(1.1)
	2	25.0(0.00)	26.3(0.8)	27.0(1.1)	27.0(1.3)	27.0(1.1)
	DirE	-	0.001	0.004		0.004
10%	1	75.0(0.00)	73.9(0.9)	68.9(1.2)	73.2(1.3)	68.9(1.2)
	2	25.0(0.00)	26.1(0.9)	31.1(1.2)	26.8(1.3)	31.1(1.2)
	DirE	-	0.001	0.005		0.005
20%	1	75.0(0.00)	73.5(1.2)	65.1(1.0)	72.5(1.7)	65.1(1.0)
	2	25.0(0.00)	26.5(1.2)	34.9(1.0)	27.5(1.7)	34.9(1.0)
	DirE	-	0.001	0.009		0.009
30%	1	75.0(0.00)	73.8(1.0)	62.4(0.9)	73.0(1.4)	62.4(0.9)
	2	25.0(0.00)	26.2(1.0)	37.6(0.9)	27.0(1.4)	37.6(0.9)
	DirE	-	0.001	0.003		0.003
40%	1	75.0(0.00)	74.0(0.8)	60.3(1.6)	73.1(1.4)	60.3(1.6)
	2	25.0(0.00)	26.0(0.8)	39.7(1.6)	26.9(1.4)	39.7(1.6)
	DirE	-	0.002	0.008		0.008

The real relative importances ('True') by generation are provided in the third column.

From all simulated tests (Tables 1, 2, 3 and 4), we see that the the classical method and 'RobP' show the closest relative importances of the principal components to the true geometric variability in the data. Moreover, both of these approaches show a very stable behavior, and the results for the relative importances do not change that much, even when a high number of missing data is present. The results for the other two approaches, the basic robust and 'RobR', on the other hand, are much less stable, and particularly the basic robust procedure starts to underestimate the relative importances of the major components when the amount of missing data increases.

The directions remain stable for all the simulated test cases, even when a large amount of missing data is present. Over all the simulated tests, the 'RobP' with the original robust covariance bears the closest resemblance to the true directions. It can tolerate more noise compared to 'Clas', as shown in Table 3. We also conclude that the missing data do not affect the results of the PCA procedures as much as the noise. Tables 3 and 4 show that, for a large noise, the increase in sparsity can actually improve the performance of the robust method

Table 2. Results for $\sigma = [4\ 2\ 1]$

Missing	PC	True(Std)	Clas(Std)	Rob(Std)	RobP(Std)	RobR(Std)
0%	1	57.1(0.00)	56.3(0.7)	58.6(1.3)	55.8(1.0)	58.6(1.3)
	2	28.6(0.00)	28.6(0.8)	28.9(1.2)	28.7(1.0)	28.9(1.2)
	3	14.3(0.00)	15.2(0.3)	12.6(0.4)	15.5(0.4)	12.6(0.4)
	DirE	-	0.005	0.017		0.017
10%	1	57.1(0.00)	56.3(0.8)	55.7(1.3)	55.8(1.0)	54.5(1.6)
	2	28.6(0.00)	28.6(0.9)	30.0(1.2)	28.6(0.9)	30.7(1.2)
	3	14.3(0.00)	15.1(0.4)	14.3(0.6)	15.5(0.5)	14.8(0.8)
	DirE	-	0.008	0.015		0.015
20%	1	57.1(0.00)	56.2(0.8)	51.7(1.4)	55.6(1.1)	51.7(1.4)
	2	28.6(0.00)	28.6(0.9)	30.7(1.2)	28.8(1.3)	31.5(1.5)
	3	14.3(0.00)	15.2(0.3)	17.6(0.8)	15.6(0.5)	16.7(0.7)
	DirE	-	0.005	0.020		0.014
30%	1	57.1(0.00)	56.0(0.7)	49.2(0.7)	55.3(0.9)	50.9(0.7)
	2	28.6(0.00)	28.8(0.8)	31.6(0.8)	29.0(0.9)	32.1(1.3)
	3	14.3(0.00)	15.2(0.4)	19.2(1.0)	15.7(0.5)	17.1(1.3)
	DirE	-	0.006	0.013		0.012
40%	1	57.1(0.00)	56.2(0.9)	46.2(1.4)	55.8(1.2)	49.9(1.6)
	2	28.6(0.00)	28.7(1.2)	32.0(1.7)	28.5(1.3)	32.5(1.7)
	3	14.3(0.00)	15.1(0.4)	21.8(1.1)	15.6(0.7)	17.6(1.0)
	DirE	-	0.010	0.014		0.013

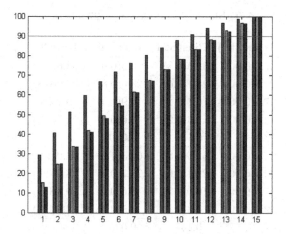

Fig. 1. Cumulative sum of the relative importances for the classical PCA using variance, the classical PCA, and the robust PCA using percentiles (from left to right).

because it decreases the absolute number of noisy observations. Interestingly, as can be seen from Table 4, the geometric variability was estimated accurately, even if the directions were wrong.

3.3   Results for PISA Data Set

Next, we apply the different PCA methods tested in the previous section to a large educational data set, namely the latest data from the Programme for International Student Assessment[1] (PISA 2012). The data contain 485490 observations, and as variables we use the 15 scale indices [24] that are known to explain the student performance in mathematics, the main assessment area in PISA 2012. The scale indices are derived variables that summarize information from student background questionnaires [22], and are scaled so that their mean is zero with a standard deviation of one. Due to the rotated design of PISA (each student answers only one of the three different background questionnaires), this data set has 33.24% of missing data by design, a special case of MCAR.

In Table 5, the relative importances $\{RI_k\}$ are depicted. The table also shows the variance-based view for the classical method, denoted as 'ClsVar'. As can be seen from the table, the first principal component is much higher for 'ClsVar' than for the other approaches. In consequence, fewer principal components would be selected with 'ClsVar' when a certain threshold of how much the principal components should account for is given. As illustrated in Fig. 1, if the threshold is set to 90%, we would select 11 components with 'ClsVar' but 13 for both the classical PCA and for the 'RobP'.

---

[1] Available at http://www.oecd.org/pisa/pisaproducts/.

Table 3. Results for $\sigma = [27\ 9\ 3\ 1]$

Missing	PC	True(Std)	Clas(Std)	Rob(Std)	RobP(Std)	RobR(Std)
0%	1	67.5(0.00)	62.5(0.8)	66.9(0.9)	63.6(1.1)	66.9(0.9)
	2	22.5(0.00)	22.1(0.6)	23.6(0.8)	22.9(0.8)	23.6(0.8)
	3	7.5(0.00)	10.1(0.2)	6.9(0.4)	9.1(0.4)	6.9(0.4)
	4	2.5(0.00)	5.3(0.2)	2.6(0.2)	4.5(0.2)	2.6(0.2)
	DirE	-	0.168	0.080		0.080
10%	1	67.5(0.00)	62.5(0.8)	62.3(1.3)	64.0(1.2)	60.3(2.3)
	2	22.5(0.00)	22.1(0.6)	25.7(0.9)	23.0(0.9)	26.9(1.7)
	3	7.5(0.00)	10.0(0.2)	8.6(0.5)	9.0(0.4)	9.2(0.6)
	4	2.5(0.00)	5.3(0.2)	3.5(0.3)	4.0(0.2)	3.6(0.4)
	DirE	-	0.157	0.045		0.046
20%	1	67.5(0.00)	62.5(1.0)	56.9(1.2)	64.2(1.2)	58.3(1.7)
	2	22.5(0.00)	22.1(0.7)	27.4(1.0)	22.9(0.9)	28.1(1.1)
	3	7.5(0.00)	10.1(0.3)	11.0(0.6)	8.9(0.5)	9.8(1.0)
	4	2.5(0.00)	5.4(0.3)	4.7(0.4)	3.9(0.2)	3.7(0.4)
	DirE	-	0.164	0.032		0.031
30%	1	67.5(0.00)	62.7(0.8)	52.1(1.6)	64.4(0.9)	57.7(1.3)
	2	22.5(0.00)	22.0(0.6)	28.2(1.4)	23.2(0.6)	28.2(1.5)
	3	7.5(0.00)	10.0(0.4)	13.2(1.0)	8.6(0.4)	10.2(0.5)
	4	2.5(0.00)	5.3(0.3)	6.5(0.5)	3.8(0.2)	3.9(0.4)
	DirE	-	0.177	0.023		0.038
40%	1	67.5(0.00)	62.7(0.8)	46.9(0.8)	64.2(1.4)	55.9(1.2)
	2	22.5(0.00)	22.2(0.6)	28.7(1.0)	23.5(1.3)	29.8(1.5)
	3	7.5(0.00)	9.9(0.3)	15.7(0.5)	8.8(0.3)	10.8(1.1)
	4	2.5(0.00)	5.2(0.3)	8.6(0.8)	3.6(0.4)	3.5(0.5)
	DirE	-	0.189	0.016		0.040

In Fig. 2, the loadings of the first two principal components are visualized for the classical and for the robust version. We see that for both versions, the three scale indices ANXMAT, FAILMAT, and ESCS are the most distinct from the others. However, the robust version is able to distinguish this finding more clearly. That *index of economic, social and cultural status* (ESCS) accounts for much of the variability in the data, being the "strongest single factor associated with performance in PISA" [21], is always highlighted in PISA documentations and can be clearly seen in Fig. 2, especially from the robust PC 1.

Table 4. Results for $\sigma = [32\ 16\ 8\ 4\ 2\ 1]$

Missing	PC	True(Std)	Clas(Std)	Rob(Std)	RobP(Std)	RobR(Std)
0%	1	50.8(0.00)	48.1(0.5)	55.3(1.0)	48.6(0.8)	55.3(1.0)
	2	25.4(0.00)	24.3(0.4)	26.7(0.6)	24.4(0.4)	26.7(0.6)
	3	12.7(0.00)	12.6(0.2)	10.6(0.4)	12.7(0.3)	10.6(0.4)
	4	6.3(0.00)	7.5(0.2)	4.7(0.3)	7.1(0.2)	4.7(0.3)
	5	3.2(0.00)	4.4(0.1)	1.6(0.1)	4.3(0.1)	1.6(0.1)
	6	1.6(0.00)	3.2(0.1)	1.0(0.1)	2.8(0.2)	1.0(0.1)
	DirE	-	0.298	0.374		0.374
10%	1	50.8(0.00)	48.0(0.6)	51.6(1.1)	48.5(1.1)	51.0(1.9)
	2	25.4(0.00)	24.3(0.5)	27.5(0.8)	24.8(0.6)	28.1(1.1)
	3	12.7(0.00)	12.6(0.2)	12.0(0.5)	12.8(0.4)	12.0(0.9)
	4	6.3(0.00)	7.5(0.2)	5.5(0.2)	7.0(0.2)	5.5(0.4)
	5	3.2(0.00)	4.4(0.1)	2.2(0.2)	4.2(0.1)	2.2(0.2)
	6	1.6(0.00)	3.2(0.2)	1.3(0.1)	2.7(0.2)	1.3(0.2)
	DirE	-	0.318	0.277		0.358
20%	1	50.8(0.00)	48.2(0.5)	48.6(1.0)	48.9(1.6)	51.3(1.4)
	2	25.4(0.00)	24.2(0.5)	27.3(0.8)	24.6(0.6)	27.3(0.7)
	3	12.7(0.00)	12.7(0.3)	13.2(0.8)	13.0(0.6)	12.3(1.2)
	4	6.3(0.00)	7.4(0.2)	6.4(0.3)	7.0(0.3)	5.5(0.6)
	5	3.2(0.00)	4.4(0.2)	2.7(0.3)	4.0(0.2)	2.2(0.2)
	6	1.6(0.00)	3.2(0.2)	1.7(0.2)	2.4(0.2)	1.3(0.3)
	DirE	-	0.372	0.090		0.137
30%	1	50.8(0.00)	48.1(0.6)	43.8(1.2)	48.6(1.4)	49.4(2.4)
	2	25.4(0.00)	24.3(0.5)	27.5(0.8)	25.0(0.8)	28.5(1.7)
	3	12.7(0.00)	12.6(0.1)	15.0(0.6)	12.9(0.5)	12.5(0.8)
	4	6.3(0.00)	7.5(0.2)	7.6(0.5)	7.1(0.4)	5.8(0.8)
	5	3.2(0.00)	4.3(0.1)	3.8(0.5)	4.0(0.2)	2.2(0.2)
	6	1.6(0.00)	3.2(0.2)	2.3(0.3)	2.4(0.2)	1.5(0.4)
	DirE	-	0.335	0.092		0.468
40%	1	50.8(0.00)	48.0(0.6)	39.7(1.5)	48.3(1.7)	50.2(2.9)
	2	25.4(0.00)	24.3(0.4)	26.6(1.0)	25.1(1.1)	28.3(2.4)
	3	12.7(0.00)	12.6(0.3)	15.8(1.0)	13.0(0.7)	11.9(1.3)
	4	6.3(0.00)	7.5(0.2)	9.5(0.8)	7.3(0.3)	6.0(0.8)
	5	3.2(0.00)	4.4(0.3)	5.1(0.5)	3.9(0.3)	2.2(0.3)
	6	1.6(0.00)	3.1(0.2)	3.3(0.4)	2.3(0.2)	1.3(0.2)
	DirE	-	0.516	0.078		0.518

Table 5. Results for PISA data

	RI_1	RI_2	RI_3	RI_4	RI_5	RI_6	RI_7	RI_8	RI_9	RI_{10}	RI_{11}	RI_{12}	RI_{13}	RI_{14}	RI_{15}
ClsVar	29.5	11.4	10.4	8.6	6.8	5.0	4.4	4.1	3.8	3.7	3.2	3.0	2.8	2.0	1.3
Cls	15.3	9.5	9.1	8.3	7.3	6.3	5.9	5.7	5.5	5.4	5.0	4.8	4.7	4.0	3.3
RobP	13.1	11.9	8.6	7.5	7.2	6.5	6.5	5.9	5.9	5.2	4.8	4.8	4.5	3.9	3.7

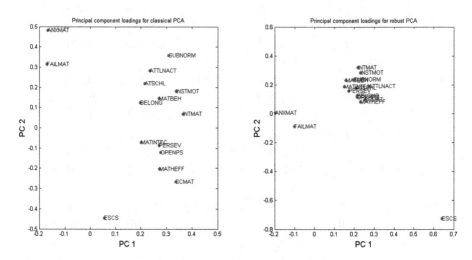

Fig. 2. Principal component loadings for PISA data for the classical (left) and robust (right) approaches.

4   Conclusions

Although PCA is one of the most widely used ML and DM techniques, systematic testing and assessment of PCA in the presence of missing data seem to still be an important topic to study. In this article, we have proposed a robust PCA method and two modifications (one using percentiles for the importance and one with a reduced covariance matrix) of this method. The testing of these three approaches was done in comparison with the classical, reference PCA for sparse data. First, we illustrated the results for carefully designed simulated data and then for a large, real educational data set.

From the simulated tests, we concluded that the percentiles-based robust method and the classical PCA showed the best results, especially when the relative importance of the principal components were compared with the true variability of the data. The basic robust approach started to underestimate the relative importance of the major components when the amount of missing data increased. The results of the simulated tests were stable, and the variance between repeated test runs was very small. Likewise, the estimated directions remained also stable even with a large amount of missing data. Tests with PISA data showed that the proposed robust methods are applicable for large, real data sets with one-third of the values missing, where the interpretation of the robust result yielded clearer known discrimination of the original variables compared to the classical PCA.

The classical PCA uses variance to estimate the importance of the principal components, which highlights (as demonstrated in Table 5 and Fig. 1) the major components. As shown by the simulated results, it is more prone to nongaussian errors in the data. These points might explain some of the difficulties the classical method faced in applications [23]. In [14], seven distinctions of the PCA problem in the presence of missing values were listed: (1) no analytical solution since even the estimation of the data covariance matrix is nontrivial, (2) the optimized cost function typically has multiple local minima, (3) no analytical solution even for the location estimate, (4) standard approaches can lead to overfitting, (5) algorithms may require heavy computations, 6) the concept of the PCA basis in the principal subspace is not easily generalized, and (7) the choice of the dimensionality of the principal subspace is more difficult than in classical PCA. We conclude that the proposed robust methods successfully addressed all these distinctions: (1) well-defined covariance matrix, (2) being positive semidefinite, (3) a unique location estimate in the form of the spatial median, (4) resistance to noise due to robustness, (5) the same linear algebra as in the classical approach, and (6)–(7) a geometrically consistent definition of the principal subspace and its dimension related to the data variability.

Acknowledgments. The authors would like to thank Professor Tuomo Rossi for many helpful discussions on the contents of the paper.

References

1. Alpaydin, E.: Introduction to Machine Learning, 2nd edn. The MIT Press, Cambridge, MA, USA (2010)
2. Äyrämö, S.: Knowledge Mining Using Robust Clustering: volume 63 of Jyväskylä Studies in Computing. University of Jyväskylä, Jyväskylä (2006)
3. Bednar, J., Watt, T.: Alpha-trimmed means and their relationship to median filters. IEEE Trans. Acoust. Speech Sig. Process. **32**(1), 145–153 (1984)
4. Bishop, C.M.: Neural Networks for Pattern Recognition. Oxford University Press, Oxford (1995)
5. Croux, C., Ollila, E., Oja, H.: Sign and rank covariance matrices: statistical properties and application to principal components analysis. In: Dodge, Y. (ed.) Statistical data analysis based on the L1-norm and related methods, pp. 257–269. Springer, Basel (2002)
6. d'Aspremont, A., Bach, F., Ghaoui, L.E.: Optimal solutions for sparse principal component analysis. J. Mach. Learn. Res. **9**, 1269–1294 (2008)
7. Gervini, D.: Robust functional estimation using the median and spherical principal components. Biometrika **95**(3), 587–600 (2008)
8. Golub, G.H., Van Loan, C.F.: Matrix Computations, 3rd edn. Johns Hopkins University Press, Baltimore, MD, USA (1996)
9. Hampel, F.R., Ronchetti, E.M., Rousseeuw, P.J., Stahel, W.A.: Robust statistics: the approach based on influence functions, vol. 114. Wiley, New York (2011)

10. Han, J., Kamber, M., Pei, J.: Data Mining: Concepts and Techniques, 3rd edn. Morgan Kaufmann Publishers Inc., San Francisco, CA, USA (2011)
11. Hettmansperger, T.P., McKean, J.W.: Robust Nonparametric Statistical Methods. Edward Arnold, London (1998)
12. Hotelling, H.: Analysis of a complex of statistical variables into principal components. J. Educ. Psychol. 24(6), 417 (1933)
13. Huber, P.J.: Robust Statistics. Wiley, New York (1981)
14. Ilin, A., Raiko, T.: Practical approaches to principal component analysis in the presence of missing values. J. Mach. Learn. Res. 11, 1957–2000 (2010)
15. Jolliffe, I.: Principal Component Analysis. Wiley Online Library, New York (2005)
16. Kärkkäinen, T., Heikkola, E.: Robust formulations for training multilayer perceptrons. Neural Comput. 16, 837–862 (2004)
17. Kärkkäinen, T., Toivanen, J.: Building blocks for odd-even multigrid with applications to reduced systems. J. Comput. Appl. Math. 131, 15–33 (2001)
18. Little, R.J.A., Rubin, D.B.: Statistical Analysis with Missing Data, vol. 4. Wiley, New York (1987)
19. Locantore, N., Marron, J.S., Simpson, D.G., Tripoli, N., Zhang, J.T., Cohen, K.L., Boente, G., Fraiman, R., Brumback, B., Croux, C., et al.: Robust principal component analysis for functional data. Test 8(1), 1–73 (1999)
20. Milasevic, P., Ducharme, G.R.: Uniqueness of the spatial median. Ann. Stat. 15(3), 1332–1333 (1987)
21. OECD: PISA Data Analysis Manual: SPSS and SAS, 2nd edn. OECD Publishing, Paris (2009)
22. OECD: PISA: Results: Ready to Learn - Students' Engagement, Drive and Self-Beliefs. OECD Publishing, Paris (2013)
23. Ringberg, H., Soule, A., Rexford, J., Diot, C.: Sensitivity of PCA for traffic anomaly detection. In: ACM SIGMETRICS Performance Evaluation Review, vol. 35, pp. 109–120. ACM (2007)
24. Saarela, M., Kärkkäinen,T.: Discovering gender-specific knowledge from Finnish basic education using PISA scale indices. In: Proceedings of the 7th International Conference on Educational Data Mining, pp. 60–68 (2014)
25. Saarela, M., Kärkkäinen, T.: Analysing student performance using sparse data of core bachelor courses. JEDM-J. Educ. Data Min. 7(1), 3–32 (2015)
26. Stigler, S.M.: Do robust estimators work with real data? Ann. Stat. 5, 1055–1098 (1977)
27. Van Ginkel, J.R.. Kroonenberg, P.M., Kiers, H.A.: Missing data in principal component analysis of questionnaire data: a comparison of methods. J. Stat. Comput. Simul. 1–18 (2013) (ahead-of-print)
28. Visuri, S., Koivunen, V., Oja, H.: Sign and rank covariance matrices. J. Stat. Plann. Infer. 91(2), 557–575 (2000)

Association and Sequential
Rule Mining

Efficient Mining of High-Utility Sequential Rules

Souleymane Zida[1], Philippe Fournier-Viger[1(✉)], Cheng-Wei Wu[2],
Jerry Chun-Wei Lin[3], and Vincent S. Tseng[2]

[1] Department of Computer Science, University of Moncton, Moncton, Canada
esz2233@umoncton.ca, philippe.fournier-viger@umoncton.ca
[2] Department of Computer Science, National Chiao Tung University,
Hsinchu, Taiwan
silvemoonfox@gmail.com, vtseng@cs.nctu.edu.tw
[3] School of Computer Science and Technology, Harbin Institute of Technology
Shenzhen Graduate School, Shenzhen, China
jerrylin@ieee.org

Abstract. High-utility pattern mining is an important data mining task
having wide applications. It consists of discovering patterns generating
a high profit in databases. Recently, the task of high-utility sequential
pattern mining has emerged to discover patterns generating a high profit
in sequences of customer transactions. However, a well-known limitation
of sequential patterns is that they do not provide a measure of the confidence or probability that they will be followed. This greatly hampers
their usefulness for several real applications such as product recommendation. In this paper, we address this issue by extending the problem of
sequential rule mining for utility mining. We propose a novel algorithm
named HUSRM (High-Utility Sequential Rule Miner), which includes
several optimizations to mine high-utility sequential rules efficiently. An
extensive experimental study with four datasets shows that HUSRM is
highly efficient and that its optimizations improve its execution time by
up to 25 times and its memory usage by up to 50 %.

Keywords: Pattern mining · High-utility mining · Sequential rules

1 Introduction

Frequent Pattern Mining (FPM) is a fundamental task in data mining, which
has many applications in a wide range of domains [1]. It consists of discovering
groups of items appearing together frequently in a transaction database. However, an important limitation of FPM is that it assumes that items cannot appear
more than once in each transaction and that all items have the same importance
(e.g. weight, unit profit). These assumptions do not hold in many real-life applications. For example, consider a database of customer transactions containing
information on quantity of purchased items and their unit profit. If FPM algorithms are applied on this database, they may discover many frequent patterns
generating a low profit and fail to discover less frequent patterns that generate

© Springer International Publishing Switzerland 2015
P. Perner (Ed.): MLDM 2015, LNAI 9166, pp. 157–171, 2015.
DOI: 10.1007/978-3-319-21024-7_11

a high profit. To address this issue, the problem of FPM has been redefined as High-utility Pattern Mining (HUPM) [2,5,7,9,12,13]. However, these work do not consider the sequential ordering of items in transactions. High-Utility Sequential Pattern Mining (HUSP) was proposed to address this issue [14,15]. It consists of discovering sequential patterns in sequences of customer transactions containing quantity and unit profit information. Although, this definition was shown to be useful, an important drawback is that it does not provide a measure of confidence that patterns will be followed. For example, consider a pattern ⟨{milk}, {bread}, {champagne}⟩ meaning that customers bought milk, then bread, and then champagne. This pattern may generate a high profit but may not be useful to predict what customers having bought milk and bread will buy next because milk and bread are very frequent items and champagne is very rare. Thus, the probability or confidence that milk and bread is followed by champagne is very low. Not considering the confidence of patterns greatly hampers their usefulness for several real applications such as product recommendation.

In FPM, a popular alternative to sequential patterns that consider confidence is to mine sequential rules [3]. A sequential rule indicates that if some item(s) occur in a sequence, some other item(s) are likely to occur afterward with a given confidence or probability. Two main types of sequential rules have been proposed. The first type is rules where the antecedent and consequent are sequential patterns [10,11]. The second type is rules between two unordered sets of items [3,4]. In this paper we consider the second type because it is more general and it was shown to provide considerably higher prediction accuracy for sequence prediction in a previous study [4]. Moreover, another reason is that the second type has more applications. For example, it has been applied in e-learning, manufacturing simulation, quality control, embedded systems analysis, web page prefetching, anti-pattern detection, alarm sequence analysis and restaurant recommendation (see [4] for a survey). Several algorithms have been proposed for sequential rule mining. However, these algorithms do not consider the quantity of items in sequences and their unit profit. But this information is essential for applications such as product recommendation and market basket analysis. Proposing algorithms that mine sequential rules while considering profit and quantities is thus an important research problem.

However, addressing this problem raises major challenges. First, algorithms for utility mining cannot be easily adapted to sequential rule mining. The reason is that algorithms for HUPM and HUSP mining such as USpan [14], HUI-Miner [8] and FHM [5] use search procedures that are very different from the ones used in sequential rule mining [3,4]. A distinctive characteristics of sequential rule mining is that items may be added at any time to the left or right side of rules to obtain larger rules and that confidence needs to be calculated. Second, the proposed algorithm should be efficient in both time and memory, and have an excellent scalability.

In this paper, we address these challenges. Our contributions are fourfold. First, we formalize the problem of high-utility sequential rule mining and its properties. Second, we present an efficient algorithm named HUSRM (High-Utility Sequential Rule Miner) to solve this problem. Third, we propose several optimizations

Table 1. A sequence database

SID	Sequences
s_1	$\langle\{(a,1)(b,2)\}(c,2)(f,3)(g,2)(e,1)\rangle$
s_2	$\langle\{(a,1)(d,3)\}(c,4),(b,2),\{(e,1)(g,2)\}\rangle$
s_3	$\langle(a,1)(b,2)(f,3)(e,1)\rangle$
s_4	$\langle\{(a,3)(b,2)(c,1)\}\{(f,1)(g,1)\}\rangle$

Table 2. External utility values

Item	a	b	c	d	e	f	g
Profit	1	2	5	4	1	3	1

to improve the performance of HUSRM. Fourth, we conduct an extensive experimental study with four datasets. Results show that HUSRM is very efficient and that its optimizations improve its execution time by up to 25 times and its memory usage by up to 50 %.

The rest of this paper is organized as follows. Sections 2, 3, 4 and 5 respectively presents the problem definition and related work, the HUSRM algorithm, the experimental evaluation and the conclusion.

2 Problem Definition and Related Work

We consider the definition of a sequence database containing information about quantity and unit profit, as defined by Yin et al. [14].

Definition 1 (Sequence Database). Let $I = \{i_1, i_2, ..., i_l\}$ be a set of items (symbols). An *itemset* $I_x = \{i_1, i_2, ..., i_m\} \subseteq I$ is an unordered set of distinct items. The *lexicographical order* $\succ_{lex}$ is defined as any total order on I. Without loss of generality, it is assumed in the following that all itemsets are ordered according to $\succ_{lex}$. A *sequence* is an ordered list of itemsets $s = \langle I_1, I_2, ..., I_n \rangle$ such that $I_k \subseteq I$ $(1 \leq k \leq n)$. A *sequence database* SDB is a list of sequences $SDB = \langle s_1, s_2, ..., s_p \rangle$ having sequence identifiers (SIDs) $1, 2...p$. Note that it is assumed that sequences cannot contain the same item more than once. Each item $i \in I$ is associated with a positive number $p(i)$, called its *external utility* (e.g. unit profit). Every item i in a sequence s_c has a positive number $q(i, s_c)$, called its *internal utility* (e.g. purchase quantity).

Example 1. Consider the sequence database shown in Table 1, which will be the running example. It contains four sequences having the SIDs 1, 2, 3 and 4. Each single letter represents an item and is associated with an integer value representing its internal utility. Items between curly brackets represent an itemset. When an itemset contains a single item, curly brackets are omitted for brevity. For example, the first sequence s_1 contains five itemsets. It indicates that items a and b occurred at the same time, were followed by c, then f, then g and lastly e. The internal utility (quantity) of a, b, c, e, f and g in that sequence are respectively 1, 2, 2, 3, 2 and 1. The external utility (unit profit) of each item is shown in Table 2.

The problem of sequential rule mining is defined as follows [3,4].

Definition 2 (Sequential Rule). A sequential rule $X \rightarrow Y$ is a relationship between two unordered itemsets $X, Y \subseteq I$ such that $X \cap Y = \emptyset$ and $X, Y \neq \emptyset$. The interpretation of a rule $X \rightarrow Y$ is that if items of X occur in a sequence, items of Y will occur afterward in the same sequence.

Definition 3 (Sequential Rule Size). A rule $X \rightarrow Y$ is said to be of size $k * m$ if $|X| = k$ and $|Y| = m$. Note that the notation $k * m$ is not a product. It simply means that the sizes of the left and right parts of a rule are respectively k and m. Furthermore, a rule of size $f * g$ is said to be larger than another rule of size $h * i$ if $f > h$ and $g \geq i$, or alternatively if $f \geq h$ and $g > i$.

Example 2. The rules $r = \{a, b, c\} \rightarrow \{e, f, g\}$ and $s = \{a\} \rightarrow \{e, f\}$ are respectively of size $3 * 3$ and $1 * 2$. Thus, r is larger than s.

Definition 4 (Itemset/Rule Occurrence). Let $s = \langle I_1, I_2 \ldots I_n \rangle$ be a sequence. An itemset I occurs or is contained in s (written as $I \sqsubseteq s$) iff $I \subseteq \bigcup_{i=1}^{n} I_i$. A rule $r = X \rightarrow Y$ occurs or is contained in s (written as $r \sqsubseteq s$) iff there exists an integer k such that $1 \leq k < n$, $X \subseteq \bigcup_{i=1}^{k} I_i$ and $Y \subseteq \bigcup_{i=k+1}^{n} I_i$. Furthermore, let $seq(r)$ and $ant(r)$ respectively denotes the set of sequences containing r and the set of sequences containing its antecedent, i.e. $seq(r) = \{s | s \in SDB \wedge r \sqsubseteq s\}$ and $ant(r) = \{s | s \in SDB \wedge X \sqsubseteq s\}$.

Definition 5 (Support). The *support* of a rule r in a sequence database SDB is defined as $sup_{SDB}(r) = |seq(r)| / |SDB|$.

Definition 6 (Confidence). The *confidence* of a rule $r = X \rightarrow Y$ in a sequence database SDB is defined as $conf_{SDB}(r) = |seq(r)| / |ant(r)|$.

Definition 7 (Problem of Sequential Rule Mining). Let $minsup, minconf \in [0, 1]$ be thresholds set by the user and SDB be a sequence database. A sequential rule r is *frequent* iff $sup_{SDB}(r) \geq minsup$. A sequential rule r is *valid* iff it is frequent and $conf_{SDB}(r) \geq minconf$. The *problem of mining sequential rules* from a sequence database is to discover all valid sequential rules [3].

We adapt the problem of sequential rule mining to consider the utility (e.g. generated profit) of rules as follows.

Definition 8 (Utility of an Item). The utility of an item i in a sequence s_c is denoted as $u(i, s_c)$ and defined as $u(i, s_c) = p(i) \times q(i, s_c)$.

Definition 9 (Utility of a Sequential Rule). Let be a sequential rule r : $X \rightarrow Y$. The utility of r in a sequence s_c is defined as $u(r, s_c) = \sum_{i \in X \cup Y} u(i, s_c)$ iff $r \sqsubseteq s_c$. Otherwise, it is 0. The utility of r in a sequence database SDB is defined as $u_{SDB}(r) = \sum_{s \in SDB} u(r, s)$ and is abbreviated as $u(r)$ when the context is clear.

Example 3. The itemset $\{a, b, f\}$ is contained in sequence s_1. The rule $\{a, c\} \rightarrow \{e, f\}$ occurs in s_1, whereas the rule $\{a, f\} \rightarrow \{c\}$ does not, because item c does not occur after f. The profit of item c in sequence s_1 is $u(c, s_1) = p(c) \times q(c, s_1) =$

$5 \times 2 = 10$. Consider a rule $r = \{c, f\} \to \{g\}$. The profit of r in s_1 is $u(r, s_1) = u(c, s_1) + u(f, s_1) + u(g, s_1) = (5 \times 2) + (3 \times 3) + (1 \times 2) = 18$. The profit of r in the database is $u(r) = u(r, s_1) + u(r, s_2) + u(r, s_3) + u(r, s_4) = 18 + 0 + 0 + 9 = 27$.

Definition 10. (Problem of High-Utility Sequential Rule Mining). Let $minsup, minconf \in [0, 1]$ and $minutil \in \mathbf{R}^+$ be thresholds set by the user and SDB be a sequence database. A rule r is a *high-utility sequential rule* iff $u_{SDB}(r) \geq minutil$ and r is a valid rule. Otherwise, it is said to be a low utility sequential rule. The *problem of mining high-utility sequential rules* from a sequence database is to discover all high-utility sequential rules.

Example 4. Table 3 shows the sequential rules found for $minutil = 40$ and $minconf = 0.65$. In this example, we can see that rules having a high-utility and confidence but a low support can be found (e.g. r_7). These rules may not be found with regular sequential rule mining algorithms because they are rare, although they are important because they generate a high profit.

Table 3. Sequential rules found for $minutil = 40$ and $minconf = 0.65$

ID	Sequential rule	Support	Confidence	Utility
r_1	$\{a, b, c\} \to \{e\}$	0.50	0.66	42
r_2	$\{a, b, c\} \to \{e, g\}$	0.50	0.66	46
r_3	$\{a, b, c\} \to \{f, g\}$	0.50	0.66	42
r_4	$\{a, b, c\} \to \{g\}$	0.75	1.0	57
r_5	$\{a, b, c, d\} \to \{e, g\}$	0.25	1.0	40
r_6	$\{a, c\} \to \{g\}$	0.75	1.0	45
r_7	$\{a, c, d\} \to \{b, e, g\}$	0.25	1.0	40
r_8	$\{a, d\} \to \{b, c, e, g\}$	1.0	1.0	40
r_9	$\{b, c\} \to \{e\}$	0.50	0.66	40
r_{10}	$\{b, c\} \to \{e, g\}$	0.50	0.66	44
r_{11}	$\{b, c\} \to \{g\}$	0.75	1.0	52
r_{12}	$\{c\} \to \{g\}$	0.75	1.0	40

Algorithms for mining sequential rules explore the search space of rules by first finding frequent rules of size $1 * 1$. Then, they recursively append items to either the left or right sides of rules to find larger rules [3]. The *left expansion* of a rule $X \to Y$ with an item $i \in I$ is defined as $X \cup \{i\} \to Y$, where $i \succ_{lex} j, \forall j \in X$ and $i \notin Y$. The *right expansion* of a rule $X \to Y$ with an item $i \in I$ is defined as $X \to Y \cup \{i\}$, where $i \succ_{lex} j, \forall j \in Y$ and $i \notin X$. Sequential rule mining algorithms prune the search space using the support because it is anti-monotonic. However, this is not the case for the utility measure, as we show thereafter.

Property 1. (antimonotonicity of utility). Let be a rule r and a rule s, which is a left expansion of r. It follows that $u(r) < u(s)$ or $u(r) \geq u(s)$. Similarly, for a rule t, which is a right expansion of r, $u(r) < u(t)$ or $u(r) \geq u(t)$.

Example 5. The rule $\{a\} \rightarrow \{b\}$, $\{a\} \rightarrow \{b,g\}$ and $\{a\} \rightarrow \{b,e\}$ respectively have a utility of 10, 7 and 12.

In high-utility pattern mining, to circumvent the problem that utility is not anti-monotonic, the solution has been to use anti-monotonic upper-bounds on the utility of patterns to be able to prune the search space. Algorithms such as Two-Phase [9], IHUP [2] and UP-Growth [12] discover patterns in two phases. During the first phase, an upper-bound on the utility of patterns is calculated to prune the search space. Then, during the second phase, the exact utility of remaining patterns is calculated by scanning the database and only high-utility patterns are output. However, an important drawback of this method is that too many candidates may be generated and may need to be maintained in memory during the first phase, which degrades the performance of the algorithms. To address this issue, one-phase algorithms have been recently proposed such as FHM [5], HUI-Miner [8] and USpan [14] to mine high-utility patterns without maintaining candidates. These algorithms introduces the concept of *remaining utility*. For a given pattern, the remaining utility is the sum of the utility of items that can be appended to the pattern. The main upper-bound used by these algorithms to prune the search space is the sum of the utility of a pattern and its remaining utility. Since one-phase algorithms were shown to largely outperform two-phase algorithms, our goal is to propose a one-phase algorithm to mine high-utility sequential rules.

3   The HUSRM Algorithm

In the next subsections, we first present important definitions and data structures used in our proposal, the HUSRM algorithm. Then, we present the algorithm. Finally, we describe additional optimizations.

3.1   Definitions and Data Structures

To prune the search space of sequential rules, the HUSRM algorithm adapts the concept of sequence estimated utility introduced in high-utility sequential pattern mining [14] as follows.

Definition 11 (Sequence Utility). The *sequence utility* (SU) of a sequence s_c is the sum of the utility of items from s_c in s_c. i.e. $SU(s_c) = \sum_{\{x\} \sqsubseteq s_c} u(x, s_c)$.

Example 6. The sequence utility of sequences s_1, s_2, s_3 and s_4 are respectively 27, 40, 15 and 16.

Definition 12 (Sequence Estimated Utility of an Item). The *sequence estimated utility* (SEU) of an item x is defined as the sum of the sequence utility of sequences containing x, i.e. $SEU(x) = \sum_{s_c \in SDB \wedge \{x\} \sqsubseteq s_c} SU(s_c)$.

Definition 13 (Sequence Estimated Utility of a Rule). The *sequence estimated utility* (SEU) of a sequential rule r is defined as the sum of the sequence utility of sequences containing r, i.e. $SEU(r) = \sum_{s_c \in seq(r)} SU(s_c)$.

Example 7. The SEU of rule $\{a\} \rightarrow \{b\}$ is $SU(s_1) + SU(s_2) + SU(s_3) = 27 + 40 + 15 = 82$.

Definition 14 (Promising Item). An item x is *promising* iff $SEU(x) \geq minutil$. Otherwise, it is *unpromising*.

Definition 15 (Promising Rule). A rule r is *promising* iff $SEU(r) \geq minutil$. Otherwise, it is *unpromising*.

The SEU measure has three important properties that are used to prune the search space.

Property 2 (Overestimation). The SEU of an item/rule w is higher or equal to its utility, i.e. $SEU(w) \geq u(w)$.

Property 3 (Pruning Unpromising Items). Let x be an item. If x is unpromising, then x cannot be part of a high-utility sequential rule.

Property 4 (Pruning unpromising rules). Let r be a sequential rule. If r is unpromising, then any rule obtained by transitive expansion(s) of r is a low utility sequential rule.

We also introduce a new structure called *utility-table* that is used by HUSRM to quickly calculate the utility of rules and prune the search space. Utility-tables are defined as follows.

Definition 16 (Extendability). Let be a sequential rule r and a sequence s. An item i *can extend* r *by left expansion* in s iff $i \succ_{lex} j, \forall j \in X, i \notin Y$ and $X \cup \{i\} \rightarrow Y$ occurs in s. An item i *can extend* r *by right expansion* in s iff $i \succ_{lex} j, \forall j \in Y, i \notin X$ and $X \rightarrow Y \cup \{i\}$ occurs in s. Let $onlyLeft(r,s)$ denotes the set of items that can extend r by left expansion in s but not by right expansion. Let $onlyRight(r,s)$ denotes the set of items that can extend r by right expansion in s but not by left expansion. Let $leftRight(r,s)$ denotes the set of items that can extend r by left and right expansion in s.

Definition 17 (Utility-Table). The *utility-table* of a rule r in a database SDB is denoted as $ut(r)$, and defined as a set of tuples such that there is a tuple $(sid, iutil, lutil, rutil, lrutil)$ for each sequence s_{sid} containing r (i.e. $\forall s_{sid} \in seq(r)$). The $iutil$ element of a tuple is the utility of r in s_{sid}. i.e., $u(r, s_{sid})$. The $lutil$ element of a tuple is defined as $\sum u(i, s_{sid})$ for all item i such that i can extend r by left expansion in s_{sid} but not by right expansion, i.e. $\forall i \in onlyLeft(r, s_{sid})$. The $rutil$ element of a tuple is defined as $\sum u(i, s_{sid})$ for all item i such that i can extend r by right expansion in s_{sid} but not by left expansion, i.e. $\forall i \in onlyRight(r, s_{sid})$. The $lrutil$ element of a tuple is defined as $\sum u(i, s_{sid})$ for all item i such that i can extend r by left or right expansion in s_{sid}, i.e. $\forall i \in leftRight(r, s_{sid})$.

Example 8. The utility-table of $\{a\} \rightarrow \{b\}$ is $\{(s_1, 5, 12, 3, 20), (s_2, 5, 0, 10, 0)\}$. The utility-table of $\{a\} \rightarrow \{b, c\}$ is $\{(s_1, 25, 12, 3, 0)\}$. The utility-table of the rule $\{a, c\} \rightarrow \{b\}$ is $\{(s_1, 25, 12, 3, 0)\}$.

The proposed *utility-table* structure has the following nice properties to calculate the utility and support of rules, and for pruning the search space.

Property 5. Let be a sequential rule r. The utility $u(r)$ is equal to the sum of *iutil* values in $ut(r)$.

Property 6. Let be a sequential rule r. The support of r in a database SDB is equal to the number of tuples in the utility-table of r, divided by the number of sequences in the database, i.e. $sup_{SDB}(r) = |ut(r)|/|SDB|$.

Property 7. Let be a sequential rule r. The sum of *iutil*, *lutil*, *rutil* and *lrutil* values in $ut(r)$ is an upper bound on $u(r)$. Moreover, it can be shown that this upper bound is tighter than $SEU(r)$.

Property 8. Let be a sequential rule r. The utility of any rule t obtained by transitive left or right expansion(s) of r can only have a utility lower or equal to the sum of *iutil*, *lutil*, *rutil* and *lrutil* values in $ut(r)$.

Property 9. Let be a sequential rule r. The utility of any rule t obtained by transitive left expansion(s) of r can only have a utility lower or equal to the sum of *iutil*, *lutil* and *lrutil* values in $ut(r)$.

Now, an important question is how to construct utility-tables. Two cases need to be considered. For sequential rules of size $1 * 1$, utility-tables can be built by scanning the database once. For sequential rules larger than size $1 * 1$, it would be however inefficient to scan the whole database for building a utility-table. To efficiently build a utility-table for a rule larger than size $1 * 1$, we propose the following scheme.

Consider the left or right expansion of a rule with an item i. The utility-table of the resulting rule r' is built as follows. Tuples in the utility-table of r are retrieved one by one. For a tuple $(sid, iutil, lutil, rutil, lrutil)$, if the rule r' appears in sequence s_{sid} (i.e. $r \sqsubseteq s_{sid}$), a tuple $(sid, iutil', lutil', rutil', lrutil')$ is created in the utility-table of r'. The value $iutil'$ is calculated as $iutil + u(\{i\}, s_{sid})$. $lutil'$ is calculated as $lutil - u(j, s_{sid}) \forall j \notin onlyLeft(r', s_{sid}) \land j \in onlyLeft(r, s_{sid}) - [u(i, s_{sid})$ if $i \in onlyLeft(r, s_{sid})]$. The value $rutil'$ is calculated as $rutil - u(j, s_{sid}) \forall j \notin onlyRight(r', s_{sid}) \land j \in onlyRight(r, s_{sid}) - [u(i, s_{sid})$ if $i \in onlyRight(r, s_{sid})]$. Finally, the value $lrutil'$ is calculated as $lrutil - u(j, s_{sid}) \forall j \notin leftRight(r', s_{sid}) \land j \in leftRight(r, s_{sid}) - [u(i, s_{sid})$ if $i \in leftRight(r, s_{sid})]$. This procedure for building utility-tables is very efficient since it requires to scan each sequence containing the rule r at most once to build the utility-table of r' rather than scanning the whole database.

Example 9. The utility-table of $r : \{a\} \rightarrow \{e\}$ is $\{(s_1, 2, 14, 0, 11), (s_2, 2, 36, 2, 0), (s_3, 2, 4, 0, 9)\}$. By adding the *iutil* values of this table, we find that $u(r) = 6$ (Property 5). Moreover, by counting the number of tuples in the utility-table and dividing it by the number of sequences in the database, we find that $sup_{SDB}(r) = 0.75$ (Property 6). We can observe that the sum of *iutil*, *lutil*, *rutil* and *lrutil* values is equal to 82, which is an upper bound on $u(r)$ (Property 7). Furthermore, this value tells us that transitive left/right expansions of r may generate high-utility sequential rules (Property 8). And more particularly, because the sum of $iutil$, $lutil$ and $lrutil$ values is equal to 80, transitive left expansions of r may generate high-utility sequential rules (Property 9). Now, consider the rule $r' : \{a, b\} \rightarrow \{e\}$. The utility-table of r' can be obtained from the utility-table of r using the afore-mentioned procedure. The result is $\{(s_1, 6, 10, 0, 11), (s_2, 6, 32, 2, 0), (s_3, 6, 0, 0, 9)\}$. This table, can then be used to calculate utility-tables of other rules such as $\{a, b, c\} \rightarrow \{e\}$, which is $\{(s_1, 16, 0, 0, 11), (s_2, 26, 12, 2, 0)\}$.

Up until now, we have explained how the proposed utility-table structure is built, can be used to calculate the utility and support of rules and can be used to prune the search space. But a problem remains. How can we calculate the confidence of a rule $r : X \rightarrow Y$? To calculate the confidence, we need to know $|seq(r)|$ and $|ant(r)|$, that is the number of sequences containing r and the number of sequences containing its antecedent X. $|seq(r)|$ can be easily obtained by counting $|ut(r)|$. However, $|ant(r)|$ is more difficult to calculate. A naive solution would be to scan the database to calculate $|ant(r)|$. But this would be highly inefficient. In HUSRM, we calculate $|ant(r)|$ efficiently as follows. HUSRM first creates a bit vector for each single item appearing in the database. The *bit vector* $bv(i)$ of an item i contains $|SDB|$ bits, where the j-th bit is set to 1 if $\{i\} \sqsubseteq s_j$ and is otherwise set to 0. For example, $bv(a) = 1111$, $bv(b) = 1011$ and $bv(c) = 1101$. Now to calculate the confidence of a rule r, HUSRM intersects the bit vectors of all items in the rule antecedent, i.e. $\bigwedge_{i \in X} bv(i)$. The resulting bit vector is denoted as $bv(X)$. The number of bits set to 1 in $bv(X)$ is equal to $|ant(r)|$. By dividing the number of lines in the utility-table of the rule $|ut(r)|$ by this number, we obtain the confidence. This method is very efficient because intersecting bit vectors is a very fast operation and bit vectors does not consume much memory. Furthermore, an additional optimization is to reuse the bit vector $bv(X)$ of rule r to more quickly calculate $bv(X \cup \{i\})$ for any left expansions of r with an item i (because $bv(X \cup \{i\}) = bv(X) \wedge bv(\{i\})$).

3.2 The Proposed Algorithm

HUSRM explores the search space of sequential rules using a depth-first search. HUSRM first scans the database to build all sequential rules of size $1 * 1$. Then, it recursively performs left/right expansions starting from those sequential rules to generate larger sequential rules. To ensure that no rule is generated twice, the following ideas have been used.

First, an important observation is that a rule can be obtained by different combinations of left and right expansions. For example, consider the rule r : $\{a, b\} \rightarrow \{c, d\}$. By performing a left and then a right expansion of $\{a\} \rightarrow \{c\}$, one can obtain r. But this rule can also be obtained by performing a right and then a left expansion of $\{a\} \rightarrow \{c\}$. A simple solution to avoid this problem is to not allow performing a left expansion after a right expansion but to allow performing a right expansion after a left expansion. Note that an alternative solution is to not allow performing a left expansion after a right expansion but to allow performing a right expansion after a left expansion.

Second, another key observation is that a same rule may be obtained by performing left/right expansions with different items. For example, consider the rule $r_9 : \{b, c\} \rightarrow \{e\}$. A left expansion of $\{b\} \rightarrow \{e\}$ with item c results in r_9. But r_9 can also be found by performing a left expansion of $\{c\} \rightarrow \{e\}$ with item b. To solve this problem, we chose to only add an item to a rule by left (right) expansion if the item is greater than each item in the antecedent (consequent) according to the total order $\succ_{lex}$ on items. By using this strategy and the previous one, no rules is considered twice.

Figure 1 shows the pseudocode of HUSRM. The HUSRM algorithm takes as parameters a sequence database SDB, and the *minutil* and *minconf* thresholds. It outputs the set of high-utility sequential rules. HUSRM first scans the database once to calculate the sequence estimated utility of each item and identify those that are promising. Then, HUSRM removes unpromising items from the database since they cannot be part of a high-utility sequential rule (Property 3). Thereafter, HUSRM only considers promising items. It scans the database to create the bit vectors of those items, and calculate $seq(r)$ and $SEU(r)$ for each rule r of size $1 * 1$ appearing in the database. Then, for each promising rule r, HUSRM scans the sequences containing r to build its utility-table $ut(r)$. If r is a high-utility sequential rule according to its utility-table and the bit-vector of its antecedent, the rule is output. Then, Property 8 and 9 are checked using the utility-table to determine if left and right expansions of r should be considered. Exploring left and right expansions is done by calling the *leftExpansion* and *rightExpansion* procedures.

The *leftExpansion* procedure (Algorithm 2) takes as input a sequential rule r and the other parameters of HUSRM. It first scans sequences containing the rule r to build the utility-table of each rule t that is a left-expansion of r. Note that the utility-table of r is used to create the utility-table of t as explained in Sect. 3.1. Then, for each such rule t, if t is a high-utility sequential rule according to its utility-table and the bit-vector of its antecedent, the rule is output. Finally, the procedure *leftExpansion* is called to explore left-expansions of t if Property 9 is verified. The *rightExpansion* procedure (Algorithm 3) is very similar to *left-Expansion* and is thus not described in details here. The main difference is that *rightExpansion* considers right expansions instead of left expansions and can call both *leftExpansion* and *rightExpansion* to search for larger rules.

3.3   Additional Optimizations

Two additional optimizations are added to HUSRM to further increase its efficiency. The first one reduces the size of utility-tables. It is based on the observations that in the *leftExpansion* procedure, (1) the *rutil* values of utility-tables are never used and (2) that *lutil* and *lrutil* values are always summed. Thus, (1) the *rutil* values can be dropped from utility-tables in *leftExpansion* and (2) the sum of *lutil* and *lrutil* values can replace both values. We refer to the resulting utility-tables as *Compact Utility-Tables* (CUT). For example, the CUT of

---

Algorithm 1. The HUSRM algorithm

---

input : SDB: a sequence database, $minutil$ and $minconf$: the two
user-specified thresholds

output: the set of high-utility sequential rules

1 Scan SDB to calculate the sequence estimated utility of each item $i \in I$;
2 $I^* \leftarrow \{i | i \in I \wedge SEU(i) \geq minutil\}$;
3 Remove from SDB each item $j \in I$ such that $j \notin I^*$;
4 Scan SDB to calculate the bit vector of each item $i \in I^*$;
5 Scan SDB to calculate R, the set of rules of the form $r : i \rightarrow j(i, j \in I^*)$
appearing in SDB and calculate $SEU(r)$ and $seq(r)$;
6 $R^* \leftarrow \{r | r \in R \wedge SEU(r) \geq minutil\}$;
7 **foreach** *rule* $r \in R^*$ **do**
8   Calculate $ut(r)$ by scanning $seq(r)$;
9   **if** $u(r) \geq minutil$ according to $ut(r)$ and $conf_{SDB}(r) \geq minconf$ **then**
        output r;
10   **if** r respects Property 8 according to $ut(r)$ **then** rightExpansion $(r, SDB,$
        $minutil, minconf)$;
11   **if** r respects Property 9 according to $ut(r)$ **then** leftExpansion $(r, SDB,$
        $minutil, minconf)$;
12 **end**

$\{a, b\} \rightarrow \{e\}$ and $\{a, b, c\} \rightarrow \{e\}$ are respectively $\{(s_1, 6, 21), (s_2, 6, 32), (s_3, 6, 9)\}$ and $\{(s_1, 16, 11), (s_2, 26, 12)\}$. CUT are much smaller than utility-tables since each tuple contains only three elements instead of five. It is also much less expensive to update CUT.

The second optimization reduces the time for scanning sequences in the *leftExpansion* and *rightExpansion* procedures. It introduces two definitions. The *first occurrence* of an itemset X in a sequence $s = \langle I_1, I_2, ...I_n \rangle$ is the itemset $I_k \in s$ such that $X \subseteq \bigcup_{i=1}^{k} I_i$ and there exists no $g < k$ such that $X \subseteq \bigcup_{i=1}^{g} I_i$. The *last occurrence* of an itemset X in a sequence $s = \langle I_1, I_2, ...I_n \rangle$ is the itemset $I_k \in s$ such that $X \subseteq \bigcup_{i=k}^{n} I_i$ and there exists no $g > k$ such that $X \subseteq \bigcup_{i=g}^{n} I_i$. An important observation is that a rule $X \rightarrow Y$ can only be expanded with items appearing after the first occurrence of X for a right expansion, and occurring before the last occurrence of Y for a left expansion. The optimization consists of keeping track of the first and last occurrences of rules and to use this information to avoid scanning sequences completely when searching for items to expand a rule. This can be done very efficiently by first storing the first and last occurrences of rules of size $1 * 1$ and then only updating the first (last) occurrences when performing a left (right) expansion.

4 Experimental Evaluation

We performed experiments to evaluate the performance of the proposed algorithm. Experiments were performed on a computer with a fourth generation 64 bit core i7 processor running Windows 8.1 and 16 GB of RAM. All memory measurements were done using the Java API.

Algorithm 2. The leftExpansion procedure

---

input : r: a sequential rule $X \to Y$, SDB, $minutil$ and $minconf$

1 $rules \leftarrow \emptyset$;
2 **foreach** $sequence\ s \in seq(r)$ according to $ut(r)$ **do**
3   **foreach** $rule\ t : X \cup \{i\} \to Y | i \in leftRight(t, s) \cup onlyLeft(t, s)$ **do**
        $rules \leftarrow rules \cup \{t\}$; Update $ut(t)$;
4 **end**
5 **foreach** $rule\ r \in rules$ **do**
6   **if** $u(r) \geq minutil$ according to $ut(r)$ and $conf_{SDB}(r) \geq minconf$ **then**
        output r;
7   **if** r respects Property 9 according to $ut(r)$ **then** leftExpansion (r, SDB, $minutil$, $minconf$);
8 **end**

Algorithm 3. The rightExpansion procedure

---

input : r: a sequential rule $X \to Y$, SDB, $minutil$ and $minconf$

1 $rules \leftarrow \emptyset$;
2 **foreach** $sequence\ s \in seq(r)$ according to $ut(r)$ **do**
3   **foreach** $rule\ t$ of the form $X \cup \{i\} \to Y$ or $X \to Y \cup \{i\}$
    $| i \in leftRight(t, s) \cup onlyLeft(t, s) \cup onlyRight(t, s)$ **do**
        $rules \leftarrow rules \cup \{t\}$; Update $ut(t)$;
4 **end**
5 **foreach** $rule\ r \in rules$ **do**
6   **if** $u(r) \geq minutil$ according to $ut(r)$ and $conf_{SDB}(r) \geq minconf$ **then**
        output r;
7   **if** r respects Property 8 according to $ut(r)$ **then** rightExpansion (r, SDB, $minutil$, $minconf$);
8   **if** r respects Property 9 according to $ut(r)$ **then** leftExpansion (r, SDB, $minutil$, $minconf$);
9 **end**

Experiments were carried on four real-life datasets commonly used in the pattern mining literature: *BIBLE*, *FIFA*, *KOSARAK* and *SIGN*. These datasets have varied characteristics and represents the main types of data typically encountered in real-life scenarios (dense, sparse, short and long sequences). The characteristics of datasets are shown in Table 4), where the $|SDB|$, $|I|$ and $avgLength$ columns respectively indicate the number of sequences, the number of distinct items and the average sequence length. *BIBLE* is moderately dense and contains many medium length sequences. *FIFA* is moderately dense and contains many long sequences. *KOSARAK* is a sparse dataset that contains short sequences and a few very long sequences. *SIGN* is a dense dataset having very long sequences. For all datasets, external utilities of items are generated between 0 and 1,000 by using a log-normal distribution and quantities of items are generated randomly between 1 and 5, similarly to the settings of [2, 8, 12].

Table 4. Dataset characteristics

Dataset	$|SDB|$	$|I|$	$avgLength$	Type of data
BIBLE	36,369	13,905	21.64	book
FIFA	573,060	13,749	45.32	click-stream
KOSARAK	638,811	39,998	11.64	click-stream
SIGN	730	267	93.00	sign language

Because HUSRM is the first algorithm for high-utility sequential rule mining, we compared its performance with five versions of HUSRM where optimizations had been deactivated (HUSRM$_1$, HUSRM$_{1,2}$, HUSRM$_{1,2,3}$, HUSRM$_{1,2,3,4}$ and HUSRM$_{1,2,3,4,5}$). The notation HUSRM$_{1,2,\ldots n}$ refers to HUSRM without optimizations O_1, O_2 ... O_n. Optimization 1 (O_1) is to ignore unpromising items. Optimization 2 (O_2) is to ignore unpromising rules. Optimization 3 (O_3) is to use bit vectors to calculate confidence instead of lists of integers. Optimization 4 (O_4) is to use compact utility-tables instead of utility-tables. Optimization 5 (O_5) is to use Property 9 to prune the search space for left expansions instead of Property 8. The source code of all algorithms and datasets can be downloaded as part of the SPMF data mining library at http://goo.gl/qS7MbH [6].

We ran all the algorithms on each dataset while decreasing the *minutil* threshold until algorithms became too long to execute, ran out of memory or a clear winner was observed. For these experiments, we fixed the *minconf* threshold to 0.70. However, note that results are similar for other values of the *minconf* parameter since the confidence is not used to prune the search space by the compared algorithms. For each dataset and algorithm, we recorded execution times and maximum memory usage.

Execution times. Fig. 1 shows the execution times of each algorithm. Note that results for HUSRM$_{1,2,3,4,5}$ are not shown because it does not terminate in less than $10,000s$ for all datasets. HUSRM is respectively up to 1.8, 1.9, 2, 3.8 and 25 times faster than HUSRM$_1$, HUSRM$_{1,2}$, HUSRM$_{1,2,3}$, HUSRM$_{1,2,3,4}$ and HUSRM$_{1,2,3,4,5}$. It can be concluded from these results that HUSRM is the fastest on all datasets, that its optimizations greatly enhance its performance, and that O_5 is the most effective optimization to reduce execution time. In this experiment, we have found up to 100 rules, which shows that mining high-utility sequential rules is very expensive. Note that if we lower *minutil*, it is possible to find more than 10,000 rules using HUSRM.

Memory usage. Table 4 shows the maximum memory usage of the algorithms for the BIBLE, FIFA, KOSARAK and SIGN datasets. Results for HUSRM$_{1,2,3,4,5}$ are not shown for the same reason as above. It can be observed that HUSRM always consumes less memory and that this usage is up to about 50 % less than that of HUSRM$_{1,2,3,4}$ on most datasets. The most effective optimization to reduce memory usage is O_4 (using compact utility-tables), as shown in Table 5.

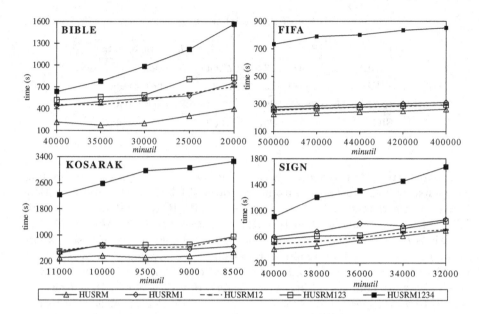

Fig. 1. Comparison of execution times (seconds)

Table 5. Comparison of maximum memory usage (megabytes)

Dataset	HUSRM	HUSRM$_1$	HUSRM$_{1,2}$	HUSRM$_{1,2,3}$	HUSRM$_{1,2,3,4}$
BIBLE	**1,022**	1,177	1,195	1,211	1,346
FIFA	**1,060**	1,089	1,091	1,169	1,293
KOSARAK	**502**	587	594	629	1,008
SIGN	**1,053**	1,052	1,053	1053	1,670

5   Conclusion

To address the lack of confidence measure in high-utility sequential pattern mining, we defined the problem of high-utility sequential rule mining and studied its properties. We proposed an efficient algorithm named HUSRM (High-Utility Sequential Rule Miner) to mine these rules. HUSRM is a one-phase algorithm that relies on a new data structure called compact utility-table and include several novel optimizations to mine rules efficiently. An extensive experimental study with four datasets shows that HUSRM is very efficient and that its optimizations respectively improve its execution time by up to 25 times and its memory usage by up to 50 %.

Acknowledgement. This work is financed by a National Science and Engineering Research Council (NSERC) of Canada research grant.

References

1. Agrawal, R., Srikant, R.: Fast algorithms for mining association rules in large databases. In: Proceedings of International Conference on Very Large Databases, pp. 487–499 (1994)
2. Ahmed, C.F., Tanbeer, S.K., Jeong, B.-S., Lee, Y.-K.: Efficient Tree Structures for High-utility Pattern Mining in Incremental Databases. IEEE Trans. Knowl. Data Eng. **21**(12), 1708–1721 (2009)
3. Fournier-Viger, P., Wu, C.-W., Tseng, V.S., Cao, L., Nkambou, R.: Mining Partially-Ordered Sequential Rules Common to Multiple Sequences. IEEE Trans. Knowl. Data Eng. (preprint). doi:10.1109/TKDE.2015.2405509
4. Fournier-Viger, P., Gueniche, T., Zida, S., Tseng, V.S.: ERMiner: sequential rule mining using equivalence classes. In: Blockeel, H., van Leeuwen, M., Vinciotti, V. (eds.) IDA 2014. LNCS, vol. 8819, pp. 108–119. Springer, Heidelberg (2014)
5. Fournier-Viger, P., Wu, C.-W., Zida, S., Tseng, V.S.: FHM: faster high-utility itemset mining using estimated utility co-occurrence pruning. In: Andreasen, T., Christiansen, H., Cubero, J.-C., Raś, Z.W. (eds.) ISMIS 2014. LNCS, vol. 8502, pp. 83–92. Springer, Heidelberg (2014)
6. Fournier-Viger, P., Gomariz, A., Gueniche, T., Soltani, A., Wu, C., Tseng, V.S.: SPMF: a java open-source pattern mining library. J. Mach. Learn. Res. **15**, 3389–3393 (2014)
7. Lin, C.-W., Hong, T.-P., Lu, W.-H.: An effective tree structure for mining high utility itemsets. Expert Syst. Appl. **38**(6), 7419–7424 (2011)
8. Liu, M., Qu, J.: Mining High Utility Itemsets without Candidate Generation. In: Proceedings of 22nd ACM International Conference on Information on Knowledge and Management, pp. 55–64 (2012)
9. Liu, Y., Liao, W., Choudhary, A.K.: A two-phase algorithm for fast discovery of high utility itemsets. In: Ho, T.-B., Cheung, D., Liu, H. (eds.) PAKDD 2005. LNCS (LNAI), vol. 3518, pp. 689–695. Springer, Heidelberg (2005)
10. Lo, D., Khoo, S.-C., Wong, L.: Non-redundant sequential rules - theory and algorithm. Inf. Syst. **34**(4–5), 438–453 (2009)
11. Pham, T.T., Luo, J., Hong, T.P., Vo, B.: An efficient method for mining non-redundant sequential rules using attributed prefix-trees. Eng. Appl. Artif. Intell. **32**, 88–99 (2014)
12. Tseng, V.S., Shie, B.-E., Wu, C.-W., Yu, P.S.: Efficient algorithms for mining high utility itemsets from transactional databases. IEEE Trans. Knowl. Data Eng. **25**(8), 1772–1786 (2013)
13. Tseng, V., Wu, C., Fournier-Viger, P., Yu, P.: Efficient algorithms for mining the concise and lossless representation of closed+ high utility itemsets. IEEE Trans. Knowl. Data Eng. **27**(3), 726–739 (2015)
14. Yin, J., Zheng, Z., Cao, L.: USpan: an efficient algorithm for mining high utility sequential patterns. In: Proceedings of 18th ACM SIGKDD International Conference on Knowledge Discovery and Data Mining, pp. 660–668 (2012)
15. Yin, J., Zheng, Z., Cao, L., Song, Y., Wei, W.: Efficiently mining top-k high utility sequential patterns. In: IEEE 13th International Conference on Data Mining, pp. 1259–1264 (2013)

MOGACAR: A Method for Filtering Interesting Classification Association Rules

Diana Benavides Prado$^{(\boxtimes)}$

Systems and Computing Engineering Department,
Universidad de Los Andes, Bogota, Colombia
dk.benavides20@uniandes.edu.co

Abstract. Knowledge Discovery process is intended to provide valid, novel, potentially useful and finally understandable patterns from data. An interesting research area concerns the identification and use of interestingness measures, in order to rank or filter results and provide what might be called better knowledge. For association rules mining, some research has been focused on how to filter itemsets and rules, in order to guide knowledge acquisition from the user's point of view, as well as to improve efficiency of the process. In this paper, we explain MOGACAR, an approach for ranking and filtering association rules when there are multiple technical and business interestingness measures; MOGACAR uses a multi-objective optimization method based on genetic algorithm for classification association rules, with the intention to find the most interesting, and still valid, itemsets and rules.

Keywords: Knowledge discovery in databases · Data mining · Interestingness measures · Genetic algorithms · Classification association rules

1 Introduction

Knowledge Discovery, a research area enclosing Data Mining (DM), is defined as the non-trivial process of identifying patterns in data. These patterns should be valid, novel, potentially useful and understandable in order to gain new knowledge within some application domain [7]. Association rules (AR) is one of the most commonly used techniques in DM, allowing to obtain groups of things that generally go together. One of the main challenges with AR is how to determine the best set of rules that represent best patterns for some user or application area, having into account that typically this technique produces a large number of rules, not necessarily novel or useful.

Interestingness measures for association rules and other DM techniques is an exciting research field. In general, interestingness measures can be classified as those which evaluate performance of the DM algorithm (technical measures) and those which assess utility, usability or some other more problem-related quality attributes (business measures) [5]. By using them together, there might be more criteria for identifying more useful and understandable patterns, from the user and problem's point of view, that are still valid and correct from the point of view of data mining techniques.

With respect to association rules, the most common algorithm Apriori [1] tends to involve a costly process. In its first phase, finding frequent sets of items or itemsets, all

© Springer International Publishing Switzerland 2015
P. Perner (Ed.): MLDM 2015, LNAI 9166, pp. 172–183, 2015.
DOI: 10.1007/978-3-319-21024-7_12

of the possible combinations of items on the available data are obtained, which are finally filtered by using a statistical measure (support). One of the main concerns in AR mining is how to provide more potentially useful, understandable and interesting rules, having into account that typically a large set of rules is obtained, making it difficult for the user of the DM solution to use them. In this paper, a proposal for obtaining more interesting classification association rules is presented, by including other kinds of measures for filtering itemsets, which might be particular to some business or application area; for accomplishing the task of selecting interesting itemsets by evaluating them from the point of view of multiple measures altogether, a heuristic approach based on multi-objective optimization is proposed.

This paper continues as follows: Sect. 2 presents background around DM and AR, Sect. 3 identifies state-of-the-art about interestingness measures and classification association rules ranking and filtering, Sect. 4 presents an approach for more interesting classification association rules by using multi-objective optimization, Sect. 5 presents this proposal validation and results, Sect. 6 presents some final remarks and ideas for future work.

2 Data Mining and Association Rules

DM is defined as the exploration and analysis of large amounts of data to discover interesting patterns and rules [3].

AR is one of the most commonly used techniques in DM. This technique allows grouping things which generally go together [3], and Apriori is its most common algorithm [1]. An association rule is defined as an implication of the form $X \rightarrow Y$, where X is a set of some items (itemset) in a set of attributes I, and Y is a single item in I that is not in X (thus $X \cap Y = \varnothing$). The rule $X \rightarrow Y$ is satisfied in a set of transactions T with the confidence factor $0 <= c <= 1$ if at least c % of transactions in T that satisfy X also satisfy Y.

Classification association rules (CAR) [11], are intended to restrict the search space in Y, in order to obtain association rules when the consequence (i.e. Y) is previously defined. In CAR, frequent itemsets are instead called ruleitems.

3 Interest of Patterns and Rules

Several theoretical frameworks have been proposed in order to include interestingness measures such as validity, novelty, potential utility and understandability of patterns, as criteria for providing final results to some user. Some previous works classify interestingness measures into three main groups: objective, subjective and semantic measures; it is typical that the last two groups are merged into subjective or problem-specific measures [8]. Another more recent point of view classify interestingness measures in two major categories: technical measures and business measures [5]. Technical measures include those which can automatically assess whether discovered knowledge is valid or not, using the data itself; business measures include concepts such as potential capacity of action or potential decisions some business or application may take, based on discovered knowledge.

Interest of a pattern p can be established by taking into account technical as well as business interestingness measures, objective (process-defined) or subjective (user-defined) [5].

$$Int(p) = I(to(p), ts(p), bo(p), bs(p)) \tag{1}$$

For technical interestingness measures in AR, there have been some proposals for appropriately ranking and filtering rules: some of them have been focused on how to rank rules based on its dominance relationships by using technical interestingness measures [4], another ones on defining additional technical measures for ranking [14] and still others on adaptation of approaches of other computer science research fields, such as information retrieval [17].

For business interestingness measures, past research has been mainly focused on actionability or potential utility of the obtained knowledge: KEFIR [13] proposes measuring the impact of an obtained pattern in health domain, by having into account business metrics impacted, and its relation with data. Profit Mining [16] proposes maximization of a business objective in retail sector (profit), taking into account the probability of a customer to purchase a product and a taxonomy of products; here, business interest is a formally defined formula, and obtained patterns are ranked according to this formula and pruned according to some projected benefit of the pattern over future customers. In OOApriori [15], an utility interestingness measure over an attribute of a database might be an additional parameter for obtaining association rules, which may be configured subjectively in the same manner as support and confidence; this allows determining profit of patterns according to whether they obey to utility parameter. In D3M [5], some business interestingness measures are defined in terms of risk and cost of a pattern around debt in social security domain. In Value-Added [6], addition of semantic or value attributes such as price, profit, security level, to values of regular attributes of the data set is addressed; this leads to the fact that frequent itemsets generated in the first phase of AR have the same characterization in terms of some semantic value.

Those solutions are mostly business or application-oriented on the one hand, and pure technical on the other; as a result, a research opportunity for defining how to appropriately combine technical and business interestingness measures, independently of the application area and having into account that user may be at the same time interested in valid and potentially useful knowledge, is open.

A method for filtering interesting classification association rules by ranking them according to technical and business interestingness measures altogether, and filtering or pruning not interesting ones, with independence of the problem or application area, is described next. For this method, a set of business interestingness measures is available and appropriately integrated into DM process [2]. A multi-objective optimization method is used, for iteratively determining the most interesting ruleitems from the point of view of the involved measures; we provide a method for ranking and filtering ruleitems, according to its interest, in order to improve knowledge acquisition as well as leveraging existence of other metrics for potentially reducing the cost of the process.

4 MOGACAR: Multi-objective Optimization Approach for Filtering Interesting Classification Association Rules

Since a set of technical and business interestingness measures is available, a definition of how to leverage them together might be first established; in this proposal, those measures are used for determining potential interest of a ruleitem, in order to improve first phase of Apriori algorithm (frequent itemset or ruleitem search) as well as to guarantee that, lately, only valid and more interesting classification association rules are obtained. In the most general sense, an interesting ruleitem p is defined as the one that, given a set of measures M, for all of its measures m it is interesting (denoted as Int(p)).

Definition 1 (Interesting ruleitem). A ruleitem p is interesting with respect to a set of technical and business interest measures M if $\forall m \in M(Int(p))$.

When we have multiple measures m and ruleitems p, Definition 1 should be accompanied by a more specific way of combining the set of measures M and determining whether some ruleitem p is more interesting than other, in terms of those measures.

Consequently, for each ruleitem p, we first need to define what it means that it is interesting from the point of view of a measure m. For a given measure m, a ruleitem p may have some current value v, i.e. value that its items, when together, take for the specific measure. On the other hand, each measure may have a previously established goal value g, i.e. its desired value.

For support technical measure, for example, some goal g may be 90 %; it means that more interesting ruleitems (from the technical point of view) will be those which value v for support is closer to 90 %; for a company in finance sector, for example, some business measure may be cost over income ratio, and its associated goal g could be, for example, 50 %; it means that ruleitems with a value v for cost over income ratio closer to 50 % will be more interesting. Consequently, we define the **potential** (interest) **of a ruleitem** as the set of differences between the values v that is takes for each of the measures m and those measures goals g:

Definition 2 (Potential of a ruleitem). The potential of a ruleitem is defined by the set of differences between a set of measures goals and the values that the ruleitem takes for each of those measures; each of those differences is defined as $g - v$.

Now, we need to take into account that various technical and business interestingness measures may be involved. Since, in some application areas, it might be difficult to discriminate them by their importance, or combine them into a single measure, we need to treat them equally, i.e. give them the same importance. For that purpose, we translate our problem into a multi-objective optimization problem, since its theoretical foundation relies on optimizing a collection of objective functions [12] and we are trying to do something similar with each of our measures m.

In the spectrum of possible solutions for multi-objective optimization problems, three approaches are typical. First, combining multiple objective functions, by a weighting or summing approach; this approach requires previous articulation of preferences (i.e. previously defining the importance of each of comparison criteria and/or how are they going to be summed up). Second, an approach with post-articulation of preferences, that requires defining importance in post-processing phase of data mining. Finally, an approach with no articulation of preferences, which may help to treat different objective functions independently. Since intention is to provide a solution that might be useful for any application area, including those where it may be difficult to establish preference between business measures (for example, cost over income ratio and risk percentage, in finance-related problems), no previous articulation of preferences should be used.

From a theoretical point of view, for making no articulation of preferences real, Pareto-frontier and dominance concepts [12] come into play. For obtaining more interesting classification association rules, referring to Pareto dominance concept will allow us to establish whether some ruleitem is more interesting than other, according to a set of technical and business interestingness measures, thus helping us to find a frontier where the most interesting ruleitems and classification association rules can be obtained.

Definition 3 (Pareto dominance). A pattern P1 is more interesting than a pattern P2 according to a set of measures M, with value v and goal g, if:

$$(\forall m \in M(g(m, P1) - v(m, P1) \leq g(m, P2) - v(m, P2))^{\wedge} \exists m(g(m, P1) - v(m, P2) < \\ g(m, P2) - v(m, P2))$$

In our case, a pattern P1 or P2 is a ruleitem. What Definition 3 implies is that, for some ruleitem P1 to be more interesting than a second ruleitem P2, P1 should be at least as good as P2 in terms of the differences between goals and current values (g and v) of the involved measures – its potential, as established by Definition 2 - and, also, there should exist at least one measure for which P1 is better than P2, i.e. its difference between goal and current value –potential - for this measure is shorter. In this case, P1 should be called a better (dominant) solution than P2, while P2 should be called a worse (dominated) solution with respect to P1.

For implementing Pareto-frontier, a genetic algorithm approach is selected, since it can be effective regardless of the nature of involved criteria/measures, as well as can converge to a global solution. Among the offers of genetic algorithms approaches for determining Pareto-frontier, and since it has been proved for a wide range of solutions, SPEA2 [18] is selected. Accordingly, a solution that allows generating classification association rules by involving SPEA2 approach into the process itself might be provided.

5 Apriori with SPEA2

In general, Apriori with SPEA2 approach allows to iteratively obtain sets of best solutions (k-ruleitems potentials) according to the number of individuals (k-ruleitems potentials) it dominates. It is important to note that SPEA2 allows determining interest of each ruleitem according to how many itemsets it is better than (i.e. for how many ruleitems is its potential better than, according to Definition 3), and to how good are those dominated ruleitems too (i.e. their own potentials). Also, by using SPEA2 density of the domination relation, i.e. how strong is domination relation of one ruleitem over the others, a stronger criterion for finding those best ruleitems can be reached.

With our approach, as is typical with Apriori, 1 to n-ruleitems are iteratively obtained and filtered by using support, guarantying that we take into account only statistically valid ruleitems. Then, at each iteration, we perform a three-step process: first, we determine the potential of each ruleitem, it is the difference between the value it takes for each of the measures and each measure goal or desired value, as established by Definition 2.

Second, once potential of each ruleitem at iteration i has been obtained, a ranking of those ruleitems should be established, determining which ones are more interesting and which ones are not, according to what was established in Definition 3; it is accomplished by making each of the ruleitems potentials to become a member of the initial population that SPEA2 approach will use in order to rank them, and thus obtaining best and worst solutions (i.e. best and worst ruleitems potentials). For step 2, SPEA2 we will then need to follow the next steps:

(1) Define an archive size (i.e. a subset of current k-ruleitems potentials to begin with).
(2) Define the number of generations to iterate through.
(3) Convert each k-ruleitem potential into its binary representation, in order to allow it to be processed by the genetic algorithm approach.
(4) Find dominated and non-dominated k-ruleitems potentials for the current population, by comparing them using Definition 3. According to SPEA2 [18], non-dominated solutions will have more chance of surviving next iteration, while dominated ones will not.
(5) If there are still dominated and non-dominated k-ruleitems potentials, then:
   a. Calculate fitness for dominated and non-dominated ruleitems potentials by using fitness and density formulas as specified in SPEA2. For density, Euclidean distance is used.
   b. Perform environmental selection; this selects current non-dominated k-ruleitems potentials and copies them to the archive, for the next iteration.
   c. Make a binary tournament.
   d. Apply operations of mutation and crossover for obtaining new population for the next iteration. This new population will be represented by a set of ruleitems potentials that has been obtained by the mutation and crossover operations.

As stated before, by using SPEA2, a subset of best solutions (best k-ruleitems potentials) is obtained, as well as a subset of worst solutions (worst k-ruleitems potentials); since those solutions derive from mutation and crossover operations, and might not represent actual k-ruleitems potentials, our third step allows to establish the real interest of each of the original k-ruleitems according to the position of the nearest best-solution found by SPEA2, i.e., the rank of the current k-ruleitem potential.

Criterion for determining closeness between a k-ruleitem potential and some best (or worst)-solution found by SPEA2 is based on Euclidean distance between their potentials. If some worst-solution is found for the current k-ruleitem potential, which is closer to the k-ruleitem potential than some best-solution found by SPEA2, k-ruleitem will be pruned. If some best-solution is found for the current k-ruleitem potential, which is closer to the k-ruleitem potential than some worst-solution found by SPEA2, k-ruleitem will be kept as an interesting ruleitem. Figure 1 summarizes this three-step approach.

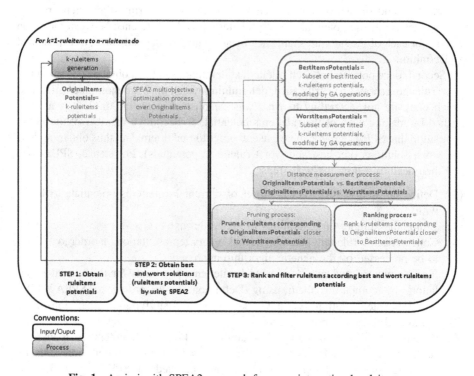

Fig. 1. Apriori with SPEA2 approach for more interesting k-ruleitems.

In this manner, it is possible to remark most interesting k-ruleitems while pruning or filtering least interesting ones, thus filtering knowledge for next iteration of the first phase of the algorithm. Apriori algorithm with SPEA2 is shown in next page.

In this algorithm, we have added to Apriori algorithm the three-step procedure mentioned in the last paragraphs: lines 1 and 17 for obtaining potentials (step 1), lines 2 and 3 and 18 and 19 for obtaining worst and best solutions by using the procedure established with SPEA2 and lines 4 and 20 for ranking and filtering ruleitems according to what has been obtained with SPEA2.

For implementing Apriori with SPEA2, an extension to Weka [10] implementation of Apriori has been developed, in order to include business interestingness measures for ranking and pruning ruleitems [2]. As part of this extension, an implementation of each of the steps proposed by Apriori with SPEA2 has been developed [9].

```
0)     F₁ = {large 1-ruleitems};
1)       PT₁= obtainPotentials(F₁);
2)       BP₁ = obtainBestPotentials(PT₁);
3)       WP₁ = obtainWorstPotentials(PT₁);
4)       F₁ = rankFilter(F₁, BP₁, WP₁ );
5)       CAR₁ = genRules(F₁);
6)       prCAR₁ = pruneRules(CAR₁);
7)       for(k=2; Fₖ₋₁ = ∅; k++) do
8)         Cₖ = candidateGen(Fₖ₋₁);
9)         for each data case d ∈ D do
10)          Cₐ = ruleSubset(Cₖ, d);
11)            for each candidate c ∈ Cₐ do
12)              c.condsupCount++
13)              if d.class = c.class then c.ruleSupCount++;
14)            end
15)        end
16)        Fₖ = {c ∈ Cₖ | c.ruleSupCount ≥ minsup};
17)        PTₖ= obtainPotentials(Fₖ);
18)        BPₖ = obtainBestPotentials(PTₖ);
19)        WPₖ = obtainWorstPotentials(PTₖ);
20)        Fₖ = rankFilter(Fₖ, BPₖ, WPₖ );
21)        CARₖ = genRules(Fₖ);
22)        prCARₖ = pruneRules(CARₖ);
23)      end
24)      CARₛ = Uₖ CARₖ;
25)      prCARₛ = UₖprCARₖ;
26)    end
```

6 Experiment and Evaluation

For validating this proposal, a case study in education domain has been chosen. In Colombia, knowledge acquired by high school students is tested with a national standard exam named SABER11. In this test, knowledge and abilities of the students in eight different basic knowledge areas are evaluated, for instance: language, mathematics, social sciences, philosophy, biology, chemistry, physics and english; an additional area, named flexible component, can be chosen by students from a set of six possibilities: environment, emphasis in biology, emphasis in social sciences, emphasis in language, emphasis in mathematics, and society and violence. ICFES, the institution administering SABER11 exam, is continuously updating and improving design of this test, in order to guarantee that all of the abilities and required knowledge are evaluated.

For our experiment, a dataset containing results of SABER11 exam in semester 2 of 2013 is used. It contains scores (from 0 to 100 %) for 547.286 students in each of the eight knowledge areas, along with flexible component and basic, family, location, social and economic information for each one of those students.

This experiment is intended to identify intrinsic relations between basic, family, location, social and economic attributes of students and flexible component, in order to determine which characteristics of students tend to go together with each of the six possible choices. Consequently, a rule will have the form $X \rightarrow Y$, where X can be any attribute characterizing a student and Y will be one of the six choices for flexible component. For ranking and filtering ruleitems that will generate those rules, each of the eight basic knowledge areas (language, mathematics, social sciences, philosophy, biology, chemistry, physics and english) will become an interestingness measure, with its respective goal value (100 %).

In the context of this case study, appropriateness of ranking and filtering approach of Apriori with SPEA2 is evaluated, by comparing obtained ruleitems with and without proposed method (only Apriori). For this experiment, SPEA2 approach is parameterized with an archive size of 10 and 5 generations to iterate through.

Our first evaluation is quantitative, by comparing the number of ruleitems obtained at each iteration:

Table 1. SPEA2 k-ruleitems reduction rate versus pure technical approach (Apriori).

Min. support	k-ruleitems	Apriori with SPEA2	Apriori	Reduction rate
35 %	1-ruleitems	2	2	0 %
	2-ruleitems	1	1	0 %
30 %	1-ruleitems	7	9	22 %
	2-ruleitems	3	7	57 %
	3-ruleitems	0	2	100 %
25 %	1-ruleitems	10	15	33 %
	2-ruleitems	14	47	70 %
	3-ruleitems	0	39	100 %
	4-ruleitems	0	9	100 %

Table 1 shows that Apriori with SPEA2 can reach up to 22 % reduction rate when generating 1-ruleitems, with a support configuration of 30 %; consequently, in next iterations of ruleitems generation, reduction rate keeps increasing (57 % for 2-ruleitems to 100 % for 3-ruleitems, for example).

Now, for qualitatively evaluating Apriori with SPEA2, a comparison of the obtained k-ruleitems with and without this method is made, in order to determine whether ranking and filtering have been performed accordingly; as an example of our results, in Table 2, details of 1-ruleitems obtained with 30 % support configuration are shown. In this table, each measure (technical, T1, representing support, and business, M1 to M8, representing each of the eight knowledge areas) is shown; for T1, we show the position that it should belong to according to potential of the ruleitem for this measure; for M1 to M8, we show the position as well as the obtained value when calculating the potential of the ruleitem.

Table 2. 1-ruleitems and its corresponding values and positions.

No.	1-ruleitem	T1	M1		M2		M3		M4		M5		M6		M7		M8	
1	COLE CALENDARIO COLEGIO=A; MEDIO-AMBIENTE	8	53,82	1	56,28	1	56,54	1	61,10	1	56,24	1	55,78	1	44,22	1	56,93	1
2	COLE GENERO POBLACION=X; MEDIO-AMBIENTE	1	54,68	5	57,18	5	57,45	5	61,98	5	57,04	5	56,87	5	43,13	5	57,99	5
3	COLE INST VLR PENSION=1; MEDIO-AMBIENTE	5	54,82	6	57,48	7	57,60	6	62,12	7	57,24	6	57,01	6	42,99	6	58,49	8
4	ECON SN CELULAR=1; MEDIO-AMBIENTE	2	54,42	4	56,90	4	57,20	4	61,72	4	56,82	4	56,58	4	43,42	4	57,61	4
5	ECON SN NEVERA=1; MEDIO-AMBIENTE	7	54,08	2	56,58	2	56,94	2	61,45	2	56,53	2	56,33	2	43,67	2	57,26	2
6	ECON SN MICROHONDAS=0; MEDIO-AMBIENTE	6	55,00	9	57,57	9	57,72	9	62,24	9	57,34	9	57,12	8	42,88	8	58,55	9
7	ECON SN AUTOMOVIL=0; MEDIO-AMBIENTE	3	54,90	7	57,52	8	57,67	8	62,18	8	57,29	8	57,13	9	42,87	9	58,44	7
8	ESTU TRABAJA=1; MEDIO-AMBIENTE	4	54,23	3	56,74	3	57,04	3	61,53	3	56,65	3	56,37	3	43,63	3	57,38	3
9	ESTU ZONA RESIDE=10; MEDIO-AMBIENTE	8	54,94	8	57,44	6	57,66	7	62,15	6	57,27	7	57,02	7	42,98	7	58,37	6

According to Table 2, best candidates for being "best" solutions ranked by Apriori with SPEA2 are those marked with color, in its respective positions; from this table one may infer that ruleitem numbered 1 should be the first ranked, since it appears in the first position except for the support measure, while the second ranked should be ruleitem number 5. Results obtained by using SPEA2 approach are shown in Table 3:

Table 3. Results for ranking using SPEA2 approach.

No.	1-ruleitem	SPEA2 rank
1	COLE CALENDARIO COLEGIO = A; MEDIO-AMBIENTE	1
2	COLE GENERO POBLACION = X; MEDIO-AMBIENTE	5
3	COLE INST VLR PENSION = 1; MEDIO-AMBIENTE	7
4	ECON SN CELULAR = 1; MEDIO-AMBIENTE	4
5	ECON SN NEVERA = 1; MEDIO-AMBIENTE	2
6	ECON SN MICROHONDAS = 0; MEDIO-AMBIENTE	Filtered
7	ECON SN AUTOMOVIL = 0; MEDIO-AMBIENTE	Filtered
8	ESTU TRABAJA = 1; MEDIO-AMBIENTE	3
9	ESTU ZONA RESIDE = 10; MEDIO-AMBIENTE	6

Ranking ruleitems using Apriori with SPEA2 shows same results as manual ranking, for this and other experiments. Also, SPEA2 filters worst ruleitems in order to provide only the most interesting knowledge, while discarding rules that appear far from contributing to reaching different measures goals.

7 Conclusion and Final Remarks

An approach for finding more interesting classification association rules has been proposed. This proposal, based on multi-objective optimization theory and implemented through a genetic algorithm following SPEA2 approach, exemplifies a simple way in which classification association rules may be ranked and filtered according to a set of equally important measures. This approach requires a set of business interestingness measures to have been previously defined and its relation to data established; though it would require an effort besides the limits of the data mining process, it would be helpful for data mining processes or techniques where interest of patterns might be necessary in order to provide useful knowledge.

This approach can be extended in different ways; first, by allowing to previously define preferences over involved measures, in order to support problems where such kind of articulation can be effectively established. Second, by allowing to define what is a better ruleitem and what is a worst ruleitem; in this approach, a better ruleitem is defined as the one which contributes the most to reaching some measures goals; however, for some business or application area, most interesting ruleitems, from the user's point of view, might be those which contribute the less to reaching those goals, because they might represent urgent actions to be taken.

Another way of extending this work could be to apply it to other kinds of data mining techniques, such as classification. In general, this approach could be used when there exist a trade-off between multiple criteria in order to determine how much valid, appropriate or interesting is knowledge; in classification using decision trees, for example, it could be used for reinforcing the criteria by which the tree is iteratively constructed, by amplifying it to other kinds of measures different to typical ones such as information gain, for naming an example.

Multiobjective optimization using SPEA2 could be also extended to other interesting research areas in data mining; to metalearning, for example, where one of the many research applications is algorithm ranking according to one or more interest criteria (accuracy, cost, computation time, among others). SPEA2 approach might be useful for providing data mining user with a more flexible way of ranking algorithms, having into account multiple criteria.

Finally, our approach could be included as part of any association rule tool for suggesting best rules; although here we were concentrated on the first phase of Apriori – frequent itemsets -, it might be used for obtaining only interesting rules too. Those tools should be parameterized with the appropriated business interestingness measures.

Acknowledgments. Special thanks to ICFES for providing data online for research purposes.

References

1. Agrawal, R., Srikant, R.: Fast algorithms for mining association rules. In: 20th International Conference on Very Large Databases Proceedings, VLDB (1994)
2. Benavides, D., Villamil, M.: KDBuss framework: knowledge discovery with association rules in the business context. In: MLDM-2013, New York, 22–25 July 2013
3. Berry, M., Linoff, G.: Data Mining Techniques: For Marketing, Sales and Customer Relationship Management, 2nd edn. Wiley, Indianapolis (2004)
4. Bouker, S., Saidi, R., Ben Yahia, S., Mephu Nguifo, E.: Ranking and selecting association rules based on dominance relationship. In: 2012 IEEE 24th International Conference on Tools with Artificial Intelligence (ICTAI) (2012)
5. Cao, L., Yu, P., Zhang, C., Zhao, Y.: Domain Driven Data Mining, 1st edn. Springer, Sydney (2010)
6. Chen, M.S., Yu, P.S., Liu, B.: Value added association rules. In: 6th Pacific-Asia Conference on Knowledge Discovery and Data Mining Proceedings PAKDD (2002)
7. Cios, K., Pedrycz, W., Swiniarski, R., Kurgan, L.: Data Mining: A Knowledge Discovery Approach, 1st edn. Springer, New York (2007)
8. Geng, L., Hamilton, H.: Interestingness measures for data mining: a survey. ACM Comput. Surv. 38(3), Article 9 (2006)
9. GitHub repository containing implementation of SPEA2. https://github.com/DianaBenavides/AprioriSPEA2
10. Hall, M., Frank, E., Holmes, G., Pfahringer, B., Reutemann, P., Witten, I.: The WEKA data mining software: an update. SIGKDD Explor. 11(1), 10–18 (2009)
11. Liu, B., Hsu, W., Ma, Y.: Integrating classification and association rule mining. In: KDD-98, New York, 27–31 Aug 1998
12. Marler, R.T., Arora, J.S.: Survey of multi-objective optimization methods for engineering. Struct. Mult. Optim. 26(6), 369–395 (2004)
13. Piatetsky-Shapiro, G., Matheus, C.: The interestingness of deviations. In: AAAI-94 Workshop on Knowledge Discovery in Databases AAAI (1994)
14. Ramaraj, E., Rameshkumar, K.: Ranking mined association rule: a new measure. Delving J. Technol. Eng. Sci. (JTES) 1(1), 57–61 (2009)
15. Shen, Y., Yang, Q., Zhang, Z.: Objective-oriented utility-based association mining. In: 2002 International Conference on Data Mining Proceedings. IEEE (2002)
16. Wang, K., Zhou, S., Han, J.: Profit mining: from patterns to actions. In: Jensen, C.S., Jeffery, K., Pokorný, J., Šaltenis, S., Bertino, E., Böhm, K., Jarke, M. (eds.) EDBT 2002. LNCS, vol. 2287, pp. 70–87. Springer, Heidelberg (2002)
17. Yang, G., Shimada, K., Mabu, S., Hirasawa, K.: A personalized association rule ranking method based on semantic similarity and evolutionary computation. In: 2008 IEEE Congress on Evolutionary Computation (CEC 2008)
18. Zitzler, E., Laumanns, M., Thiele, L.: SPEA2: Improving the strength pareto evolutionary algorithm for multiobjective optimization. In: Evolutionary methods for design, optimization, and control with applications to industrial problems, EUROGEN2011 Conference, Athens, Greece, September 19-21, 2001

Support Vector Machines

Classifying Grasslands and Cultivated Pastures in the Brazilian Cerrado Using Support Vector Machines, Multilayer Perceptrons and Autoencoders

Wanderson Costa$^{(\boxtimes)}$, Leila Fonseca, and Thales Körting

Image Processing Division, National Institute For Space Research,
São José dos Campos, SP, Brazil
{wscosta,leila,thales}@dpi.inpe.br
http://www.inpe.br

Abstract. One of the most biodiverse regions on the planet, Cerrado is the second largest biome in Brazil. Among the land changes in the Cerrado, over 500,000 km^2 of the biome have been changed into cultivated pastures in recent years. Categorizing types of land cover and its native formations is important for protection policy and monitoring of the biome. Based on remote sensing techniques, this work aims at developing a methodology to map pasture and native grassland areas in the biome. Data related to EVI vegetation indices obtained by MODIS images were used to perform image classification. Support Vector Machine, Multilayer Perceptron and Autoencoder algorithms were used and the results showed that the analysis of different attributes extracted from EVI indices can aid in the classification process. The best result obtained an accuracy of 85.96 % in the study area, identifying data and attributes required to map pasture and native grassland in Cerrado.

Keywords: Data mining · Image processing · Brazilian cerrado · Support vector machine · Multilayer perceptron · Autoenconder

1 Introduction

Issues such as population growth, climate change and the continuous demand for energy, water and food are a global threat and may cause serious environmental risks if natural resources are not used properly [1]. Within this perspective, it can be mentioned the problems caused by transformations in land use and cover in the Brazilian Cerrado, the second largest biome in the country. More than half of Cerrado's area has been changed, mostly to make room for cattle and cash crops, losing more than 1,000,000 km^2 of its original vegetation [14]. The spacial distribution of these areas is shown in Fig. 1.

Cultivated pastures cover more than 500,000 km^2 and croplands surpass 100,000 km^2, while protected areas comprise only about 33,000 km^2. The destruction of the Cerrado's vegetation is thrice larger than the amount of the deforested

© Springer International Publishing Switzerland 2015
P. Perner (Ed.): MLDM 2015, LNAI 9166, pp. 187–198, 2015.
DOI: 10.1007/978-3-319-21024-7_13

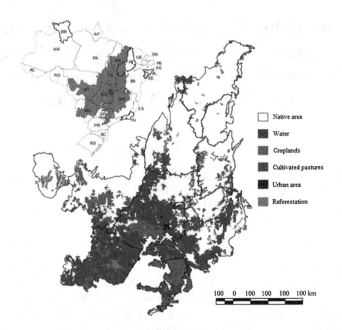

Fig. 1. Spatial distribution of changed areas in the Cerrado biome. Source: Sano et al. [18].

area in the Amazon region. This scenario represents a high environmental cost and implies vegetation degradation, water pollution, loss of biodiversity, soil erosion, instability of the carbon cycle, regional climatic changes and variations in fire events, which are typical in the biome [13].

A large number of definitions for Cerrado can be found in the literature and, as a result, there are several proposals for a topology of patterns describing native formations. Among the physiognomic types presented in the biome, there are the grassland formations, which include the physiognomies of *Clean, Dirty* and *Rocky Fields* [16]. With the identification and monitoring of these vegetation types using satellite imagery, policies can estimate the productivity of the degraded areas and promote its chemical, physical and biological integrity.

In order to promote the recovery of degraded regions and the policies that protect the Brazilian Cerrado, it is pivotal to create maps to analyze the land cover and use of the biome. However, cultivated pastures, similarly to native grasslands, may vary from formations with predominance of grasses to areas that present dominance of pioneer trees and shrub species [9]. Therefore, the mapping of pasture areas and native formations in Cerrado is a difficult task if only the spectral information obtained by satellite images is used [18].

To overcome this problem, this work proposes a methodology for mapping areas of Cultivated Pasture and Native Field (*Clean, Dirty* and *Rocky Fields*), using pattern recognition algorithms, such as Support Vector Machine (SVM), Multilayer Perceptron (MLP) and Autoencoder techniques. Data related to EVI

(Enhanced Vegetation Index) obtained by MODIS images were used to perform image classification. Time series extracted from EVI indices were used as attributes in the classification process.

The rest of the paper is organized as follows. In Sect. 2, we present a brief description of the Brazilian Cerrado, its native formations and the cultivated pastures in the biome. The methodological procedures and data used are depicted in Sect. 3. In Sect. 4, we discuss the results obtained in this work. Finally, we describe the conclusion in Sect. 5.

2   The Brazilian Cerrado

The Brazilian Cerrado has an area of approximately 2 million km^2, comprising about 24 % of the Brazilian territory [3]. Cerrado is the richest tropical savanna in the world [13] and the biome occupies the central region of Brazil, extending from the northeast coast of the Maranhão state (MA) to the north of Paraná (PR) state, as seen in Fig. 1. Generally, Cerrado can be understood as a grass field coexisting with scattered trees and shrubs [23] and, after Amazonia, it is the second largest biome of Brazil [13].

Cerrado is the Portuguese word for Brazil's plateau of savannas, grasslands, woodlands, gallery and dry forests [8]. Generally, the grassland formations refer to regions with predominance of herbaceous species and some shrubs, without the occurrence of trees in the landscape [16].

The native vegetation included in grassland comprises areas of *Clean, Dirty* and *Rocky Fields* (Fig. 2). The phytophysiognomy of *Clean Field* has a predominance of grasses interspersed with underdeveloped woody plants, without the presence of trees. On the other hand, areas of *Dirty Field* and *Rocky Field* present a physiognomic type predominantly herbaceous and shrubby. However, *Rocky Field* areas usually occupy regions of rocky outcrops at elevations above 900 m, which include micro-relief landscapes with typical species [16].

More than half of Cerrado's original vegetation has been converted into cultivated pasture areas, agriculture and other uses. In addition, studies indicate that changes in land use in the Cerrado take place with greater intensity than in the Amazon region [13,18].

Henceforth, these transformations in land cover impose substantial threats to ecosystems and species of the biome. Only 2.2 % of its area is legally protected

Fig. 2. Grassland formations in the Brazilian Cerrado. Clean Field (left), Dirty Field (center) and Rocky Field (right).

and various species of animals and plants are endangered. Besides the vegetation degradation and soil erosion, the introduction of non-native species and the use of fire to create pastures can have an adverse impact on ecosystems and cause significant loss of biodiversity in the biome [13].

2.1 Cultivated Pastures

Comprising about 500,000 km^2 of the biome, nearly 50 % of these cultivated pasture areas are severely degraded (Fig. 3), causing loss of soil fertility, increased erosion and predominance of invasive species [11]. For that reason, the recovery of these areas increase the producers' income and it can reduce the environmental impact by decreasing erosion, emission of carbon dioxide and opening new areas for cattle [6].

Fig. 3. Examples of managed (left) and degraded pastures (right). Source: EMBRAPA and INPE [9].

Different data sets can be extracted and analyzed to support the detection of these regions, such as climatic, biophysical and radiometric (vegetation indices) data from satellite images [10]. Vegetation indices produced by remote sensing data represent improved measures of surface vegetation conditions and the pattern analysis of these indices can be useful for the observation of land use and cover [21]. Additionally, wet and dry seasons are well defined in the Cerrado and, consequently, this fact may assist to identify the radiometric and biophysical characteristics of the cultivated pasture regions [11].

However, the classification of cultivated pastures is difficult because the degradation of these regions can, for instance, influence the percentage of vegetation cover and the response of vegetation indices. Misclassification may happen when pastures are managed improperly, since those areas might be infested by invasive species or even it may occur the revival of species of native shrubs and trees [11]. Among the monitoring projects of cultivated pastures in Brazil, we can highlight the GeoDegrade Project [20], which it is in final stages of development and aims the identification and monitoring of levels of degradation in cultivated pastures in Amazon, Cerrado and Atlantic Forest regions. Nevertheless, GeoDegrade project does not focus on the automatic classification of cultivated pastures and grasslands.

Therefore, in order to improve the discrimination of such targets it is necessary to use temporal and field data and also to better understand the biophysical properties of these areas [11]. Thus, the analysis using time series extracted from satellite images allows more accurate identification of temporal patterns.

3   Data and Methods

The study area covers a Cerrado's location that encompasses a region of Serra da Canastra National Park and neighboring regions, illustrated in Fig. 4. It is located in the south-central state of Minas Gerais, southeastern of Brazil. The chosen scene contains the targets of interest, both native grassland areas and cultivated pastures.

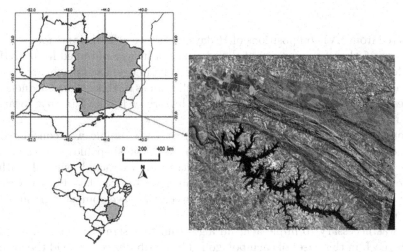

Fig. 4. Location of the study area. Crop of a Landsat-5 TM (R5B4G3) image of the study area.

Information on cultivated pasture areas, for the year 2006, was provided by the Brazilian Ministry of the Environment. The regions of native areas, for the year 2009, were obtained from the Forest Inventory of Minas Gerais State, created by the University of Lavras (UFLA), which ranks the grassland formations into two classes: Field (*Clean* and *Dirty Fields*) and *Rocky Field* [19]. These data are shapefiles with polygons categorized with the classes of interest. To evaluate the classification results, these data were merged and used as ground truth. The classes used in this work includes the following standards: Cultivated Pasture, Native Field and Others. The class Native Field combined the Field and *Rocky Field* classes into a single class, whereas the class Others covered all other native formations in the study that are neither Field nor *Rocky Field.*

Images of EVI MODIS sensor were used to collect the attributes, with atmospheric correction and spatial resolution of 250 m. Specifically, annual profiles were used as feature vectors for 2006 and 2009. Each annual profile was

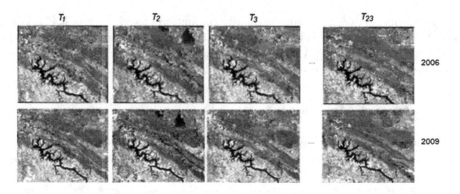

Fig. 5. Sequence of images to compose the EVI time series for 2006 and 2009.

extracted from EVI compositions of 16 days, resulting in 23 instants for each profile. Figure 5 shows a summary of the sequence of images obtained from MODIS data for the 23 instants of 2006 and 2009. In one image, each pixel corresponds to a vegetation index value (EVI), so that the values of this ordered sequence of images result a time series that stores the history of the EVI in that location, as shown in Fig. 6.

For the extraction of predictive attributes, each sample was considered as the set of information relating to a polygon from the shapefiles. For each time series, attributes related to vegetation indices were extracted using algorithms developed in TerraLib 5 [4]. TerraLib is an open source class library written in C++ that allows collaborative work between the development community of geographic applications.

It was necessary to implement an algorithm to assign the mean of all pixel values (EVI, in this case) of each polygon. The attributes correspond the average time series of the pixels inside each polygon for each EVI image at the instant $t_i (i = 1, 2, \cdots, 23)$.

To minimize the influence of edge pixels, an approach was used to define the attributes of EVI for each polygon. Using TerraLib 5, an algorithm was implemented that sets the value of the attribute for each polygon in each one of the images t_1 to t_{23}, equal to the mean of "pure" pixels of a given polygon.

For each polygon, the algorithm did not consider pixels in the border when computing the average time series. Figure 7 shows two examples of pixels in the decision-making process of the algorithm. The first pixel is "pure" and it would be considered in the calculation of attributes, whereas the second one would be disregarded.

With the generation of the time series for each polygon, the algorithm developed in this work also calculates the variables derived from the extracted time series. These variables are also used as attributes in the experiments. The variables extracted from the time series are the values of: minimum, maximum, amplitude, sum, mode, mean and standard deviation.

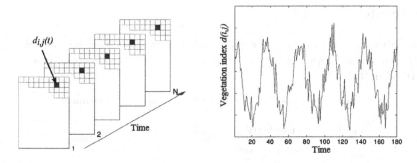

Fig. 6. Example of time series of vegetation index in the pixel d(i,j)

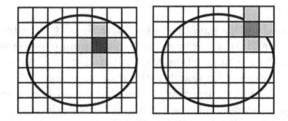

Fig. 7. For each polygon, the pixel highlighted in the figure on the left is considered "pure", while the second pixel does not follow the spatial relation defined in the algorithm and, therefore, it is not taken into consideration.

For the classification of the areas, this work used the algorithms of Support Vector Machine [2,22], Multilayer Percetron [17] and Autoencoder [15], using the tools Weka [12] and LibSVM [5]. We used the strategy of percentage division of the set of all samples, with 2/3 of the samples for the training set and 1/3 for the test set.

In evaluation process, confusion matrices were used, deriving descriptive statistics, such as the accuracy of the results, so that the confusion between classes can be clearly presented and understood [7]. In the case of SVM classification, the calculations of the confusion matrix were implemented in R and C++ algorithms.

In addition to classification with SVM and MLP using the 23 spectral attributes and the 7 attributes derived from the time series, we applied an Autoencoder trained with all 30 attributes. Two new data sets with the attributes generated by Autoencoder were used. These new data sets were then trained and classified with the SVM and MLP networks.

After using and evaluating the classifiers, it can be constructed as output of the classification process a thematic map for each of the classified experiments. Therefore, it is possible to generate an information layer by integration and processing of the input data, derived from satellite images.

4   Results

For observation and analysis of results, three experiments were performed. Each experiment uses a specific set of attributes in the classification process, as described in Table 1.

Table 1. Summary of the datasets used in each experiment.

Experiment	Predictive attributes used
1	23 attributes (time series)
2	Attributes extracted from the time series
3	Time series and attributes extracted from the time series

The classes used in this study include Native Field, Cultivated Pasture and Others. The SVM and MLP networks were used in all experiments. The RBF kernel was used in SVM classification, and the parameters C and γ were adjusted for each test using exhaustive search with cross-validation. On the other hand, results using 1 and 2 hidden layers were observed for MLP networks and values of η and *momentum* were defined as 0.3 and 0.2, respectively.

Furthermore, in the Experiment 3, classifications were also performed with SVM and MLP from the attributes generated by the Autoenconder, trained with the 30 original attributes. Two sets of data were created from the Autoencoder. One set contains 15 (half of the original attributes) and the other has 25 attributes.

Based on algorithms implemented in TerraLib 5 using the "pure" pixel approach, described in the previous section, a data set was extracted with a total of 508 samples to perform all experiments, in which each sample is related to the mean of pixels values (EVI) of a polygon. Thus, the training set had 337 samples and the test set contained 171.

In Experiment 1, the potential of time series of EVI images as unique set of attributes in the classification was evaluated, in order to analyze the contribution of time series itself. The best classification resulted in an accuracy of 84.21 % using SVM. The results of Experiment 1 are shown in Table 2.

Table 2. Experiment 1 results

Classifier	Accuracy (%)
SVM	**84.21**
MLP (1 hidden layer)	77.19
MLP (2 hidden layers)	83.04

In Experiment 2, the contribution of variables extracted from each of the time series was assessed. Seven predictive attributes were used: minimum, maximum, amplitude, sum, mode, mean and standard deviation. All of them were also extracted and implemented using TerraLib 5. Table 3 shows the accuracies obtained for Experiment 2. It was noted that using only this information did not result in good classifications, if compared with Experiment 1 accuracies. The highest accuracy was only 77.19 % using the MLP network with two hidden layers.

Table 3. Experiment 2 results

Classifier	Accuracy (%)
SVM	76.60
MLP (1 hidden layer)	71.35
MLP (2 hidden layers)	**77.19**

Lastly, in Experiment 3 it was analyzed whether the combination of the attributes of Experiment 1 and Experiment 2 would achieve better results. Table 4 shows the results obtained. From the evaluation of the results, it can be noticed a small increase in the accuracy of the classifiers regarding the other experiments applied. Experiment 3 obtained the best result of the classifications and achieved an accuracy of 85.96 % using SVM classifier. It was reached an increase in accuracy of 1.75 % compared to Experiment 1 with SVM, without providing more significant growth.

Table 4. Experiment 3 results

Classifier	Accuracy (%)
SVM	**85.96**
MLP (1 hidden layer)	77.19
MLP (2 hidden layers)	84.21
Autoencoder (15 attributes) > SVM	75.44
Autoencoder (25 attributes) > SVM	82.46
Autoencoder (15 attributes) > MLP (1 hidden layer)	75.44
Autoencoder (15 attributes) > MLP (2 hidden layers)	81.87
Autoencoder (25 attributes) > MLP (1 hidden layer)	77.19
Autoencoder (25 attributes) > MLP (2 hidden layers)	84.21

The experiments with Autoenconder showed that similar accuracies were found reducing the amount of attributes via this network. For example, the MLP with two hidden layers using the 30 original attributes obtained the same accuracy that the experiment with the 25 attributes generated by the Autoencoder, using a MLP with the same equivalent number of hidden layers.

Afterwards, the thematic map with the best result (Experiment 3 with SVM) was generated, as illustrated in Fig. 8.

Concerning the computational cost, the algorithm implemented in TerraLib 5 for extracting attributes regarding only "pure" pixels of the polygon obtained a processing time of nearly 5 hours to complete the extraction of the attributes set.

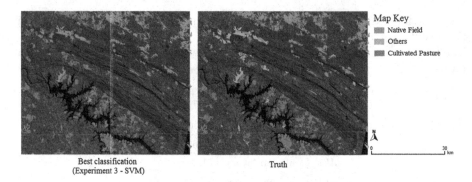

Fig. 8. Thematic map created from the best result of Experiment 3 (left) and thematic map with ground truth (right). The classification are superimposed on a Landsat-5 TM image (R5B4G3).

Observing the classification by SVM and MLP, these classifiers performed the predictions of classes in less than a second.

From an operational point of view, the creation of a classification model by SVM with the attributes of Experiment 3 showed results similar to Experiment 1, since the improvement in accuracy was small for the mapping of the three classes. Looking at the results with Autoencoder, they showed that accuracies using this network were similar to those tests that uses the 30 original attributes. Henceforth, this network can decrease the amount of attributes in the classification processs and, consequently, reduce the data dimensionality.

Furthermore, this work can contribute to the classification and integration of remote sensing data in geographic information systems from the construction of thematic maps as output of the classification process, since it was possible to generate a layer of information from the input data.

5   Conclusion

By comparing pattern recognition techniques, such as Support Vector Machines, Multilayer Perceptrons and Autoencoders, this work aimed to classify and create thematic maps of areas of native grasslands and cultivated pasture areas in the Brazilian Cerrado using times series extracted from satellite images.

With regard to the processing time, SVM and MLP networks showed a minimum computational cost for classification of results. The SVM algorithm showed better results in separation between classes Native Field and Cultivated Pasture. The best result obtained an accuracy of 85.96 % accuracy by combining all the attributes related to time series and variables derived from it. The results also showed that tests using a smaller number of attributes from the Autoencoder had accuracies similar to those obtained in the tests that used all original attributes, emphasizing the ability of this network to reduce the dimensionality of the data.

References

1. Beddington, J.: Food, energy, water and the climate: a perfect storm of global events. In: Lecture to Sustainable Development UK 09 Conference, vol. 19 (2009)
2. Bovolo, F., Camps-Valls, G., Bruzzone, L.: A support vector domain method for change detection in multitemporal images. Pattern Recogn. Lett. **31**(10), 1148–1154 (2010)
3. Brossard, M., Barcellos, A.O.: Conversão do cerrado em pastagens cultivadas e funcionamento de latossolos. Cadernos de Ciência & Tecnologia **22**(1), 153–168 (2005)
4. Câmara, G., Vinhas, L., Ferreira, K.R., De Queiroz, G.R., De Souza, R.C.M., Monteiro, A.M.V., De Carvalho, M.T., Casanova, M.A., De Freitas, U.M.: Terralib: an open source gis library for large-scale environmental and socio-economic applications. In: Brent Hall, G., Leahy, M.G. (eds.) Open source approaches in spatial data handling. Advances in Geographic Information Science, vol. 2, pp. 247–270. Springer, Heidelberg (2008)
5. Chang, C.C., Lin, C.J.: Libsvm: a library for support vector machines. ACM Trans. Intell. Syst. Technol. (TIST) **2**(3), 27 (2011)
6. Chaves, J.M., Moreira, L., Sano, E.E., Bezerra, H.S., Feitoza, L., Krug, T., Fonseca, L.M.G.: Uso da técnica de segmentação na identificação dos principais tipos de pastagens cultivadas do cerrado. In: Anais, pp. 31–33. INPE, São José dos Campos (2001)
7. Congalton, R.G., Green, K.: Assessing the Accuracy of Remotely Sensed Data: Principles and Practices. CRC Press, Boca Raton (2008)
8. Eiten, G.: Delimitação do conceito de cerrado. Arquivos do Jardim Botânico (1977)
9. EMBRAPA, INPE: Survey information of use and land cover in the amazon. Executive summary, TerraClass, Brasília (2011)
10. Ferreira, L.G., Fernandez, L.E., Sano, E.E., Field, C., Sousa, S.B., Arantes, A.E., Araújo, F.M.: Biophysical properties of cultivated pastures in the brazilian savanna biome: an analysis in the spatial-temporal domains based on ground and satellite data. Remote Sens. **5**(1), 307–326 (2013)
11. Ferreira, L.G., Sano, E.E., Fernandez, L.E., Araújo, F.M.: Biophysical characteristics and fire occurrence of cultivated pastures in the brazilian savanna observed by moderate resolution satellite data. Int. J. Remote Sens. **34**(1), 154–167 (2013)
12. Hall, M., Frank, E., Holmes, G., Pfahringer, B., Reutemann, P., Witten, I.H.: The weka data mining software: an update. ACM SIGKDD Explor. Newsl. **11**(1), 10–18 (2009)
13. Klink, C.A., Machado, R.B.: Conservation of the brazilian cerrado. Conserv. Biol. **19**(3), 707–713 (2005)
14. Machado, R.B., Ramos Neto, M.B., Pereira, P.G.P., Caldas, E.F., Gonçalves, D.A., Santos, N.S., Tabor, K., Steininger, M.: Estimativas de perda da área do cerrado brasileiro. Conservation International do Brasil, Brasília (2004)
15. Ng, A.: Sparse autoencoder. CS294A Lecture notes p. 72 (2011)
16. Ribeiro, J.F., Walter, B.M.T.: As principais fitofisionomias do bioma cerrado. In: Cerrado: Ecologia e Flora, vol. 1, pp. 152–212. EMBRAPA, Braslia (2008)
17. Ruck, D.W., Rogers, S.K., Kabrisky, M., Oxley, M.E., Suter, B.W.: The multilayer perceptron as an approximation to a bayes optimal discriminant function. IEEE Trans. Neural Netw. **1**(4), 296–298 (1990)
18. Sano, E.E., Rosa, R., Brito, J.L.S., Ferreira, L.G.: Mapeamento semidetalhado do uso da terra do bioma cerrado. Pesquisa Agropecuária Brasileira **43**(1), 153–156 (2008)

19. Scolforo, J.: Inventário florestal de Minas Gerais: cerrado: florística, estrutura, diversidade, similaridade, distribuição diamética e de altura, volumetria, tendências de crescimento e áreas aptas para manejo florestal. UFLA (2008)

20. da Silva, G.B.S., de Araújo Spinelli, L., Nogueira, S.F., Bolfe, E.L., de Castro Victoria, D., Vicente, L.E., Grego, C.R., Andrade, R.G.: Sistema de informação geográfica (sig) e base de dados geoespaciais do projeto geodegrade. In: Embrapa Monitoramento por Satélite-Artigo em anais de congresso (ALICE). In: SIMPÓSIO BRASILEIRO DE SENSORIAMENTO REMOTO, 16, 2013, Foz do Iguaçú. Anais... São José dos Campos: INPE (2013)

21. Tucker, C.J.: Red and photographic infrared linear combinations for monitoring vegetation. Remote Sens. Environ. 8(2), 127–150 (1979)

22. Vapnik, V.N.: The Nature of Statistical Learning Theory. Springer, New York (1995)

23. Walter, B.M.T.: Fitofisionomias do bioma Cerrado: síntese terminológica e relações florísticas. Universidade de Brasília, Doutorado em ecologia (2006)

Hybrid Approach for Inductive Semi Supervised Learning Using Label Propagation and Support Vector Machine

Aruna Govada[✉], Pravin Joshi, Sahil Mittal, and Sanjay K. Sahay

BITS, Pilani, K.K. Birla Goa Campus, Zuarinagar 403726, Goa, India
garuna@goa.bits-pilani.ac.in

Abstract. Semi supervised learning methods have gained importance in today's world because of large expenses and time involved in labeling the unlabeled data by human experts. The proposed hybrid approach uses SVM and Label Propagation to label the unlabeled data. In the process, at each step SVM is trained to minimize the error and thus improve the prediction quality. Experiments are conducted by using SVM and logistic regression(Logreg). Results prove that SVM performs tremendously better than Logreg. The approach is tested using 12 datasets of different sizes ranging from the order of 1000s to the order of 10000s. Results show that the proposed approach outperforms Label Propagation by a large margin with F-measure of almost twice on average. The parallel version of the proposed approach is also designed and implemented, the analysis shows that the training time decreases significantly when parallel version is used.

Keywords: Semi-supervised learning · Data mining · Support vector machine · Label propagation

1 Introduction

Semi supervised learning methods are mainly classified into two broad classes: Inductive and Transductive. In both Inductive and transductive the learner has both labeled training data set $(x_i, y_i)_{i=1...l} \sim p(x, y)$ and unlabeled training data set $(x_i)_{i=l+1...l+u}$ $p(x)$, where $l \ll u$. The inductive learner learns a predictor f : X → Y , f ∈ F where F is the hypothesis space, $x \in X$ is an input instance, $y \in Y$ its class label. The predictor learns in such a way that it predicts the future test data better than the predictor learned from the labeled data alone. In transductive learning, it is expected to predict the unlabeled data $(x_i)_{i=l+1...l+u}$ without any expectations of generalizing the model to future test data. Gaussian processes, transductive SVM and graph-based methods fall in the latter category. On the other hand, the former models are based on joint distribution and examples include Expectation Maximization. In many real world scenarios, it is easy to collect a large amount of unlabeled data {x}. For example, the catalogue of celestial objects can be obtained from sky surveys, Geo spatial data

© Springer International Publishing Switzerland 2015
P. Perner (Ed.): MLDM 2015, LNAI 9166, pp. 199–213, 2015.
DOI: 10.1007/978-3-319-21024-7_14

can be received from satellites, the documents can be browsed through the web. However, their corresponding labels {y} for the prediction, such as classification of the galaxies, prediction of the climate conditions, categories of documents often requires expensive laboratory experiments, human expertise and a lot of time. This labeling obstruction results in an insufficiency in labeled data with an excess of unlabeled data left over. Utilizing this unlabeled data along with the limited labeled data in constructing the generalized predictive models is desirable. Semi supervised learning model can either be built by first training the model on unlabeled data and using labeled data to induce class labels or vice versa. The proposed inductive approach labels the unlabeled data using a hybrid model which involves both Label Propagation and SVM. At every step in the process, it fits the model to minimize error and thus improve the prediction quality. The rest of the paper is organized as follows. The related work is discussed in Sect. 2. Label propagation, SVM are discussed in Sect. 3 and the proposed approach is presented in Sect. 4. Section 5 contains the experimental results and comparison of the proposed approach with the Label Propagation algorithm followed by Conclusion and future work in Sect. 6.

2   Related Work

A decent amount of work has been done in the field of semi-supervised learning in which major role has been played by the unlabeled data which is in huge amount as compared to the labeled data.Castelli et al. [1,2] and Ratsaby et al. [3] showed that unlabeled data can predict better if the model assumption is correct. But if the model assumption is wrong, unlabeled data may actually hurt accuracy. Cozman et al. [4] provide theoretical analysis of deterioration in performance with an increase in unlabeled data and argue that bias is adversely affected in such situations. Another technique that can be used to get the model correct is to down weight the unlabeled data by Corduneanu et al. [5]. Callison-Burch et al. [6] used the down-weighing scheme to estimate word alignment for machine translation.

A lot of algorithms have been designed to make use of abundant unlabeled data. Nigam et al. [7] apply the Expectation Maximization [8] algorithm on mixture of multinomial for the task of text classification and showed that the resulting classifiers predict better than classifier trained only on labeled data.

Clustering has also been employed over the years to make use of unlabeled data along with the labeled data.The dataset is clustered and then each cluster is labeled with the help of labeled data. Demiriz et al. [9] and Dara et al. [10] used this cluster and label approach successfully to increase prediction performance.

A commonly used technique for semi-supervised learning is self-training. In this, a classifier is initially trained with the small quantity of labeled data. The classifier is then used to classify the unlabeled data and unlabeled points which are classified with most confidence are added to the training set. The classifier is re-trained and the procedure is repeated. Word sense disambiguation is successfully achieved by Yarowsky et al. [11] using self-training. Subjective nouns are identified by Riloff et al. [12]. Parsing and machine translation is also done with the help of self-training methods as shown by Rosenberg et al. [13] in detection of object systems from images.

A method which sits apart from all already mentioned methods in the field of semi supervised learning is Co-training. Co-training [14] assumes that (i) features can be split into two sets; (ii) Each sub-feature set is sufficient to train a good classifier; (iii) the two sets are conditionally independent given the class. Balcan et al. [15] show that co-training can be quite effective and that in the extreme case only one labeled point is needed to learn the classifier.

A very effective way to combine labeled data with unlabeled data is described by Xiaojin Zhu et al. [16] which propagated labels from labeled data points to unlabeled ones. An approach based on a linear neighborhood model is discussed by Fei Wang et al. [17] which can propagate the labels from the labeled points to the whole data set using these linear neighborhoods with sufficient smoothness. Graph based method is proposed in [18] wherein vertices represent the labeled and unlabeled records and edge weights denote similarity between them. Their extensive work involves using the label propagation to label the unlabeled data, role of active learning in choosing labeled data, using hyper parameter learning to learn good graphs and handling scalability using harmonic mixtures.

3 Label Propagation

Let $(x_1, y_1)......(x_l, y_l)$ be labeled data, where $(x_1.......x_l)$ are the instances, $Y = (y_1.....y_l) \in (1....C)$ are the corresponding class labels. Let $(x_{l+1}, y_{l+1}).....$ (x_{l+u}, y_{l+u}) be unlabeled data where $Y_U = (y_{l+1}.....y_u)$ are unobserved. Y_U has to be estimated using X and Y_L, where $X = (x_1......x_l........x_{l+u})$

$$F : L \cup U \longrightarrow R$$

w_{ij} is similarity between i and j, where $i, j \in X$ F should minimize the energy function

$$(f) = \frac{1}{2} \sum_{i,j} w_{ij}(f(i) - f(j))^2 = f^T \triangle f$$

and f_i, F_j should be similar for a high w_{ij}.

Label propagation assumes that the number of class labels are known, and all classes present in the labeled data.

3.1 Support Vector Machine

The learning task in binary SVM [19] can be represented as the following

$$min_w = \frac{\| w \|^2}{2}$$

subject to $y_i(w.x_i + b) \geq 1$, $i = 1, 2, ...N$ where w and b are the parameters of the model for total N number of instances.

Using Legrange multiplier method the following equation to be solved,

$$L_p = \frac{\parallel w \parallel^2}{2} - \sum_{i=1...N} \lambda_i(y_i(w.x_i + b) - 1), \lambda_i$$

are called legrange multipliers. By solving the following partial derivatives, we will be able to derive the decision boundary of the SVM.

$$\frac{\partial p}{\partial w} = 0 \implies w = \sum_{i=1....N} \lambda_i y_i x_i = 0$$

$$\frac{\partial p}{\partial b} = 0 \implies w = \sum_{i=1....N} \lambda_i y_i = 0$$

4   Proposed Approach

The proposed Algorithm is an inductive semi supervised learning, an iterative approach in which at each iteration the following steps are executed. In the first step label propagation is run on the training data and the probability matrix is computed for the unlabeled data. The second step is to train the SVM on the available labeled data. Now in the third step, the classes for unlabeled data are predicted using SVM and class probabilities computed in step 1 are compared with the threshold. In step 4, all the records for which both label propagation and SVM agree on the class label, are labeled with corresponding class.This continues till all the records are labeled or no new records are labeled in an iteration. One-one multi class SVM is used as the data consists of multi classes.

The records of each of the data sets are shuffled, 70 % is considered for training and the rest is considered for testing. 80 % of the training data is unlabeled.

The algorithm is implemented in both the serial and parallel versions.

4.1   Serial Version

Input: Classifier, Threshold
Output: F-measure

1. (labeled_records, unlabeled_records) = select_next_train_folds()
   # Each fold of data is split into labeled and unlabeled records with 20:80 ratio
   # unlabeled_records have their class field set to -1
2. test_records = select_next_test_fold()
   # Concatenate labeled and unlabeled records to get train_records
3. train_records = labeled_records + unlabeled_records
4. newly_labeled = 0
5. while len(labeled_records) < len (train_records):
   5.1 lp_probability_matrix = run_lp(labeled_records + unlabeled_records)
   5.2 model = fit_classifier(classifier, labeled_records)
   5.3 labeled_atleast_one = False

5.4 for record in unlabeled_records:

    i. classifier_out = model.predict_class(record.feature_vector)

    # *Test for LP and classifier agreement*

    ii. if lp_probability_matrix[record.feature_vector][classifier_out] ≥ threshold:

        a. unlabeled_records.remove(record)

        b. record.class_label = classifier_out #label the record

        c. labeled_records.add(record) #add the newly labeled record to set of labeled records

        d. newly_added += 1

        # *Set labeled_atleast_one flag to True if at least one new record is labeled in current iteration of while loop*

        e. labeled_atleast_one = True

        # *Break the loop if no new record is labeled in current iteration of while loop*

5.5 if labeled_atleast_one == False:

    5.5.1 break

    # *Compute F-measure of constructed model*

6. test_records_features = test_records.get_feature_vectors()

7. test_records_labels = test_records.get_labels()

8. predicted_labels = model.predict(test_records_features)

9. f-measure = compute_fmeasure(predicted_labels, test_records_labels)

4.2 Parallel Version

Input: Classifier, Threshold, No_of_tasks #Number of parallel processes
Output: F-measure

1. newly_labeled = 0

2. while len(labeled_records) < len(train_records):

    2.1 lp_train_records = labeled_records + unlabeled_records

    2.2 lp_probability_matrix = []; classifier_out = []

    2.3 lp_process = new_process(target = run_lp, args = (lp_train_records, lp_probability_matrix))

    2.4 lp_process.start()

    2.5 classifier_process = new_process(target = fit_classifier, args = (classifier, labeled _records, unlabeled_records, classifier_all_out))

    2.6 classifier_process.start()

    2.7 lp_process.join()

    2.8 classifier_process.join()

    2.9 atleast_one_labeled = False

    2.10 chunk_size = len(unlabeled_records) / No_of_tasks

    2.11 all_pids = []

    2.12 None_initialize(labeled_lists, No_of_tasks)

    2.13 None_initialize(unlabeled_copies, No_of_tasks)

    2.14 for i in range(len(labeled_lists)):

     i. start = i * chunk_size
    ii. end = (i+1) * chunk_size
   iii. unlabeled_copies = unlabeled_records[start : end]
   iv. lp_probabilities = lp_probability_matrix[start : end]
    v. classifier_outs = classifier_all_outs[start : end]
   vi. label_records_process = new_process(func = label_data, args = (unlabeled_copies[i], labeled_lists[i], lp_probabilities, classifier_outs, threshold))
  vii. label_records_process.start()
 viii. all_pids.append(label_records_process)

2.15 unlabeled_records = []

2.16 done_processes = []

2.17 while len(done_pids) < len(all_pids):
    i. for i in range(len(all_pids)):
       if not all_pids[i].is_alive() and (i not in done_pids):
       a. done_processes.append(i)
       b. unlabeled_records += unlabeled_copies[i]
       c. labeled_records += labeled_lists[i]

2.18 if atleast_one_labeled == False:
  2.18.1 break
       # Compute F-measure of constructed model

3. predicted_labels = [ ]

4. test_records_features = test_records.get_feature_vectors()

5. test_records_labels = test_records.get_labels()

6. run_parallel_classifier(predicted_labels, labeled_records, test_records_features, classifier, no_of_tasks)

7. f-measure = compute_fmeasure(predicted_labels, test_records_labels)

5 Experimental Section

In our experimental analysis, we considered 12 different datasets. The datasets along with their number of attributes (excluding the class label) and number of instances are as follows (in the format *dataset: (no of attributes, no of records)*): Vowel: (10, 528), Letter: (16, 10500), Segment: (18, 2310), Iris scale random: (4, 149), Satimage: (36, 1331), 10000 SDSS: (7, 10000), 1000 SDSS: (7, 1000), Glass scale random: (9, 214), Letter 1: (16, 4500), Mfeat: (214, 2000), Pendigits: (16, 7494) and Shuttle: (9, 12770). In our experimental analysis, the following comparisons are made.

1. Serial version of our hybrid approach with Zhu et al. [16].
2. Serial version of our hybrid approach with supervised learning classifier SVM.
3. Parallelization of our algorithm with our own serial implementation.

Before performing the above comparisons,the serial version of our hybrid approach is examined on the following aspects. The algorithm is run for different values of threshold of the probability matrix of label propagation and percentage of initially labeled data in each of the training data set. We observed the values of percentage of increase in labeled records for each iteration, percentage of labeled data at the final iteration and finally the training time and F-measure. An alternative classifier Logreg is also implemented and its performance is compared with SVM based on factors like F-measure and training time.

As it can be seen from Fig. 1 varying the threshold of the probability matrix of label propagation has little impact on the F-measure. Considering only label propagation increase in threshold would lead to stricter labeling resulting in increase in precision and decrease in recall. Similarly, the decrease in probability threshold results in an increase in number of unlabeled records being considered for labeling. Hence, it should increase the labeling rate of the records. But when SVM is used with label propagation precision and recall are not allowed to vary significantly because a record can be labeled only when SVM and Label Propagation agree on the class label. So, unlabeled records marked by label propagation for labeling with low confidence are discarded by output of SVM. Thus the percentage of labeled data at the end of the final iteration fluctuates very little for all the thresholds (as it is shown in next graph Fig. 2). So change in thresholds, has little effect on F-measure of the model.

To see the effect of the dimension and the number of records in the dataset, In Fig. 3, we plotted Training time with respect to the cube of number of records in the dataset for the best performing classifier and threshold. The axis were chosen by keeping in mind the O $(dim*N^3)$ complexities of both SVM and Logreg. The graph tends to show polynomial increase in training time as N increases (instead of linear). This may be the effect of neglecting lower order terms in the complexity expression of SVM and Logreg.

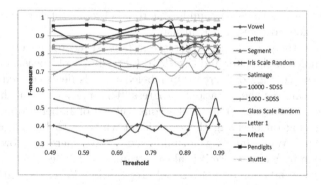

Fig. 1. F-measure of the datasets by varying the threshold of the probability matrix

The analysis of Figs. 1 and 2 explains the following feature of the algorithm. For the datasets: 1000 records Sloan Digital Sky survey (SDSS), 10000 records SDSS, Mfeat, Pendigits and Shuttle,Percentage of labeled data is very low 0–20%. But F-measures are reasonably high between 0.67 to 0.9. This shows that their high F-measure does not always require high amount of unlabeled data to

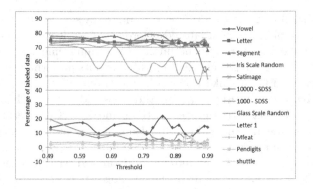

Fig. 2. Percentage of the labeled data by the final iteration

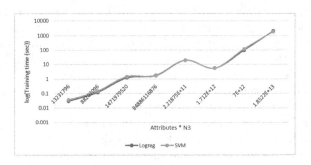

Fig. 3. Training time of the model by using different classifier SVM and Logreg

be labeled. As long as the algorithm is able to label representative records in the dataset, it is likely to give good F-measure.

We observed the percentage of increase in labeled records for every iteration for the best choice of classifier and threshold (according to F-measure). In Fig. 4, it shows that percentage of increase in labeled records decreases exponentially (note the log scale) as the iterations progress. This means not much data gets labeled as loop progresses. This is the consequence of Label propagation and SVM not agreeing on deciding the class label which is to be assigned to the unlabeled record. While labeling an unlabeled record, there is a low chance of misclassification by SVM , since it is always trained on labeled data. This means that the quality of labeling done by Label propagation decreases significantly as the iterations progresses. This deterioration in Label Propagations quality has very little effect on algorithms overall prediction quality because of the right predictions done by SVM at every step while labeling the unlabeled records leading to better performance than Label propagation.

The performance of the proposed approach is compared with the label propagation algorithm [16] for all the datasets and shown in Fig. 5. F-measures of the proposed approach were noted for best choice of classifier and threshold. For label propagation, all the unlabeled records were labeled according to the

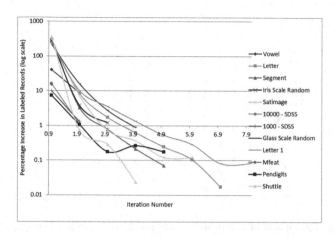

Fig. 4. Percentage of increase in labeled records for every iteration

class corresponding to highest label propagation Probability. In all the cases, the proposed approach outperforms label propagation by a large margin. This high quality of performance can be attributed use of SVM together with Label Propagation to label the unlabeled examples. No unlabeled example is labeled without the agreement of both SVM and label propagation. This significantly reduces the pitfalls caused by label propagation increasing the prediction quality of overall approach.

The analysis of our approach(semi supervised learning) with the supervised SVM is also studied. Results in Fig. 6 show that the F-measure of our approach is comparable.

Figures 7, 8, 9, 10, 11 and 12 are plotted for different percentages of initially unlabeled data for the data sets Vowel, Irisscalerandom, Satimage, 1000-SDSS, Glassscalrandom, Mfeat respectively considering best choice of Threshold (as per the F-measure). As can be seen, F-measure of the model tends to fall as percentage of initially unlabeled data increases. This is intuitive. As the amount of labeled data in the initial data increases, the algorithm is able to learn the pattern present in a representative sample of the dataset. Thus it can successfully generalize the pattern to the test set leading to a overall increase in F-measure.

Finally, The parallel version of the proposed approach is also implemented. As the results can be seen from Fig. 13, parallelizing the algorithm helps to improve training time of the algorithm. Two of the most expensive steps in the algorithm are training SVM and label propagation on the data. Doing these two in parallel reduces training time significantly (note that training time is in log scale).

The analysis is done for SDSS dataset for samples of different sizes.The results are shown in Fig. 14. For each sample, we ran different number of parallel tasks and training time is observed. Results show that number of parallel tasks have reasonable effect on training time only when dataset size exceeds a certain threshold (around 60000 records in this case). Further, for each dataset, there is an

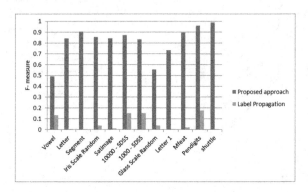

Fig. 5. The comparison of the proposed approach with the label propagation [16]

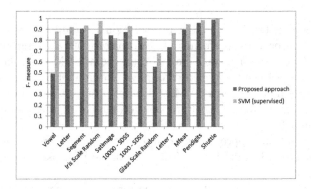

Fig. 6. The comparison of proposed approach with supervised one-one SVM

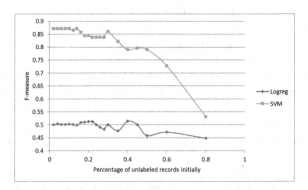

Fig. 7. F-measure of Vowel data set by varying the percentage of initially unlabeled records

optimum number of parallel tasks which yields minimum training time. If number of parallel tasks is above this optimum level, the cost of maintaining these parallel tasks exceeds the gain earned by parallel computation. On the other hand, if number of parallel tasks is set to a value less than optimum level, system resources are poorly utilized. So it is necessary to set number of parallel tasks to optimum value for maximum benefit.

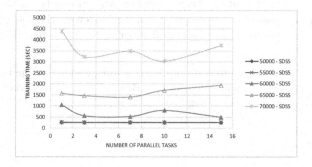

Fig. 14. The number of parallel tasks w.r.t. training time

The proposed approach is tested on skewed datasets also. The proportion of class labels in the skewed dataset is at least 1:8. F-measure of 10000-SDSS drops by approximately 0.1 when it has skewed class proportions. But skewed version of Shuttle shows exceptional behavior. Its F-measure remains almost the same. This shows that skewness of the data has little or no effect on F-measure of algorithm. It can be inferred that the distribution of the data plays a major role in performance of semi-supervised algorithm. The results are shown in Fig. 15.

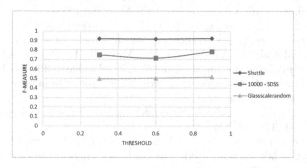

Fig. 15. The analysis of F-measure of skewed data sets

6   Conclusion and Future Work

The proposed approach uses SVM along with Label Propagation algorithm to yield a very high overall prediction quality. It can use a small amount of labeled data along with a large quantity of unlabeled data to yield a high F-measure on test data. It has a very small margin for error since it labels the unlabeled data using consent of both - SVM and Label Propagation. On testing both the algorithms on 12 different datasets we can conclude that the proposed approach performs much better than label propagation [16] alone. It yields F-measure values which are almost twice as compared to Label Propagation. Further, we designed the parallel version of the approach and were able to decrease the training time significantly. In future, the parallel algorithm can be further enhanced to yield linear or super linear scale up. Further research on the role of supervised algorithms in the field of semi supervised learning could be beneficial.

Acknowledgments. We are thankful for the support provided by the Department of Computer Science and Informations Systems, BITS, Pilani, K.K. Birla Goa Campus to carry out the experimental analysis.

References

1. Castelli, V., Cover, T.: The exponential value of labeled samples. Pattern Recogn. Lett. **16**, 105–111 (1995)
2. Castelli, V., Cover, T.: The relative value of labeled and unlabeled samples in pattern recognition with an unknown mixing parameter. IEEE Trans. Inf. Theory **42**, 2101–2117 (1996)
3. Ratsaby, J., Venkatesh, S.: Learning from a mixture of labeled and unlabeled examples with parametric side information. In: Proceedings of the Eighth Annual Conference on Computational Learning Theory, pp. 412–417 (1995)
4. Cozman, F.G., Cohen, I., Cirelo, M.C.: Semi-supervised learning of mixture models. In: ICML (2003)
5. Corduneanu, A., Jaakkola, T.: Stable mixing of complete and incomplete information, Technical report AIM-2001-030, MIT AI Memo (2001)
6. Callison-Burch, C., Talbot, D., Osborne, M.: Statistical machine translation with word- and sentence-aligned parallel corpora. In: Proceedings of the ACL (2004)
7. Nigam, K., McCallum, A.K., Thrun, S., Mitchell, T.: Text classification from labeled and unlabeled documents using EM. Mach. Learn. **39**, 103–134 (2000)
8. Dempster, A., Laird, N., Rubin, D.: Maximum likelihood from incomplete data via the EM algorithm. J. Roy. Stat. Soc. B, 1–38 (1977)
9. Bennett, K., Demiriz, A.: Semi-supervised support vector machines. Adv. Neural Inf. Proc. Syst. **11**, 368–374 (1999)
10. Dara, R., Kremer, S., Stacey, D.: Clustering unlabeled data with SOMs improves classification of labeled real-world data. In: Proceedings of the World Congress on Computational Intelligence (WCCI) (2002)
11. Yarowsky, D.: Unsupervised word sense disambiguation rivaling supervised methods. In: Proceedings of the 33rd Annual Meeting of the Association for Computational Linguistics, pp. 189–196 (1995)
12. Riloff, E., Wiebe, J., Wilson, T.: Learning subjective nouns using extraction pattern bootstrapping. In: Proceedings of the Seventh Conference on Natural Language Learning (CoNLL-2003) (2003)
13. Rosenberg, C., Hebert, M., Schneiderman, H.: Semi-supervised self training of object detection models. In: Seventh IEEE Workshop on Applications of Computer Vision (2005)

14. Blum, A., Mitchell, T.: Combining labeled and unlabeled data with co-training. In: Proceedings of the Workshop on Computational Learning Theory, COLT (1998)
15. Balcan, M.-F., Blum, A.: An augmented PAC model for semi-supervised learning. In: Chapelle, O., Sch Olkopf, B., Zien, A. (eds.) Semi-supervised learning. MIT Press, Cambridge (2006)
16. Zhu, X., Ghahramani, Z.: Learning from labeled and unlabeled data with label propagation. Technical report CMU-CALD-02-107, Carnegie Mellon University (2002)
17. Wang, F., Zhang, C.: Label propagation through linear neighborhoods. IEEE Trans. Knowl. Data Eng. **20**(1), 55–67 (2008)
18. Zhu, X., Lafferty, J., Rosenfeld, R.: Semi-supervised learning with graphs. Diss. Carnegie Mellon University, Language Technologies Institute, School of Computer Science (2005)
19. Hearst, M.A., et al.: Support vector machines. IEEE Intel. Syst. Appl. **13**(4), 18–28 (1998)

Frequent Item Set Mining
and Time Series Analysis

Optimizing the Data-Process Relationship for Fast Mining of Frequent Itemsets in MapReduce

Saber Salah, Reza Akbarinia, and Florent Masseglia$^{(\boxtimes)}$

Inria and LIRMM, Zenith Team, University of Montpellier, Montpellier, France
{saber.salah,reza.akbarinia,florent.masseglia}@inria.fr

Abstract. Despite crucial recent advances, the problem of frequent itemset mining is still facing major challenges. This is particularly the case when: (i) the mining process must be massively distributed and; (ii) the minimum support (*MinSup*) is very low. In this paper, we study the effectiveness and leverage of specific data placement strategies for improving parallel frequent itemset mining (PFIM) performance in MapReduce, a highly distributed computation framework. By offering a clever data placement and an optimal organization of the extraction algorithms, we show that the itemset discovery effectiveness does not only depend on the deployed algorithms. We propose ODPR (Optimal Data-Process Relationship), a solution for fast mining of frequent itemsets in MapReduce. Our method allows discovering itemsets from massive datasets, where standard solutions from the literature do not scale. Indeed, in a massively distributed environment, the arrangement of both the data and the different processes can make the global job either completely inoperative or very effective. Our proposal has been evaluated using real-world data sets and the results illustrate a significant scale-up obtained with very low *MinSup*, which confirms the effectiveness of our approach.

1 Introduction

With the availability of inexpensive storage and the progress that has been made in data capture technology, several organizations have set up very large databases, known as Big Data. This includes different data types, such as business or scientific data [1], and the trend in data proliferation is expected to grow, in particular with the progress in networking technology. The manipulation and processing of these massive data have opened up new challenges in data mining [2]. In particular, frequent itemset mining (FIM) algorithms have shown several flaws and deficiencies when processing large amounts of data. The problem of mining huge amounts of data is mainly related to the memory restrictions as well as the principles and logic behind FIM algorithms themselves [3].

In order to overcome the above issues and restrictions in mining large databases, several efficient solutions have been proposed. The most significant

Saber Salah—This work has been partially supported by the Inria Project Lab Hemera.

© Springer International Publishing Switzerland 2015
P. Perner (Ed.): MLDM 2015, LNAI 9166, pp. 217–231, 2015.
DOI: 10.1007/978-3-319-21024-7_15

solution required to rebuild and design FIM algorithms in a parallel manner relying on a specific programming model such as MapReduce [4]. MapReduce is one of the most popular solutions for big data processing [5], in particular due to its automatic management of parallel execution in clusters of commodity machines. Initially proposed in [6], it has gained increasing popularity, as shown by the tremendous success of Hadoop [7], an open-source implementation.

The idea behind MapReduce is simple and elegant. Given an input file, and two map and reduce functions, each MapReduce job is executed in two main phases. In the first phase, called map, the input data is divided into a set of splits, and each split is processed by a map task in a given worker node. These tasks apply the map function on every key-value pair of their split and generate a set of intermediate pairs. In the second phase, called reduce, all the values of each intermediate key are grouped and assigned to a reduce task. Reduce tasks are also assigned to worker machines and apply the reduce function on the created groups to produce the final results.

Although MapReduce refers as an efficient setting for FIM implementations, most of parallel frequent itemset mining (PFIM) algorithms have brought same regular issues and challenges of their sequential implementations. For instance, invoking such best PFIM algorithm with very low minimum support ($MinSup$) could exceed available memory. Unfortunately, dealing with massive datasets (up to terabytes of data) implies working with very low supports since data variety lowers item frequencies. Furthermore, if we consider a FIM algorithm which relies on a candidate generation principle, its parallel version would remain carrying the same issues as in its sequential one. Therefore, covering the problem of FIM algorithms does not only involve the distribution of computations over data, but also should take into account other factors.

Interestingly and to the best of our knowledge, there has been no focus on studying data placement strategies for improving PFIM algorithms in MapReduce. However, as we highlight in this work, the data placement strategies have significant impacts on PFIM performance. In this work, we identify, investigate and elucidate the fundamental role of using such efficient strategies for improving PFIM in MapReduce. In particular, we take advantage of two data placement strategies: Random Transaction Data Placement (RTDP) and Similar Transaction Data Placement (STDP). In the context of RTDP, we use a random placement of data on a distributed computational environment without any data constraints, to be consumed by a particular PFIM algorithm. However, in STDP, we use a similarity-based placement for distributing the data around the nodes in the distributed environment. By leveraging the data placement strategies, we propose ODPR (Optimal Data-Process Relationship), a new solution for optimizing the global extraction process. Our solution takes advantage of the best combination of data placement techniques and the extraction algorithm.

We have evaluated the performance of our solution through experiments over ClueWeb and Wikipedia datasets (the whole set of Wikipedia articles in English). Our results show that a careful management of the parallel processes along with adequate data placement, can dramatically improve the performance and make a big difference between an inoperative and a successful extraction.

The rest of this paper is organized as follows. Section 2 gives an overview of FIM problem and Sect. 3 gives the necessary background on MapReduce and some basic FIM algorithms. In Sect. 4, we propose our techniques of data placement for an efficient execution of PFIM algorithms. Section 5 reports on our experimental validation over synthetic and real-world data sets. Section 6 discusses related work, and Sect. 7 concludes.

2   Problem Definition

The problem of frequent itemset mining has been initially proposed in [8], and then numerous algorithms have been proposed to solve it. Here we adopt the notations used in [8].

> **Itemset:** Let $I = \{i_1, i_2, ..., i_n\}$ be a set of literals called *items*. An *Itemset* X is a set of items from I, i.e. $X \subseteq I$. The *size* of the itemset X is the number of items in it.
> **Transaction:** A transaction T is a set of elements such that $T \subseteq I$ and $T \neq \emptyset$. A transaction T supports the item $x \in I$ if $x \in T$. A transaction T supports the *itemset* $X \subseteq I$ if it supports any item $x \in X$, i.e. $X \subseteq T$.
> **Database:** A database D is a set of transactions.
> **Support:** The *support* of the *itemset* X in the database D is the number of transactions $T \in D$ that contain X.
> **Frequent Itemset:** An *itemset* $X \subseteq I$ is *frequent* in D if its *support* is equal or higher than a ($MinSup$) threshold.

The goal of FIM is as follows: given a database D and a user defined minimum support $MinSup$, return all frequent itemsets in D.

Example 1. Let us consider database D with 4 transactions as shown in Table 1. With a minimum support of 3, there will be no frequent items (and no frequent itemsets). With a minimum support of 2, there will be 6 frequents itemsets:$\{(a), (b), (e), (f), (ab), (ef)\}$.

Table 1. Database D

TID	Transaction
T_1	a, b, c
T_2	a, b, d
T_3	e, f, g
T_4	d, e, f

In this paper, we consider the specific problem of PFIM, where the data set is distributed over a set of computation nodes. We consider MapReduce as a programming framework to illustrate our approach, but we believe that our proposal would allow to obtain good performance results in other parallel frameworks too.

3 Requirements

In this section, we first describe briefly MapReduce and its working principles. Then, we introduce some basic FIM algorithmic principles which we use in our PFIM algorithms.

3.1 MapReduce and Job Execution

Each MapReduce job includes two functions: map and reduce. For executing the job, we need a master node for coordinating the job execution, and some worker nodes for executing the map and reduce tasks. When a MapReduce job is submitted by a user to the cluster, after checking the input parameters, e.g., input and output directories, the input *splits* (blocks) are computed. The number of input splits can be personalized, but typically there is one split for each 64 MB of data. The location of these splits and some information about the job are submitted to the master. The master creates a job object with all the necessary information, including the map and reduce tasks to be executed. One map task is created per input split.

When a worker node, say w, becomes idle, the master tries to assign a task to it. The map tasks are scheduled using a locality-aware strategy. Thus, if there is a map task whose input data is kept on w, then the scheduler assigns that task to w. If there is no such task, the scheduler tries to assign a task whose data is in the same rack as w (if any). Otherwise, it chooses any task.

Each map task reads its corresponding input split, applies the map function on each input pair and generates *intermediate key-value* pairs, which are firstly maintained in a buffer in main memory. When the content of the buffer reaches a threshold (by default 80 % of its size), the buffered data is stored on the disk in a file called spill. Once the map task is completed, the master is notified about the location of the generated intermediate key-values.

In the reduce phase, each intermediate key is assigned to one of the reduce workers. Each reduce worker retrieves the values corresponding to its assigned keys from all the map workers, and merges them using an external merge-sort. Then, it groups pairs with the same key and calls the reduce function on the corresponding values. This function will generate the final output results. When, all tasks of a job are completed successfully, the client is notified by the master.

During the execution of a job, there may be idle nodes, particularly in the reduce phase. In Hadoop, these nodes may be used for *speculative* task execution, which consists in replicating the execution of incomplete slow tasks in those nodes. When one of the replicated tasks gets complete, its results are kept and the rest of copies are stopped and their results discarded.

3.2 PFIM

One of the primordial FIM algorithms is Apriori [8]. This algorithm starts mining the database D by figuring out frequent items of size one, say L_1. Then, builds the potential frequent itemsets of size two C_2 by joining items in L_1. The algorithm tests the *support* of each C_2 element in D, and returns a list of frequent itemsets L_2. The mining process is carried out until there is no more frequent itemset in D. The main drawback of Apriori is the size of intermediate itemsets that need to be generated. Actually, with itemsets having a maximum length of n, Apriori needs to compute n generation of candidates, each being supersets of the previous frequent itemsets. Usually, the number of intermediate itemsets grows follows a normal distribution according to the generation number. In other words, the number of candidates reaches its higher number in the middle of the process. A straightforward implementation of this algorithm in MapReduce is very easy since each database scan is replaced by a MapReduce job for candidate support counting. However, the performances are very bad mainly because intermediate data have to be communicated to each mapper.

In the context of investigating PFIM in MapReduce and the effect of data placement strategies, we need to briefly describe the SON [9] algorithm that simplifies the mining process by dividing the FIM problem into two steps, and this makes it very suitable for being used in MapReduce. The steps of SON are as following:

> **Step 1:** It divides the input database D into $|P| = n$ chunks where $P = \{p_1, p_2, ..., p_n\}$. Then, it mines each data chunk in the memory, based on a *localMinSup* and given FIM algorithm. Thus, the first step of SON algorithm is devoted to determine a list of local frequent itemsets (LFI).
>
> **Step 2:** Based on the first step, the algorithm filters the list of LFI by comparing them against the entire database D using a *globalMinSup*. Then, it returns a list of global frequent itemsets (GFI) which is a subset of LFI.

As stated in the first step of SON, a specific FIM algorithm can be applied to mine each data chunk. In this work, we have implemented and tested different algorithms for this step. The first one is Apriori as described above. The second one is CDAR [10], which relies on the following mining principle:

> **Step 1:** The algorithm divides the database D into $|P| = n$ data partitions, $P = \{p_1, p_2, ..., p_i, ...p_n\}$. Each partition p_i in P only holds transactions whose length is i, where the length of a transaction is the number of items in it.
>
> **Step 2:** Then, CDAR starts mining the data partitions according to trans- action lengths in decreasing order. A transaction in each partition accounts for an itemset. If a transaction T is frequent in partition p_{i+1} then, it will be stored in a list of frequent itemsets L, otherwise, CDAR stores T in a temporary data structure $Temp$. Then, after checking the frequency of all T in p_{i+1}, CDAR generates i subsets of all T in $Temp$ and adds them to par- tition p_i. The same mining process is carried out until visiting all partitions $p_i \subset D$. Before, counting the *support* of a transaction T, CDAR checks its inclusion in L, and if it is included, then CDAR does not consider T, as it is already in L which is considered as frequent.

4   Optimal Data-Process Relationship

Let us now introduce our PFIM architecture, called Parallel Two Steps (P2S), which is designed for data mining in MapReduce. From the mining point of view, P2S is inspired from SON algorithm [9]. The main reason behind opting SON as a reference to P2S is that a parallel version of the former algorithm does not require costly overhead between mappers and reducers. However, as illustrated by our experiments in Sect. 5, a straightforward implementation of SON in MapReduce would not be the best solution for our research problem. Therefore, with P2S, we propose new solutions for PFIM mining, within the "two steps" architecture.

The principle of P2S is drawn from the following observation. Dividing a database D into n partitions $p_1, p_2, ..., p_n$, where $\cup p_i = D$, $i = 1...n$

$$GFI \subseteq \cup LFI \tag{1}$$

where GFI denotes global frequent itemsets and LFI refers to local frequent itemsets. This particular design allows it to be easily parallelized in two teps as follow:

Job 1: Each mapper takes a data split, and performs particular FIM algorithm. Then, it emits a list of local frequent itemsets to the reducer
Job 2: Takes an entire database D as input, and filters the global frequent itemsets from the list of local frequent itemsets. Then, it writes the final results to the reducer.

P2S thus divides the mining process into two steps and uses the dividing principle mentioned above. As one may observe from its pseudo-code, given by Algorithm 1, P2S is very well suited for MapReduce.

The first MapReduce job of P2S consists of applying specific FIM algorithm at each mapper based on a local minimum support ($localMinSup$), where the latter is computed at each mapper based on $MinSup$ δ percentage and the number of transactions of the split being processed. At this stage of P2S, the job execution performance mainly depends on a particular data placement strategy (i.e. RDTP or STDP). This step is done only once and the resulting placement remains the same whatever the new parameters given to the mining process (*e.g. MinSup* δ, local FIM algorithm, etc.). Then P2S determines a list of local frequent itemsets LFI. This list includes the local results of all data splits found by all mappers. The second step of P2S aims to deduce a global frequent itemset GFI. This step is carried out relying on a second MapReduce job. In order to deduce a GFI list, P2S filters the LFI list by performing a global test of each local frequent itemset. At this step, each mapper reads once the list of local frequent itemset stored in Hadoop Distributed Cache. Then, each mapper takes a transaction at a time and checks the inclusion of its itemsets in the list of the local frequent itemset. Thus, at this map phase of P2S algorithm, each mapper emits all local frequent itemsets with their complete occurrences in the whole database (i.e. key: itemset, value: 1). The reducer of the second P2S step, simply computes the sum of the count values of each key (i.e. local frequent itemset) by iterating over the value list of each key. Then, the reducer compares the number of occurrences of each local frequent itemset to $MinSup$ δ, if it is greater or equal to δ, then, the local frequent itemset is considered as a global frequent

---

Algorithm 1. P2S

    Input: Database D and $MinSup$ δ
    Output: Frequent Itemsets
1 //**Map Task 1**
2 **map**($key{:}Null : \mathcal{K}_1$, $value =$ Whole Data Split: $\mathcal{V}_1$)
3   - Determine a local $MinSup$ ls from $\mathcal{V}_1$ based on δ
4   - Perform a complete FIM algorithm on $\mathcal{V}_1$ using ls
5   **emit** *(key: local frequent itemset, value: Null)*

6 //**Reduce Task 1**
7 **reduce**(*key:local frequent itemset, list(values)*)
8   **emit** *(key,Null)*

9 //**Map Task 2**
10 Read the list of local frequent itemsets from Haddop Distributed Cache LFI
    once   **map**(*key:line offset* : $\mathcal{K}_1$, $value =$ Database Line: $\mathcal{V}_1$)
11   **if** *an itemseti* $\in LFI$ *and* $i \subseteq \mathcal{V}_1$ **then**
12        $key \leftarrow i$
13   **emit** *(key:i, value: 1)*

14 //**Reduce Task 2**
15 **reduce**(*key:i, list(values)*)
16      $sum \leftarrow 0$ **while** *values.hasNext()* **do**
17        $sum+ = values.next().get()$
18   **if** $sum >= \delta$ **then**
19   **emit** *(key:i, value: Null)*

---

itemset and it will be written to the Hadoop distributed file system. Otherwise, the reducer discards the key (i.e. local frequent itemset).

Theoretically, based on the inner design principles of P2S algorithm, different data placements would have significant impacts on its performance behavior. In particular, the performance of P2S algorithm at its first MapReduce job, and specifically at the mapper phase, strongly depends on RDTP or STDP used techniques. That is due to the sensitivity of the FIM algorithm being used at the mappers towards its input data.

The goal of this paper is to provide the best combination of both data placement and local algorithm choice in the proposed architecture. In Sect. 4.1, we develop two data placement strategies and explain more their role in the overall performances.

4.1 Data Placement Strategies

The performance of PFIM algorithms in MapReduce may strongly depend on the distribution of the data among the workers. In order to illustrate this issue, consider an example of a PFIM algorithm which is based on a candidate generation approach. Suppose that most of the workload including candidate generation is being done on the mappers. In this case, the data split or partition that holds most lengthy frequent itemsets would take more execution time. In the worst case, the job given to that specific mapper would not complete, making the global extraction process impossible. Thus, despite the fairly automatic data distribution by Hadoop, the computation would depend on the design logic of PFIM algorithm in MapReduce.

Actually, in general, FIM algorithms are highly susceptible to the data sets nature. Consider, for instance, the Apriori algorithm. If the itemsets to be extracted are very long, it will be difficult for this algorithm to perform the extraction. And in case of very long itemsets, it is even impossible. This is due to the fact that Apriori has to enumerate each subset of each itemset. The longer the final itemset, the larger the number of subsets (actually, the number of subsets grows exponentially). Now let us consider Job 1, mentioned above. If a mapper happens to contain a subset of D that will lead to lengthy local frequent itemsets, then it will be the bottleneck of the whole process and might even not be able to complete. Such a case would compromise the global process.

On the other hand, let us consider the same mapper, containing itemsets with the same size, and apply the CDAR algorithm to it. Then CDAR would rapidly converge since it is best suited for long itemsets. Actually, the working principle of CDAR is to first extract the longest patterns and try to find frequent subsets that have not been discovered yet. Intuitively, grouping similar transactions on mappers, and applying methods that perform best for long itemsets seems to be the best choice. This is why a placement strategy, along with the most appropriate algorithm, should dramatically improve the performances of the whole process.

From the observations above, we claim that optimal performances depend on a particular care of massive distribution requirements and characteristics, calling for particular data placement strategies. Therefore, in order to boost up the efficiency of some data sensitive PFIM algorithms, P2S uses different data placement strategies such as *Similar Transaction Data Placement (STDP)* and *Random Transaction Data Placement (RTDP)*, as presented in the rest of this section.

RTDP Strategy. RTDP technique merely refers to a random process for choosing bunch of transactions from a database D. Thus, using RTDP strategy, the database is divided into n data partitions $p_1, p_2, ..., p_n$ where $\cup p_i = D$, $i = 1...n$. This data placement strategy does not rely on any constraint for placing such bunch of transactions in same partition p.

STDP Strategy. Unlike RTDP data placement strategy, STDP relies on the principle of similarity between chosen transactions. Each bucket of similar transactions is mapped to the same partition p. Therefore, the database D is split into n partitions and $\cup p_i = D$, $i = 1...n$.

In STDP, each data split would be more homogeneous, unlike the case of using RDTP. More precisely, by creating partitions that contain similar transactions, we increase the chance that each partition will contain frequent local itemset of high length.

4.2   Data Partitioning

In STDP, data partitioning using similarities is a complex problem. A clustering algorithm may seem appropriate for this task. However, we propose a graph data partitioning mechanism that will allow a fast execution of this step, thanks to existing efficient algorithms for graphs partitioning such as Min-Cut [11]. In the following, we describe how transaction data can be transformed into graph data for doing such partitioning.

- First, for each unique *item* in D, we determine the list of transactions L that contain it. Let D' be the set of all transaction lists L.
- Second, we present D' as a graph $G = (V, E)$, where V denotes a set of vertices and E is a set of edges. Each transaction $T \in D$ refers to a vertex $v_i \in G$ where $i = 1...n$. The weight w of an edge that connects a pair of vertices $p = (v_i, v_j)$ in G equals to the number of common items between the transactions representing v_i and v_j.
- Then, after building the graph G, a Min-Cut algorithm is applied in order to partition D'.

In the above approach, the similarity of two transactions is considered as the number of their common items, i.e. the size of their intersection. In order to illustrate our graph partitioning technique, let us consider a simple example as follows.

Example 2. Let us consider D, the database from Table 1. We start by mapping each item in D to its transactions holder. As illustrated in the table of Fig. 1, T_1 and T_2 have 2 common items, likewise, T_3 and T_4 have 2 common items, while the intersection of T_2 and T_3 is one. The intersection of transactions in D' refers to the weight of their edges. In order to partition D', we first build a graph G from D' as shown in Fig. 1. Then, the algorithm Min-Cut finds a minimum cut in G (red line in Fig. 1), which refers to the minimum capacity in G. In our example, we created two partitions: $Partition_1 =< T_1, T_2 >$ and $Partition_2 =< T_3, T_4 >$.

We have used a particular graph partitioning tool namely PaToH [12] in order to generate data partitions. The reason behind opting for Patoh lies in its set of configurable properties, e.g. the number of partitions and the partition load balance factor.

Based on the architecture of P2S and the data placement strategies we have developed and efficiently designed two FIM mining algorithms. Namely Parallel Two Steps CDAR (P2SC) and Parallel Two Steps Apriori (P2SA) depending on the itemset mining algorithm implemented for itemset mining on the mapper, in the first step of P2S. These two algorithms are highly data-sensitive PFIM algorithms.

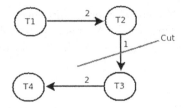

Fig. 1. Transactions of a database (left) & Graph representation of the database (right)(Color figure online)

For instance, if we consider P2SC as a P2S algorithm with STDP strategy, its performance would not be the same as we feed it with RDTP. Because relying on STDP, each split of data fed to such a mapper holds similar transactions, thus, there is less generation of transaction subsets. These expectations correspond to the intuition given in Sect. 4.1. The impact of different data placement strategies will be better observed and illustrated through out experimental results as shown in Sect. 5.

As shown by our experimental results in Sect. 5, P2S has given the best performance when instanciated with CDAR along with STDP strategy.

5   Experiments

To assess the performance of our proposed mining approach and investigate the impact of different data placement strategies, we have done an extensive experimental evaluation. In Sect. 5.1, we depict our experimental setup, and in Sect. 5.2 we investigate and discuss the results of our experiments.

5.1   Experimental Setup

We implemented our P2S principle and data placement strategies on top of Hadoop-MapReduce, using Java programming language. As mining algorithms on the mappers, we implemented Apriori as well as CDAR. For comparison with PFP-Growth [13], we adopted the default implementation provided in the Mahout [14] machine learning library (Version 0.7). We denote by P2Sx-R and P2Sx-S the use of our P2S principle with STPD (P2Sx-S) or RTPD (P2Sx-R) strategy for data placement, where local frequent itemsets are extracted by means of the 'x' algorithm. For instance, P2SA-S means that P2S is executed on data arranged according to STPD strategy, with Apriori executed on the mappers for extracting local frequent itemsets. MR-Apriori is the straightforward implementation of Apriori in MapReduce (one job for each length of candidates, and database scans for support counting are replaced by MapReduce jobs). PApriori does not use any particular data placement strategy. To this end, we just opted to test the algorithm with a RTDP data placement strategy for a comparison sake. Eventually, the instance of P2S architecture with Apriori

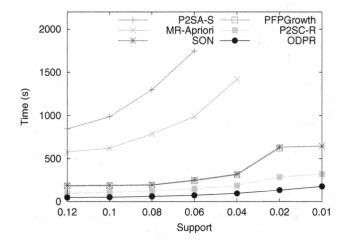

Fig. 2. All algorithms executed on the whole set of wikipedia articles in English

exploited for local frequent itemset mining on the mappers and data arranged according to the RTPD strategy has to be considered as a straightforward implementation of SON. Therefore, we consider this version of P2S being the original version of SON in our experiments.

We carry out all our experiments based on the Grid5000 [15] platform, which is a platform for large scale data processing. We have used a cluster of 16 and 48 machines respectively for Wikipedia and ClueWeb data set experiments. Each machine is equipped with Linux operating system, 64 Gigabytes of main memory, Intel Xeon $X3440$ 4 core CPUs, and 320 Gigabytes SATA II hard disk.

To better evaluate the performance of ODPR and the impact of data placement strategies, we used two real-world datasets. The first one is the 2014 English wikipedia articles [16] having a total size of 49 Gigabytes, and composed of 5 millions articles. The second one is a sample of ClueWeb English dataset [17] with size of 240 Gigabytes and having 228 millions articles. For each dataset we performed a data cleaning task, by removing all English stop words from all articles and obtained a dataset where each article accounts for a transaction (where items are the corresponding words in the article) to each invoked PFIM algorithm in our experiments.

We performed our experiments by varying the $MinSup$ parameter value for each algorithm along with particular data placement strategy. We evaluate each algorithm based on its response time, in particular, when $MinSup$ is very low.

5.2 Performance Results

Figures 2 and 3 report our results on the whole set of Wikipedia articles in English. Figure 2 gives a complete view on algorithms performances for a support varying from 0.12 % to 0.01 %. We see that MR-Apriori runtime grows exponentially, and gets quickly very high compared to other presented PFIM

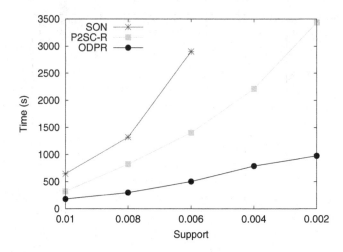

Fig. 3. A focus on algorithms that scale on wikipedia articles in English

algorithms. In particular, this exponential runtime growth reaches its high-est value with 0.04 % threshold. Below this threshold, MR-Apriori needs more resources (e.g. memory) than what exists in our tested machines, so it is impos-sible to extract frequent patterns with this algorithm. Another interesting obser-vation is that P2SA-S, i.e. the two step algorithm that use Apriori as a local mining solution, is worse that MR-Apriori. This is an important result, since it confirms that a bad choice of data-process relationship compromises a com-plete analytics process and makes it inoperative. Let us now consider the set of four algorithms that scale. The less effective are PFPGrowth and P2SA-R. It is interesting to see that two very different algorithmic schemes (PFPGrowth is based on the pattern tree principle and P2SA-R is a two steps principle with Apriori as a local mining solution with no specific care to data placement) have similar performances. The main difference being that PFPGrowth exceeds the available memory below 0.02 %. Eventually, P2SC-R and ODPR give the best performances, with an advantage for ODPR.

Figure 3 focuses on the differences between the three algorithms that scale in Fig. 2. The first observation is that P2SA-R is not able to provide results below 0.008 %. Regarding the algorithms based on the principle of P2S, we can observe a very good performance for ODPR thanks to its optimization between data and process relationship. These results illustrate the advantage of using a two steps principle where an adequate data placement favors similarity between transac-tions, and the local mining algorithm does better on long frequent itemsets.

In Fig. 4, similar experiments have been conducted on the ClueWeb dataset. We observe that the same order between all algorithms is kept, compared to Figs. 2 and 3. There are two bunches of algorithms. One, made of P2SA-S and MR-Apriori which cannot reasonably applied to this dataset, whatever the min-imum support. In the other bunch, we see that PFPGrowth suffers from the

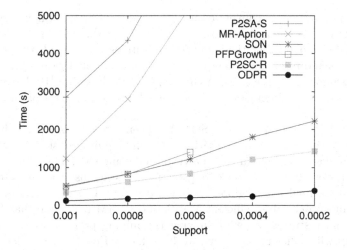

Fig. 4. Experiments on ClueWeb data

same limitations as could be observed on the Wikipedia dataset in Fig. 2, and it follows a behavior that is very similar to that of P2SA-R, until it becomes impossible to execute.

On the other hand, P2SC-R and ODPR are the two best solutions, while ODPR is the optimal combination of data placement and algorithm choice for local extraction, providing the best relationship between data and process.

6   Related Work

In data mining literature, several efforts have been made to improve the performance of FIM algorithms [18–20]. However, due to the trend of the explosive data growth, an efficient parallel design of FIM algorithms has been highly required. There have been many solutions proposed to design most of FIM algorithms in a parallel manner [9,13].

FP-Growth algorithm [20] has shown an efficient scale-up compared to other FIM algorithms, it has been worth to come up with a parallel version of FP-Growth [13] (i.e. PFP-Growth). Even though, PFP-Growth is distinguishable with its fast mining process, it has several drawbacks. In particular, with very low $MinSup$, PFP-Growth may run out of memory as illustrated by our experiments in Sect. 5. Parma algorithm [21], uses an approximation in order to determine the list of frequent itemsets. It has shown better running time and scale-up than PFP-Growth. However, it does not determine an exhaustive list of frequent itemsets, instead, it only approximates them.

A parallel version of Apriori algorithm proposed in [2] requires n MapReduce jobs, in order to determine frequent itemsets of size n. However, the algorithm is not efficient because it requires multiple database scans.

In order to overcome conventional FIM issues and limits, a novel technique, namely CDAR has been proposed in [10]. This algorithm uses a top down approach in order to determine the list of frequent itemsets. It avoids the generation of candidates and renders the mining process more simple by dividing the database into groups of transactions. Although, CDAR algorithm [10] has shown an efficient performance behavior, yet, there has been no proposed parallel version of it.

Another FIM technique, called SON, has been proposed in [9], which consists of dividing the database into n partitions. The mining process starts by searching the local frequent itemsets in each partition independently. Then, the algorithm compares the whole list of local frequent itemsets against the entire database to figure out a final list of global frequent itemsets. In this work, we inspired by SON, and proposed an efficient MapReduce PFIM technique that leverages data placement strategies for optimizing the mining process. Indeed, in order to come up with efficient solutions, we have focused on the data placement as fundamental and essential mining factor in MapReduce.

In [22], the authors proposed an algorithm for partitioning data stream databases in which the data can be appended continuously. In the case of very dynamic databases, instead of PatoH tool which we used in this paper for graph partitioning, we can use the approach proposed in [22] to perform the STDP partitioning efficiently and quickly after arrival of each new data to the database.

7   Conclusion

We have identified the impact of the relationship between data placement and process organization in a massively distributed environment such as MapReduce for frequent itemset mining. This relationship has not been investigated before this work, despite crucial consequences on the extraction time responses allowing the discovery to be done with very low minimum support. Such ability to use very low threshold is mandatory when dealing with Big Data and particularly hundreds of Gigabytes like we have done in our experiments. Our results show that a careful management of processes, along with adequate data placement, may dramatically improve performances and make the difference between an inoperative and a successful extraction.

This work opens interesting research avenues for PFIM in massively distributed environments. In general, we would like to deeply investigate a larger number of algorithms and the impact of data placement on them. More specifically, there are two main factors we want to study. Firstly, we need to better identify what algorithms can be implemented in MapReduce while avoiding to execute a large number of jobs (because the larger the number of jobs, the worse the response time). Secondly, we want to explore data placement alternatives to the ones proposed in this paper.

References

1. Hsinchun, C., Chiang, R.H.L., Storey, V.C.: Business intelligence and analytics: from big data to big impact. MIS Q **36**(4), 1165–1188 (2012)
2. Anand, R.: Mining of Massive Datasets. Cambridge University Press, Cambridge (2012)
3. Goethals, B.: Memory issues in frequent itemset mining. In: Haddad, H., Omicini, A., Wainwright, R.L., Liebrock, L.M.(eds.) Proceedings of the 2004 ACM Symposium on Applied Computing (SAC), Nicosia, Cyprus, March 14–17, 2004, pp. 530–534. ACM (2004)
4. White, T.: Hadoop : The Definitive Guide. O'Reilly, Beijing (2012)
5. Bizer, C., Boncz, P.A., Brodie, M.L., Erling, O.: The meaningful use of big data: four perspectives - four challenges. SIGMOD Rec. **40**(4), 56–60 (2011)
6. Dean, J., Ghemawat, S.: Mapreduce: simplified data processing on large clusters. Commun. ACM **51**(1), 107–113 (2008)
7. Hadoop (2014). http://hadoop.apache.org
8. Agrawal, R., Srikant, R.: Fast algorithms for mining association rules in large databases. In: Bocca, J.B., Jarke, M., Zaniolo, C. (eds.) Proceedings of International Conference on Very Large Data Bases (VLDB), pp. 487–499. Chile, Santiago de Chile (1994)
9. Savasere, A., Omiecinski, E., Navathe, S.B.: An efficient algorithm for mining association rules in large databases. In: Proceedings of International Conference on Very Large Data Bases (VLDB), pp. 432–444 (1995)
10. Tsay, Y.-J., Chang-Chien, Y.-W.: An efficient cluster and decomposition algorithm for mining association rules. Inf. Sci **160**(1–4), 161–171 (2004)
11. Even, S.: Graph Algorithms. Computer Science Press, Potomac (1979)
12. Patoh (2011). http://bmi.osu.edu/umit/PaToH/manual.pdf
13. Li, H., Wang, Y., Zhang, D., Zhang, M., Chang, E.Y.: Pfp: parallel fp-growth for query recommendation. In: Pu, P., Bridge, D.G., Mobasher, B., Ricci, F. (eds.) Proceedings of the ACM Conference on Recommender Systems (RecSys) Lausanne, Switzerland, pp. 107–114. ACM (2008)
14. Owen, S.: Mahout in Action. Manning Publications Co, Shelter Island, N.Y. (2012)
15. Grid5000. https://www.grid5000.fr/mediawiki/index.php/Grid5000:Home
16. English wikipedia articles. http://dumps.wikimedia.org/enwiki/latest
17. The clueweb09 dataset (2009). http://www.lemurproject.org/clueweb09.php/
18. Song, W., Yang, B., Zhangyan, X.: Index-bittablefi: an improved algorithm for mining frequent itemsets. Knowl.-Based Syst. **21**(6), 507–513 (2008)
19. Jayalakshmi, N., Vidhya, V., Krishnamurthy, M., Kannan, A.: Frequent itemset generation using double hashing technique. Procedia Eng. **38**, 1467–1478 (2012)
20. Han, J., Pei, J., Yin, Y.: Mining frequent patterns without candidate generation. SIGMODREC: ACM SIGMOD Record, 29 (2000)
21. Riondato, M., DeBrabant, J.A., Fonseca, R., Upfal, E.: Parma: a parallel randomized algorithm for approximate association rules mining in mapreduce. In: 21st ACM International Conference on Information and Knowledge Management (CIKM), Maui, HI, USA, pp. 85–94. ACM (2012)
22. Liroz-Gistau, M., Akbarinia, R., Pacitti, E., Porto, F., Valduriez, P.: Dynamic workload-based partitioning algorithms for continuously growing databases. In: Hameurlain, A., Küng, J., Wagner, R. (eds.) TLDKS XII. LNCS, vol. 8320, pp. 105–128. Springer, Heidelberg (2013)

Aggregation-Aware Compression of Probabilistic Streaming Time Series

Reza Akbarinia and Florent Masseglia[✉]

Inria & LIRMM, Zenith Team - Université. Montpellier,
Montpellier cedex 5, France
{reza.akbarinia,florent.masseglia}@inria.fr

Abstract. In recent years, there has been a growing interest for probabilistic data management. We focus on probabilistic time series where a
main characteristic is the high volumes of data, calling for efficient compression techniques. To date, most work on probabilistic data reduction
has provided synopses that minimize the error of representation w.r.t.
the original data. However, in most cases, the compressed data will be
meaningless for usual queries involving aggregation operators such as
SUM or AVG. We propose *PHA* (Probabilistic Histogram Aggregation),
a compression technique whose objective is to minimize the error of such
queries over compressed probabilistic data. We incorporate the aggregation operator given by the end-user directly in the compression technique, and obtain much lower error in the long term. We also adopt a
global error aware strategy in order to manage large sets of probabilistic
time series, where the available memory is carefully balanced between
the series, according to their individual variability.

1 Introduction

The abundance of probabilistic and imprecise data, in the past decade, has been
the center of recent research interest on *probabilistic data management* [8,14].
This growing production of probabilistic information is due to various reasons,
such as the increasing use of monitoring devices and sensors, the emergence of
applications dealing with moving objects, or scientific applications, with very
large sets of experimental and simulation data (so much so that Jim Gray has
identified their management and analysis as the "Fourth Paradigm" [9]). Example 1 gives an illustration of such a scientific application, where one must deal
with very large sets of probabilistic data. These data arrive as probabilistic distribution functions (pdf), where each value (*e.g.* the observation of a measure)
is associated to a level of probability.

Example 1. Phenotyping is an emerging science that bridges the gap between
genomics and plant physiology. In short, it aims to observe the interactions
between the functions of a plant (growth, temperature, etc.) and the physical
world in which it develops. Then, depending on the genomic background of the
plant, conclusions can be drawn about plant productivity, which is a major

© Springer International Publishing Switzerland 2015
P. Perner (Ed.): MLDM 2015, LNAI 9166, pp. 232–247, 2015.
DOI: 10.1007/978-3-319-21024-7_16

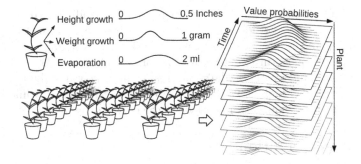

Fig. 1. A phenotyping platform. For each plant, several probabilistic values are measured at regular time intervals and the whole platform produces a large number of streaming probabilistic time series

concern from ecological, economic and societal points of view. In this context, as illustrated by Fig. 1, several measures, such as height and weight growth, or evaporation, can be performed by sensors attached to each plant. Due to device characteristics, each measure is associated to a level of probability and expressed as a pdf. The pdfs of a plant are stored at regular time intervals, leading to a probabilistic time series. In our case, we have as much probabilistic time series as plants in the platform.

There are many other possible applications involving streaming probabilistic time series, where the measures may be, for example, the amount of light received by an array of solar panels in a photovoltaic system, the number of gallons passing by a dam, the quantity of pesticides received by the plants of an observation plot, the support of probabilistic frequent itemsets over time in a probabilistic data stream, etc. All the above applications have a common characteristic: they produce time series where, for each interval of time and for each series, we have a pdf corresponding to the possible measures during this interval. In almost all these applications, the user intends to issue specific aggregation queries over the time intervals, e.g. the sum of plant growth in a sequence of time intervals.

Depending on the application, the number of probabilistic values may be very high. Compared to deterministic time series, the data size is multiplied by the range of the pdf (*e.g.* with a thousand of series, ten thousands of measures per day and a hundred of probabilistic values per measure, there are one billion probabilistic values to handle per day). When the number of values is larger than the available memory, data has to be compressed, and when that number gets even much larger, the compression cannot be without information loss. Our goal is to lower the information loss due to probabilistic data compression. Actually, *we take the view that such data are usually intended to be intensively queried, especially by aggregation queries.* In this paper, we propose a compression approach, called *PHA*, that allows high quality results to specific aggregation operators. This is done by incorporating the aggregation in the compression process itself. The idea is that when the granularity of the representation decreases, the aggregation operator tackled by the end-user is used at the core of time series compression. Our approach includes strategies for probabilistic streaming time series management by considering their tolerance to compression. PHA is distribution independent, so it can work with any kind of distributions (uniform, normal, etc.). We perform extensive evaluations to illustrate the performance of *PHA* over real and synthetic datasets. The results show it's ability to apply compression ratios as high as 95 % over probabilistic time series, while keeping extremely low (almost null) error rates w.r.t. the results of aggregation operators.

2   Problem Definition

We are interested in compressing probabilistic time series by aggregating their histograms that are representations of the data distributions in time intervals of time series.

2.1   Preliminaries

We define a *probabilistic histogram* as follows.

Definition 1. [Probabilistic histogram] *A probabilistic histogram H is an approximate representation of a pdf (probabilistic distribution function) based on partitioning the domain of values into m buckets, and assigning the same probability to all values of each bucket. Each bucket b is a pair (I, p) where $I = [s, e)$ is a value interval with s and e as start and end points respectively, and p is the probability of each value in the bucket. We assume that s and e are integer. For each value v, we denote by $H[v]$ the probability of v in the distribution. The buckets of a histogram H are denoted by $bucks(H)$, and the ith bucket by $bucks(H)[i]$. The number of buckets is $|bucks(H)|$.*

Hereafter, unless otherwise specified, when we write histogram we mean a probabilistic histogram. A probabilistic time series is defined as a sequence of histograms as follows.

Definition 2. [Probabilistic time series] *a probabilistic time series $X = \{H_1, \ldots, H_n\}$ is a sequence of histograms defined on a sequence of time intervals $T_1, \ldots, T_n$, such that histogram H_i represents the probability distribution in interval T_i. The intervals $T_1, \ldots, T_n$ are successive, i.e. there is no gap between them.*

Example 2. Let us consider a plant p in the phenotyping application of Example 1. The height growth of p has been measured by a sensor. The resulting pdfs are represented by two histograms H_1 and H_2 (see Fig. 2): $H_1 = \{([0, 1), 0), ([1, 2), 0), ([2, 3), 0), ([3, 4), 0.2), ([4, 5), 0.2), ([5, 6), 0.4), ([6, 7), 0.2), ([7, 8), 0)\}$ from 8 am to 7:59 pm, and $H_2 = \{([0, 1), 0), ([1, 2), 0.2), ([2, 3), 0.4), ([3, 4), 0.2), ([4, 5), 0.2), ([5, 6), 0), ([6, 7), 0), ([7, 8), 0)\}$ from 8 pm to 7:59 am. For instance, in H_1, the probability that the plant grew by 5 units is 40 %. Meanwhile, during the night, the most probable growth is only 2 units. The probabilistic time series of p is made of these two histograms on their respective time intervals.

Below, we define a data stream based on time series.

Definition 3. [Data stream] *a data stream* $S = \{X_1, \ldots, X_m\}$ *is a set of time series defined on a sequence of intervals* $T_1, \ldots, T_n$ *such that each time series* X_i *has its own histograms for the intervals* $T_1, \ldots, T_n$.

The aggregation of histograms can be defined using the aggregation of probability distribution functions.

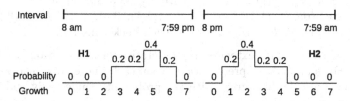

Fig. 2. A probabilistic time series, made of two probabilistic histograms on contiguous time intervals

Definition 4. [Aggregation of pdfs] *let* G_1 *and* G_2 *be two pdfs, and* f *be an aggregation function. Then, aggregating pdfs* G_1 *and* G_2 *by* f *means to generate a pdf* G *such that for any two values* $x_1 \in G_1$ *and* $x_2 \in G_2$ *there is a value* $x \in G$ *where* $x = f(x_1, x_2)$, *and the probability of* x *in* G *is equal to the probability of all combinations of values* $x_1 \in D_{G_1}$ *and* $x_2 \in D_{G_2}$ *such that* $f(x_1, x_2) = x$.

Definition 5. [Aggregation of probabilistic histograms] *given an aggregation function* f, *and two histograms* H_1 *and* H_2 *representing two pdfs* G_1 *and* G_2 *respectively, then aggregating histograms* H_1 *and* H_2 *by* f *means to make a histogram on the aggregated pdf of* G_1 *and* G_2.

We compare the probabilistic histograms by measuring their absolute difference defined as follows.

Definition 6. [Absolute difference] *the absolute difference of two histograms* H_1 *and* H_2, *denoted as* $abs\_dif(H_1, H_2)$, *is defined as the cumulative difference of their probability distributions for all possible values. In other words:*

$$abs\_dif(H_1, H_2) = \int_{-\infty}^{+\infty} (|H_1(x) - H_2(x)|) \, \mathrm{d}x.$$

We compress the histograms of time series in order to hold stream management constraints. For each compressed histogram, there is a set of original histograms which we define as follows.

Definition 7. [Original histograms] *Let* $X = \{H_1, \ldots, H_n\}$ *be a time series defined on time intervals* $T_1, \ldots, T_n$. *Assume* H' *is a histogram resulted from compressing the histograms of intervals* $T_i, \ldots, T_j$. *Then, the original histograms of* H', *denoted as* $org\_hist(H')$, *are the histograms of* X *in time intervals* $T_i, \ldots, T_j$. *In other words,* $org\_hist(H') = \{H_i, \ldots, H_j\}$.

In our algorithms, sometimes we need to extend the histograms. The extension of a histogram is defined as follows.

Definition 8. [Extending a probabilistic histogram] *the extension of a probabilistic histogram H, denoted as $Ext(H)$, is a probabilistic histogram H' such that the interval of each bucket in H' is of size one (i.e. $e - s = 1$), and for each value v, $H'(v) = H(v)$.*

Example 3. Let $H = \{([0, 2), 0.4), ([2, 4), 0.6)\}$ be a histogram with two buckets. The extension of H is $Ext(H) = \{([0, 1), 0.4), ([1, 2), 0.4), ([2, 3), 0.6), ([3, 4), 0.6)\}$. Thus, in $Ext(H)$ the size of each bucket interval is 1.

Hereafter, when we write extended representation of a histogram we mean its extension. For simplicity of presentation, when each bucket of a histogram is of size one, we write the histogram with the probability values only and we omit the start and end points.

2.2 Problem Statement

Our objective is to compress the histograms of a data stream's time series by aggregating their histograms, while respecting memory constraints on the size of histograms. Concretely, the results of each aggregation should be represented by a histogram of at most m buckets, where m is a predefined value. At the same time, we want to preserve the quality of aggregation, i.e. minimum error in the results.

Our goal can be defined as follows: given a set of given time series $S = \{X_1, \ldots, X_n\}$, an aggregation function f, and an integer number m, our objective is to compress S to a set of time series $S' = \{X'_1, \ldots, X'_n\}$ such that each time series X'_i is composed of a set of histograms $\{H'_1, \ldots, H'_k\}$, while:

1. Minimizing the error of aggregation, given by:

$$\sum\nolimits_{\forall X' \in S'} \left(\sum\nolimits_{\forall H' \in X'} abs\_dif(H', f(org\_hist(H'))) \right)$$

2. Limiting the number of buckets of each histogram to m. In other words:

$$\forall X' \in S', \forall H' \in X', |bucks(H'_i)| \leq m.$$

2.3 Background

In our underlying applications, when the number of buckets becomes larger than the available memory, we must reduce the data size. Since we are working with probabilistic time series (*i.e.* series of histograms) we have two possible dimensions for data compression. The first one is the bucket dimension. In this case, we merge several contiguous buckets into one bucket having larger granularity. Wavelets and regression [7] are two popular methods that can be adapted for this purpose. The second one is the time dimension. In this case, when the number of histograms is too large, we merge several histograms, corresponding to a number of time slots, into one aggregated histogram that correspond to a larger time slot, thus increasing the granularity of our representation.

3 Compression Approach

In this section, we propose our compression approach on histograms called PHA (Probabilistic Histogram Aggregation) and a strategy for histogram selection (for compression) in the global streaming process.

3.1 Bucket Dimension

As stated previously, there are two possible dimensions for compressing probabilistic time series: bucket and time dimensions. In this subsection, we present our approach for compression on the bucket dimension. We first describe our method for aggregating probabilistic histograms of time intervals based on the SUM aggregation operator, which is the running operator of this paper. Then, we explain the compression of aggregated histograms. It is not difficult to extend our approach to other aggregation operators such as AVG (see Appendix for AVG).

Aggregating Histograms with SUM. Given two histograms H_1 and H_2 on two time intervals T_1 and T_2 respectively, aggregating them by a function f means to obtain a histogram H such that the probability of each value k in H is equal to the cumulative probability of all cases where $k = f(x_1, x_2)$ such that $H_1(x_1) > 0$ and $H_2(x_2) > 0$. In deterministic context, given two histograms H_1 and H_2, for each value k the SUM operator returns $H(k) = H_1(k) + H_2(k)$ as output. However, this method does not work for the case of probabilistic histograms, because we need to take into account the probability of all cases where sum is equal to k. Below, we present a lemma that gives a formula for performing the SUM operator over probabilistic histograms.

Lemma 1. *Let H_1 and H_2 be two probabilistic histograms. Then, the probability of each value k in the probabilistic histogram obtained from the sum of H_1 and H_2, denoted as $SUM(H_1, H_2)[k]$, is given by $\sum_{i=0}^{k}((H_1[i] \times H_2[k-i] + H_1[k-i] \times H_2[i])$*

Proof. The probability of having a value k in $SUM(H_1, H_2)$ is equal to the probability of having values i in H_1 and j in H_2 such that $i + j = k$. Thus, we must compute the cumulative probability of all cases where the sum of two values from H_1 and H_2 is equal to k. This is done by the sigma in Lemma 1.

Example 4. Consider the histograms H_1 and H_2 in Example 2. Then, the probability of value $k = 4$ in the histogram $H = SUM(H_1, H_2)$ is computed as follows:

$$H[4] = H_1[0] \times H_2[4] + H_1[1] \times H_2[3] + H_1[2] \times H_2[2] + H_1[3] \times H_2[1] + H_1[4] \times H_2[0] = 0 + 0 + 0 + 0.2 \times 0.2 + 0 = 0.04.$$

Compressing Aggregated Histograms. An aggregated histogram is not a compressed histogram by itself (it is just a representation having a lower granularity on the time dimension). Let us consider the SUM operator on H_1

Aggregation	Value probabilities															Error
SUM(H1,H2)	0	0	0	0	0.04	0.12	0.2	0.28	0.2	0.12	0.04	0		0	0	
SUM(Ext(Wav(H1)),Ext(Wav(H2)))	0	0	0.01	0.02	0.07	0.12	0.17	0.22	0.17	0.12	0.07	0.02	0.01	0	0	0.24
SUM(Ext(Reg(H1)),Ext(Reg(H2)))	0	0	0	0.02	0.04	0.14	0.17	0.26	0.17	0.14	0.04	0.02	0	0	0	0.16
PHA(H1,H2)	0.008	0.008	0.008	0.008	0.008	0.12	0.2	0.28	0.2	0.12	0.04	0		0	0	0.064

Fig. 3. Comparing the results of SUM operator on histograms compressed using different compression techniques

and H_2 from Example 2. The result of this operator is given by Fig. 3 (line "SUM($H1, H2$)" in bold characters). Obviously, the number of buckets is not significantly reduced and the requirement of data compression when the memory budget has been reached is not met. This is due to the fact that the number of values has grown, and now corresponds to the sum of the maximum values in H_1 and H_2. Therefore, if we want to obtain a compressed version of this histogram, we need to aggregate data on the buckets dimensions. We chose the regression approach and apply it to the aggregated histogram H'. Then, we obtain an aggregation on the time dimension, combined to a compression on the buckets dimension. We adopt a bottom-up approach to calculate the regression form of our histograms. First, we build a sorted list of distances between the contiguous buckets. Then, the two most similar buckets are merged and the list of distances is updated. This process is repeated until the desired number of buckets is reached. Here, when two contiguous buckets b_1 and b_2 are merged into the resulting bucket b', then b' inherits from the start point of b_1 and the end-point of b_2. The value of b' is the weighted average of values in b_1 and b_2. The weight of a bucket b' is the number of original buckets that have been merged into b'. Our claim is that paying attention to the order of these operations (*i.e.* compression+aggregation vs. aggregation+compression) makes the difference between appropriate and irrelevant manipulation of probabilistic data. Since an aggregation operator considers the probabilistic meaning of each value in a histogram, it gives better results on the original data, compared to the result obtained on the compressed (*i.e.* damaged) data. This is illustrated in the next example.

Comparative Example. The operations reported in Fig. 3 illustrate the main motivation for our approach: *in order to choose a representation for probabilistic time series compression, we should not focus on the error w.r.t. the original histograms, but rather on the error w.r.t. the results of aggregation operators that will run on the compressed data.* The table of Fig. 3 gives the results of the SUM operator on H_1 and H_2 from Example 2 (second line, "$SUM(H_1, H_2)$"). We compare the results of this operator on the compressed versions of H_1 and H_2, using the techniques presented Sects. 2.3 and 3.1. First, we report the results of SUM on the compressed histograms on the bucket dimension, *i.e.* $SUM(Ext(Wav(H1)), Ext(Wav(H2)))$. In other words, we apply SUM on the extended histograms of the wavelet transform of H_1 and H_2 and obtain the approximate value probabilities on the time interval

$T_1, \ldots, T_2$. In the last column, we report the error (24 %) of this representation, compared to the true probability values of $SUM(H_1, H_2)$. Second, we apply SUM on the extended histograms of the regression compression on H_1 and H_2, i.e. $SUM(Ext(Reg(H1)), Ext(Reg(H2)))$ and also report the error (i.e. 16 %).

This comparison is provided with SUM operator applied to the histogram compressed on the time dimension. We apply a regression compression to the result of SUM on the original histograms. The compressed histogram is given by the "$Reg(SUM(H1, H2))$" line and its extension is given by the last line. In this case, the error is only 6.4 %.

In this illustration, we can observe the difference between the errors of each representation. Intuitively, we observe that PHA gives the best results (bottom line of Fig. 3). When the histograms are independent, the regression technique should give better results than *wavelets*.

3.2   Time Dimension

The time dimension is very important in the technique used for compressing time series. In data streams, this dimension has a major impact since it gives the organization of events records (i.e. their order of arrival). Moreover, most streams are cohesive on this dimension since the series have usually low variation from one timestamp to an other. By working on this dimension, we are thus able to lower the information loss due to the representation granularity. In this section, we study two strategies under the decaying factor point of view and propose a global optimization framework where the available space is balanced between the series in the model, in order to obtain the best possible global error.

Logarithmic Tilted Time Windows. The logarithmic Tilted Time Windows (TTW) is a management principle based on a decaying factor [5]. When new transactions arrive in the stream, older transactions are merged. The older the transactions, the larger the granularity of representation (and the size of windows). The main advantage of this representation is to give more space to recent transactions. The main drawback is to merge old transactions with a blind approach that does not optimize the available space.

Global Error Optimization. Let us consider S, the data stream of Definition 3. At each step s (a new histogram is added to each probabilistic time series), we want to update the representations and their error rates in our data structure. In order to maintain a globally satisfying error rate, we need to choose the representations that will minimize this error. The main idea is that merging the most similar histogram will have low impact on the resulting error. In other words, similar histograms have higher tolerance to merging error. Therefore, for each new histogram in S, let X_i be the time series having the most similar consecutive histograms, i.e. H_1 and H_2, then i) H_1 and H_2 are merged into H_3 ii) the distances between H_3 and its direct neighbors are calculated and iii) the sorted list of distances between consecutive histograms in S is updated. Based on this general principle, we devise the strategies described hereafter.

PHA 'a la Ward'. The goal of *PHA* is to minimize the error of a merged histogram w.r.t. the original merged histogram. It is not possible to know, in advance, which couple of histograms will give the lowest error (after merging) among all the histograms in S. Therefore, an optimal strategy, by means of *PHA*, should try all the possible combinations and pick up the best one. This principle is inspired from the Ward criterion in hierarchical clustering. However, as illustrated by our experiments, this does not always give the best results in terms of global error.

Fast PHA. We also propose a fast strategy based on *PHA*. In order to reduce the time spent on each merging operation, we propose to merge two consecutive histograms if the sum of their number of values is minimum. Surprisingly, our experiments illustrate the good quality of results obtained with this strategy, whereas no information about the histogram's content has been taken into account (only their number of probability values).

4   On Retrieving Original Values

When the histograms have been merged, the remaining question is "how to use this compressed information to retrieve the original histograms?". An ideal solution would be able to get back to the original histograms (even if some information has been lost in the compression). In this section, we first show that in general if the original histograms are different, it is not possible to find them from the compressed histogram, because of information loss in the compression process. Then, by assuming that the original histograms are identical, we propose a method to return them. Let us first study of the case where the original histograms are different. Let us assume that the size of the original histograms is n. The following lemma shows the impossibility of computing them from the merged histogram.

Lemma 2. *Suppose a histogram $H = SUM(H_1, H_2)$, and assume it has been generated from two different probabilistic histograms. Then, we cannot compute H_1 and H_2 by using H.*

Proof. Let $2n$ be the size of histogram $H = SUM(H_1, H_2)$. Let $X_i = H_1[i]$ and $Y_i = H_2[i]$ for $0 \leq i \leq n$. To find the variables X_i and Y_i for $0 \leq i \leq n$, we must be able to solve the following equation system:

$$\begin{cases} X_0 \times Y_0 = H[0] \\ X_0 \times Y_1 + X_1 \times Y_0 = H[1] \\ X_0 \times Y_i + X_1 \times Y_{i-1} + \cdots + X_i \times Y_0 = H[i] \\ \cdots \\ X_n \times Y_n = H[2n] \end{cases} \tag{1}$$

In this system, there are $2 \times n + 1$ equations, and $2 \times n + 2$ variables, i.e. X_i and Y_i for $0 < i < n$. Since the number of equations is less than the number of variables, it is not possible to find the value of variables.   □

This is usual for data streams, where decreasing the granularity of representation leads to approximate representations and information loss. However, our experiments show that our approach allows very high compression rates with no, or very few, loss for the queries performed on the stream (while traditional approaches show very important loss). Furthermore, when the original histograms are equal, the following lemma gives us a formula for retrieving their exact original values.

Lemma 3. *Suppose an aggregated histogram H, and assume it has been generated from H_1 and H_2 with the SUM operator, i.e. $H = SUM(H_1, H_2)$. Suppose that H_1 and H_2 are two equal probabilistic histograms, i.e. $H_1 = H_2$, then H_1 can be reconstructed from H using the following formulas:*

$$H_1(k) = \begin{cases} \frac{H[k] - \sum_{i=1}^{k-1} H_1[i] \times H_1[k-i]}{2 \times H_1[0]} & \text{for } k > 0; \\ \sqrt{H[0]} & \text{for } k = 0 \end{cases} \qquad (2)$$

Proof. The proof is done by using Lemma 1 and the assumption that $H_1[k] = H_2[k]$ for all k. We first prove the equation for the case of $k = 0$, and then for $k > 0$. From Lemma 1, we have : $H_1[0] \times H_2[0] = H[0]$. By replacing $H_2[0]$ with $H_1[0]$, we obtain: $H_1[0] = \sqrt{H[0]}$. To prove the second case, i.e. $k > 0$, let us write the equation of Lemma 1 as follows:
$H[k] = H_1[0] \times H_2[k] + H_1[k] \times H_2[0] + \sum_{i=1}^{k-1}(H_1[i] \times H_2[k-i] + H_1[k-i] \times H_2[i])$
Then, by replacing $H_2[i]$ with $H_1[i]$ for $0 \leq i \leq k$, we have
$H[k] = 2 \times H_1[0] \times H_1[k] + 2 \times \sum_{i=1}^{k-1}((H_1[i] \times H_1[k-i]]$
Thus, we have :
$H_1[k] = \frac{H[k] - \sum_{i=1}^{k-1} H_1[i] \times H_1[k-i]}{2 \times H_1[0]}$ □

By assuming the equality of the original histograms of a merged histogram, Algorithm 1 computes the original histograms.

Algorithm 1. Retrieving original histograms in the case where they are identical

---

Require: H: a probabilistic histogram generated by applying SUM operator on two identical histograms; $2 \times n$: highest value in the domain of H;
Ensure: H_1: the original probabilistic histogram from which H has been generated;
1: $H_1[0] = sqrt H[0]$;
2: **for** $k = 1$ to n **do**
3:     $s = 0$;
4:     **for** $i = 1$ to $k - 1$ **do**
5:         $s = s + H_1[i] \times H_1[k - i]$;
6:     **end for**
7:     $H_1[k] = (H[k] - s)/(2 \times H_1[0])$;
8: **end for**
9: **return** H_1

---

5   Experiments

We have implemented the strategies described in Sect. 3.2 with our *PHA* compression technique.

We evaluated our approach on a real-world dataset and two synthetic datasets. The real world dataset has been built over probabilistic frequent itemsets (PFI) [1] extracted from the *accident* dataset of the FMI repository[1]. The original accident dataset contains 11 millions of events, 340K transactions and 468 items. We have added an existential probability $P \in [0..1]$ to each event in these datasets, with a uniform distribution. Then, we have extracted PFIs [3] from this file. Each PFI has a probability distribution of its frequentness at regular time intervals in the file. Our interest in this dataset is due to the significant number of PFIs that can be extracted from it. Actually, we have recorded the evolution of these distributions for 15 itemsets, over 574 K timestamps, with a range of 49 possible values. For synthetic datasets, we have written a generator that builds a set of S probabilistic time series having length N of histograms having H buckets (normal or uniform distributions) as follows. For a normal distribution, we first generate a random time series of length N. Then, for each value in the time series, we consider it as the "mean" value of the distribution and we chose a random "variance". We have generated two datasets: U.S20N100K.H50 and N.S10N100K.H40. For both of these datasets, S is the number of series, N the series length and H the size of each histogram.

In our tests, we don't adopt Wavelets as a compression strategy on probability distributions because this technique does not fit the constraints of probabilistic streaming time series. Actually, Wavelets need to be applied to static datasets (they don't apply to streaming or incremental environments) and they require datasets size to be a power of two (for instance, in our context, both the number of timestamps and the number of buckets should be a power of two for a Wavelet compression).

We compare our compression approach with extensions of a competitive approach, which we call *averaging*, that compresses two given histograms by computing their average for any given value in the domain. Formally, averaging works as follows. Let H_1 and H_2 be two given histograms, then their compression means to generate a histogram H' such that for any value v in their domain, $H'(v) = AVG(H_1(v), H_2(v))$. Actually, in our context, the time dimension is very important since successive histograms are more likely to be similar than random histograms in the series. Therefore, we consider that the most adequate approach, for a comparison to *PHA*, is the *averaging* compression technique.

In the following, *TTW.AVG* is the *averaging* compression in the Tilted Time Windows strategy, *TTW.PHA* is the *PHA* compression in the TTW strategy, *GOF.PHA* is *PHA* in the Global Optimization strategy with the *Fast* approach (Sect. 3.2), *GOW.PHA* is *PHA* in the global optimization 'a la Ward' (Sect. 3.2) and *GOS.AVG* is the *averaging technique* in the global optimization strategy with SSE as a distance between histograms. When quality is measured, it is with

---

[1] http://fimi.ua.ac.be/data/.

regards to the goal defined in Sect. 2.2. For this purpose, when a stream has been summarized by means of *PHA* and a technique T, we compare the quality of the chosen operator (in our case, it is *SUM*) over the intervals built by *PHA* on the corresponding histograms in the representation of T and in the original data. Our goal is to measure the information loss of various compression techniques, compared to that of *PHA*, having the original data as a reference. Since the intervals built by *PHA* are not the same from one experiment to another, we have one comparison diagram per dataset and per couple of compression techniques. Measures are given according to a compression ratio. For instance, a compression of 90 % for a file with 1, 000, 000 histograms means that our representation is done with 100, 000 compressed histograms.

5.1   Global Optimization: Evaluation of the Synopses Techniques

During our experiments, we observed that *GOW.PHA* does not justify the difference in response time. Actually, the results in terms of quality were similar to that of *GOF.PHA* with much higher response times. This is the case for all the datasets we have tested, but due to lack of space, we don't report this result in our figures. Our first set of experiments aims to compare the synopses techniques based on *averaging*, and *PHA*. For this purpose, we use the global optimization principle described in Sect. 3.2.

Figure 4 illustrates the difference in quality of approximation according to the compression ratio on the real and synthetic datasets. We can observe that *GOF.PHA* gives very good results in terms of quality when *GOS.AVG* reaches high error levels (up to 75 % on *real* data). For instance, on the *normal* dataset (Fig. 4(b)) *GOF.PHA* is able to keep an error close to zero with compression rates up to 80 %. This is a very good result, illustrating the capability of *PHA* to keep most of the information during the compression, compared to the high error rates of *GOS.AVG*.

5.2   TTW Vs. Global Optimization

Here, our goal is first to investigate the difference between *TTW.PHA* and *TTW.AVG*. Figure 5(d) illustrates the difference between both approaches in terms of compression and error rates on the *normal* dataset. We observe similar differences, compared to our previous experiments between *GOF.PHA* and *GOS.AVG*. Then, we have measured the error rate, timestamp by timestamp, for a TTW strategy on one hand, compared to a Global Optimization, on the other hand. We expect TTW to give better results on the most recent histograms, since they are designed for this purpose. Figure 5(e) gives a detailed error comparison between *GOF.PHA* and *TTW.AVG* on the accident data set with a compression ratio of 85%. Figure 5(f) gives the same comparison on the *normal* dataset with a compression ratio of 95%. We observe that the difference is negligible for the most recent histograms, while it increases greatly for old histograms (one can observe the quality plateau due to changes of TTW in the representation).

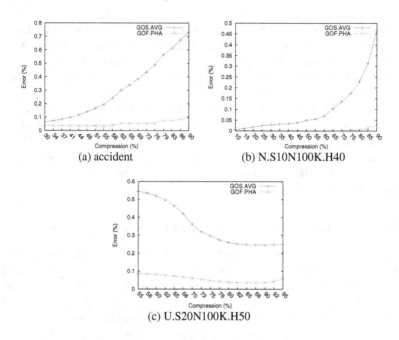

(a) accident

(b) N.S10N100K.H40

(c) U.S20N100K.H50

Fig. 4. Quality of approximation: Global Optimization strategy

6   Related Work

Today, probabilistic data management is recognized as a major concern [8], where an important challenge is to provide reliable models that will allow data querying with relevance and cohesion [14]. In this section, we review the works done on the problems that are related to the problem that we address. Aggregate query processing is an important issue of probabilistic data management. In this context, some works were devoted to developing efficient algorithms for returning the expected value of aggregate values, e.g. [4,10]. In [6,13], approximate algorithms have been proposed for probabilistic aggregate queries. The central limit theorem [12] is one of the main methods to approximately estimate the distribution of aggregation functions, e.g. SUM, for sufficiently large numbers of probabilistic values. In [11], Kanagal et al. deal with continuous aggregate queries over correlated probabilistic data streams. They assume correlated data which are Markovian and structured in nature. In [2], a dynamic programming algorithm for evaluating the SUM aggregate queries was proposed. The goal of almost all the work on probabilistic aggregate query processing is to efficiently process these queries over uncompressed probabilistic data. But, the goal of our work is to develop compression techniques such that the aggregate queries return high quality results on the compressed data. Despite the data overload and the quality issues that are at stake, we only find a few contributions on probabilistic data reduction [7,15]. Currently, the main grip for research in this domain is to consider the difference between deterministic and probabilistic data in the compression techniques.

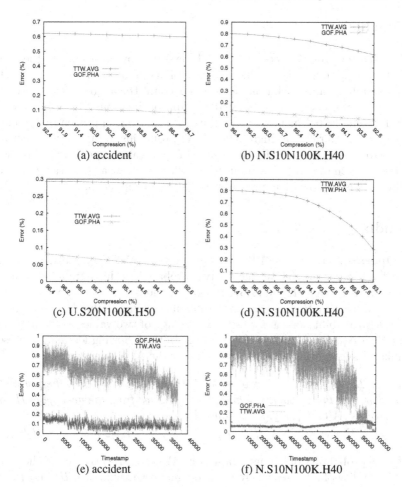

Fig. 5. Quality of approximations: TTW vs. Global Optimization (a,b,c), TTW only (d) and detailed quality of approximation, transaction by transaction (e,f)

In [7] the authors build synopses of probabilistic data represented by histograms that associate a value to its probability. In [15], a wavelet approach is designed for probabilistic time series compression. The authors work on a series of probabilistic histograms and consider two dimensions in their Wavelet algorithm. Aside from [7,15], most of existing works on probability distributions has focused on normal distributions. We didn't find solutions for probabilistic streaming time series where the numbers of series, buckets and timestamps have no constraints. and they don't consider the possible use of aggregation operations that can be made on these data.

7 Conclusions

We addressed the problem of probabilistic time series compression with the objective of allowing high quality results to aggregation operators. We proposed PHA, a new probabilistic approach that takes into account the aggregation operator in the compression process. We evaluated our approach through experiments over several datasets. Experimental results illustrate that our approach is capable to obtain compression ratios as high as 95 % over probabilistic time series, with very low (close to zero) error rates for the results of aggregation operators. This shows that PHA fits well to analytical applications where specific aggregation operators are frequently issued over compressed probabilistic time series.

Appendix

AVG Operator. To extend PHA to the average (AVG) operator, we need a method for computing the average of two probabilistic histograms. Given two probabilistic histograms H_1 and H_2, aggregating them by an AVG operator means to make a histogram H such that the probability of each value k is equal to the cumulative probability of cases where the average of two values x_1 from H_1 and x_2 from H_2 is equal to k, i.e. $k = (x_1 + x_2)/2$. In the following lemma, we present a formula for aggregating two histograms using the AVG operator.

Lemma 1. *Let H_1 and H_2 be two probabilistic histograms. Then, the probability of each value k in the probabilistic histogram obtained from the average of H_1 and H_2, denoted as $AVG(H_1, H_2)[k]$, is computed as:*

$$AVG(H_1, H_2)[k] = \sum_{i=0}^{k}((H_1[i] \times H_2[2 \times k - i] + H_1[2 \times k - i] \times H_2[i])$$

Proof. The probability of having a value k in $AVG(H_1, H_2)$ is equal to the cumulative probability of all cases where the average of two values i in H_1 and j in H_2 is equal to k. In other words, j should be equal to $2 \times k - i$. This is done by the sigma in the above equation.

References

1. Akbarinia, R., Masseglia, F.: Fast and exact mining of probabilistic data streams. In: Blockeel, H., Kersting, K., Nijssen, S., Železný, F. (eds.) ECML PKDD 2013, Part I. LNCS, vol. 8188, pp. 493–508. Springer, Heidelberg (2013)
2. Akbarinia, R., Valduriez, P., Verger, G.: Efficient evaluation of sum queries over probabilistic data. IEEE Trans. Knowl. Data Eng. **25**(4), 764–775 (2013)
3. Bernecker, T., Kriegel, H.P., Renz, M., Verhein, F., Zuefle, A.: Probabilistic frequent itemset mining in uncertain databases. In: Proceedings of the 15th ACM SIGKDD International Conference on Knowledge Discovery and Data Mining, KDD 2009, pp. 119–128. ACM (2009)
4. Burdick, D., Deshpande, P.M., Jayram, T.S., Ramakrishnan, R., Vaithyanathan, S.: OLAP over uncertain and imprecise data. VLDB J. **16**(1), 123–144 (2007)

5. Chen, Y., Dong, G., Han, J., Wah, B.W., Wang, J.: Multi-dimensional regression analysis of time-series data streams. In: Proceedings of the 28th International Conference on Very Large Data Bases, VLDB 2002, pp. 323–334. VLDB Endowment (2002)
6. Cormode, G., Garofalakis, M.: Sketching probabilistic data streams. In: Proceedings of the 2007 ACM SIGMOD International Conference on Management of Data, SIGMOD 2007, pp. 281–292 (2007)
7. Cormode, G., Garofalakis, M.: Histograms and wavelets on probabilistic data. IEEE Trans. Knowl. Data Eng. **22**(8), 1142–1157 (2010)
8. Dalvi, N., Suciu, D.: Efficient query evaluation on probabilistic databases. VLDB J. **16**(4), 523–544 (2007)
9. Hey, A.J.G., Tansley, S., Tolle, K.M. (eds.): The fourth paradigm: data-intensive scientific discovery, Microsoft Research, Redmond, Washington (2009)
10. Jayram, T.S., McGregor, A., Muthukrishnan, S., Vee, E.: Estimating statistical aggregates on probabilistic data streams. ACM Trans. Database Syst. **33**(4), 26:1–26:30 (2008)
11. Kanagal, B., Deshpande, A.: Efficient query evaluation over temporally correlated probabilistic streams. In: Proceedings of the 2009 IEEE International Conference on Data Engineering, ICDE 2009, pp. 1315–1318 (2009)
12. Rempala, G., Wesolowski, J.: Asymptotics for products of sums and u-statistics. Electron. Commun. Probab. **7**(5), 47–54 (2002)
13. Ross, R., Subrahmanian, V.S., Grant, J.: Aggregate operators in probabilistic databases. J. ACM **52**(1), 54–101 (2005)
14. Sathe, S., Jeung, H., Aberer, K.: Creating probabilistic databases from imprecise time-series data. In: Proceedings of the 2011 IEEE 27th International Conference on Data Engineering. ICDE 2011, pp. 327–338 (2011)
15. Zhao, Y., Aggarwal, C., Yu, P.: On wavelet decomposition of uncertain time series data sets. In: Proceedings of the 19th ACM International Conference on Information and Knowledge Management, CIKM 2010, pp. 129–138 (2010)

Clustering

Applying Clustering Analysis to Heterogeneous Data Using Similarity Matrix Fusion (SMF)

Aalaa Mojahed$^{(\boxtimes)}$, Joao H. Bettencourt-Silva, Wenjia Wang,
and Beatriz de la Iglesia

Norwich Research Park, University of East Anglia, Norwich, Norfolk, UK
{a.mojahed,J.Bettencourt-Da-Silva,Wenjia.Wang,b.Iglesia}@uea.ac.uk

Abstract. We define a heterogeneous dataset as a set of complex objects, that is, those defined by several data types including structured data, images, free text or time series. We envisage this could be extensible to other data types. There are currently research gaps in how to deal with such complex data. In our previous work, we have proposed an intermediary fusion approach called SMF which produces a pairwise matrix of distances between heterogeneous objects by fusing the distances between the individual data types. More precisely, SMF aggregates partial distances that we compute separately from each data type, taking into consideration uncertainty. Consequently, a single fused distance matrix is produced that can be used to produce a clustering using a standard clustering algorithm. In this paper we extend the practical work by evaluating SMF using the k-means algorithm to cluster heterogeneous data. We used a dataset of prostate cancer patients where objects are described by two basic data types, namely: structured and time-series data. We assess the results of clustering using external validation on multiple possible classifications of our patients. The result shows that the SMF approach can improved the clustering configuration when compared with clustering on an individual data type.

Keywords: Heterogeneous data · Big data · Distance measure · Intermediate data fusion · Clustering · Uncertainty

1 Introduction

Recently, the analysis of so-called "big data" has become very topical [1], but many questions remain. Big data presents many challenges because of its characteristics including large-volume, heterogeneity of data types and velocity [2]. In our research, we focus on data heterogeneity by dealing with objects described by a variety of data types, e.g. structured data, text, images, time-series, etc. In our preliminary work [3], we gave a definition of this problem and explained the context. For example, suppose that we have a dataset of animal species where

Funded by: King Abdulaziz University, Faculty of Computing and Information Technology, Jeddah, Saudi Arabia.

© Springer International Publishing Switzerland 2015
P. Perner (Ed.): MLDM 2015, LNAI 9166, pp. 251–265, 2015.
DOI: 10.1007/978-3-319-21024-7_17

each representative of a species can be characterised by descriptive properties represented as structured data, images, free text description, recording of sounds, and birth or mortality rates that are observed over a period of time and are represented as time-series. To cluster or classify such data, we believe that we should not relay on a limited interpretation that assesses our objects from a single perspectives but instead combine all the information conveyed by the different data elements that fully describe each species. Essentially, limited views concluded from each data type independently may put boundaries on revealing interesting patterns that might only be fully explored by a comprehensive understanding of the data. Thus, our aim is to seek and explore complex associations among our heterogeneous objects by aggregating multiple views using fusion techniques.

Clustering is a widely studied data mining problem with well-developed algorithms. The objective of clustering is to group objects into homogeneous clusters. Homogeneity is usually measured by dis/similarity among objects. Thus, applying appropriate measures results in more accurate configurations of the data [4]. Accordingly, several measures have been proposed and tested in the literature. These range from simple approaches that reflect the level of dis/similarity between two attribute values, to others such as those that categorize conceptual similarity [5]. However, most of the available, reliable and widely used measures can only be applied to one type of data. The extension of this work, which is to develop similarity measures for heterogeneous data, is received limited attention in the machine learning community.

We have proposed a simple fusing technique, SMF [3], using an intermediate fusion approach to calculate a single distance matrix for heterogeneous objects. In this paper, we use SMF with the clustering algorithm k-means to evaluate its potential on a dataset of cancer patients that are described by structured data and multiple time series. We evaluate whether the combined information embedded in the fused matrix is more helpful to the clustering that the distance measures for the individual data types.

The remainder of the paper is structured as follows: Sect. 2 summarises the related work in the literature. Section 3 introduces the practical work by presenting our definition of heterogeneous data. Our proposed SMF approach is outlined in Sect. 4. Section 5 discusses the experimental work and shows the results and then we conclude the paper in Sect. 6.

2   Related Work

There is no accepted single definition of data heterogeneity in the literature; instead different researchers have their own perspective as this kind of data is produced in several domains. For example, [6] describes datasets collected in scientific, engineering, medical and social applications as heterogeneous and complex. Data fusion approaches have become popular for heterogeneous data. Data fusion is referred to as the process of integration of multiple data and knowledge from the same real-world object into a consistent, accurate, and useful representation. Data fusion has been evolving for a long time in multi-sensor research [7,8] and other areas such as robotics and machine learning [9,10]. However, there has been little interaction with data mining research until recently [11].

Data fusion approaches can be classified into three categories [12]: early integration, late integration and intermediate integration according to the stage at which the fusion takes place. We propose to use a intermediate integration scheme in which the heterogeneous views of the object are integrated through the distance calculations. Though many studies (e.g. [13–15]) have examined data fusion in classification there is less work in the clustering domain. However, a similar approach the one we are proposing was investigated by Yu et al. [16] and found to be promising. We are considering here a wider number of data types, including for example multiple time series. Additionally, we intend to incorporate uncertainty into our future algorithms as this concept is central to data fusion [8].

3   Problem Definition

We define a heterogeneous dataset, H, as a set of N objects such that $H = \{O_1, O_2, ..., O_i, ..., O_N\}$. Each of this objects is defined by M elements $O_i = \{\mathcal{E}^1_{O_i}, ..., \mathcal{E}^j_{O_i}, ..., \mathcal{E}^M_{O_i}\}$, where $\mathcal{E}^j_{O_i}$ represents the data relating to $\mathcal{E}^j$ for O_i. Every i^{th} object in H, O_i, is defined by a unique *Object Identifier*, $O_i.ID$; we use the dot notation to access the identifier and other component parts of an object. Each full element, $\mathcal{E}^j$, for $1 \leq j \leq M$, may be considered as representing and storing a different data type. We begin by considering two data types, namely Structured Data (*SD*) and Time Series data (*TS*), yet the definition is extensible and allows for the introduction of further data types such as images, video, sounds, etc. We do not have any restriction on having different length *TSs* among different objects for the same time-series element. Moreover, our data permits incomplete objects where one or more of their elements are absent.

4   Similarity Matrix Fusion (SMF)

After defining a suitable data representation to describe the dataset and enable the application of suitable distance measures, we obtain the SMF by performing the following operations:

1. Calculate the DMs for each element independently such that $\forall$ O_i and O_j, we calculate distances for each zth element, $\mathcal{E}^z$, as follows:

$$DM^{\mathcal{E}^z}_{O_i,O_j} = dist(O_i.\mathcal{E}^z, O_j.\mathcal{E}^z),$$

where $dist$ represents an appropriate distance measure for the given data type. We chose in our case the Standardized Euclidean distance to measure distance between structured elements and Dynamic Time Warping using the original lengths of TSs to measure distance between TS elements. For incomplete objects where $\mathcal{E}^z$ is missing in O_i and/or O_j the value of $DM^{\mathcal{E}^z}_{O_i,O_j}$ becomes null. In addition, after computing all M DMs, we normalize the values to lie in the range $[0 - 1]$ prior to dealing with uncertainty.

254 A. Mojahed et al.

2. Fuse the DMs efficiently into one Fusion Matrix (FM) using a weighted average approach. Weights can estimated manually to allow some particular elements to have more influence on the calculations or to mask their contribution. If this is not necessary, all weights can be set to 1. The fusion weighted approach $\forall i,j \in \{1,2...N\}$ can be defined as:

$$FM_{O_i,O_j} = \frac{\sum_{z=1}^{M} w^z \times DM_{O_i,O_j}^{\mathcal{E}^z}}{\sum_{z=1}^{M} w^z}$$

where w^z is the weight given to the zth element.

3. Measure uncertainty associated with the FM due to the following situations: comparing objects with missing element(s) and/or disagreement between DMs. UFM reflects uncertainty arising from assessing incomplete objects and DFM expresses the degree of disagreement between DMs. They are calculated as follows:

$$UFM_{O_i,O_j} = \frac{1}{M}\sum_{z=1}^{M} \begin{cases} 0, & DM_{O_i,O_j}^{\mathcal{E}^z} \neq null \\ 1, & otherwise \end{cases}$$

$$DFM_{O_i,O_j} = \left(\frac{1}{M}\sum_{z=1}^{M} (DM_{O_i,O_j}^{\mathcal{E}^z} - \overline{DM_{O_i,O_j}})^2 \right)^{\frac{1}{2}},$$

where,

$$\overline{DM_{O_i,O_j}} = \frac{1}{M}\sum_{z=1}^{M} DM_{O_i,O_j}^{\mathcal{E}^z}$$

4. Apply cluster analysis using FM matrix, taking into account uncertainty computations. We choose the k-means [17] algorithm to apply the analysis and provide it with the fused distances. In this paper we only attempt to use uncertainty by removing uncertain objects/elements. However, we plan to devise a modified k-means algorithm that uses uncertainty more meaningfully.

5 Experiment

5.1 Dataset Description and Preparation

We begin with a real dataset containing data on a total of 1,904 patients diagnosed with prostate cancer during a period of seven years from 2004 to 2010 at the Norwich and Norfolk University Hospital (NNUH), UK. The dataset was created by Bettencourt-Silva et al. [18] by integrating data from nine different hospital information systems. The data represents the follow up of patients diagnosed with prostate cancer from the time of diagnosis until various points, so in some cases it covers more than 8 years after diagnosis. In addition, it also includes pre-diagnosis description data for a period that ranges between 1 day to, in some patients, more than 38 years before the day of discovering this medical condition. Each patient's data is represented by 26 attributes describing demographics (e.g. age, death indicator), disease states (e.g. Gleason score, tumor staging) and kinds of treatments that the patient has had (e.g. Hormone Therapy, radiology, surgery). In addition, 23 different blood test results (e.g. Vitamin D, MCV, Urea) are recorded over time and may be indicative of the health of the patient over the period of observation. Time is considered as 0 at the day of diagnosis and then reported as number of days from that day. Data for all blood test results before $time = 0$ is recorded with corresponding negative numbers which represent number of days before diagnosis. There are also outcome indicator attributes which record if the patient is dead or alive at the end of the study period, as well as the causes of death.

In the preparation stage we have changed the data representation and conducted some modifications. The 26 attributes are considered as forming the SD element. Note that from now on we are going to use the name of the elements as they are recorded in the dataset[1]. We have added categories in attributes with missing data in order to represent patients that fall in the same category but were left blank. For example, in the death indicator attribute, there are two values and both of them represent dead patients; 1 corresponds to patients that died because of other reasons and 2 corresponds to those that died because of prostate cancer. The values in this attribute were left blank for patients that survived to the end of the study, thus we recorded them as group 3. This type of logical replacement method was used in other cases when missing data are categorical and objects with blank values represent one possible group of patients. We end up with 100 % completed SD elements for all the 1,598 patients.

With regards to the blood test results, we put them in the form of 23 distinct TS elements. For every TS, data reported at $time < 0$ was discarded. Before starting distance calculations, we restricted ourselves to those patients for which we had data for 3 years follow up after diagnosis, in order to have comparable objects. Ideally, we would have looked at 5 years follow up, yet according to one of the dataset creators, this would have reduced the cohort considerably (we would end up with ≈600 patients only). All the data in the TSs corresponding to time >1095 (i.e. three years after diagnosis) was excluded and the death indicator attribute was modified to have the value of the third group (i.e. survived) even if he died after the 3 years to note 3 year survival. For 3 objects, we found patients that died just 1 to 3 days after the 3 years period and we includes them as died within 3 years. This is because there is usually some delay in reporting dates and that was an acceptable delay according to the dataset generator(s). So, on the one hand, we kept the original values of the death indicator attribute (1 or 2) for those that died before the 3 years period and 3 for those that were alive. On the other hand, the death indicator was changed to have the value of 4, which represents a forth group for some patients. This group represent patients with insufficient follow up (i.e. less than 3 years) but with an indicator of 3 (i.e. the patient was alive) at the last check. They represent a group of patients with

---

[1] Contact the authors for more information about the data dictionary.

an uncertain outcome. After this, z-normalization was conducted on all values remaining in the TSs before calculating the distance matrices. This was done for each TS separately, i.e. each TS then has values that have been normalised for that TS to achieve mean equal to 0 and unit variance. Also, we cleaned the data by discarding blood tests where there were mostly missing values for all patients, and removed patients which appeared to hold invalid values for some attributes, etc. At the end of this stage, we still had 1,598 patient objects with SD for 26 attributes and 22 distinct TSs.

5.2   Experimental Set up

After constructing 23 pairwise DMs where each DM reports distances between objects in relation to an individual element, at the fusion phase and by means of using these 23 DMs, we apply the next stage in our SMF approach and compute the primary fusion matrix, FM. Initially we set equal weights for all the 23 elements (i.e. $w^i = 1$ for the SD element and for every TS element) and from here on we refer to this fusion matrix using the notation FM-1. In addition, we have calculated the two primary uncertainty matrices UFM-1 and DFM-1, which report the level of uncertainty related to the fused distances in FM-1. We can also created more than one fusion matrix by alterning the weighting scheme of the components as presented in Sect. 5.3.

Next, we apply k-means clustering on the individual distance matrices and the results are presented in Sect. 5.3. We also clustered using several Fusion Matrices, FMs, independently. The performance of the SMF approach is then evaluated by comparing the results of applying the same clustering algorithm to each element separately and to the FMs. Our hypothesis is that the combined information contained in the FMs produces better clustering than the individual elements. We evaluate clustering in relation to various possible groupings of the patients in the dataset. Each experiment consists of the following steps: 1. Choose a grouping system; the possible groupings are described below; 2. Set the number of clusters, k, to be equal to the number of categories existing in the grouping system; 3. Produce a clustering solution using k-means with one of the 23 DMs; 4. Use external validation methods to evaluate the solution taking advantage of the labels; 5. Validate also using an internal method; and, 6. **Repeat** step 3 to 5 **for** each DM and for the different FMs.

The natural grouping systems for patients were suggested by the data donors. They are as follows:

– **NICE system for risk stratification:** There are a number of approaches used to classify risk groups for prostate cancer patients. A widely used system is a composite risk score. It uses three data variables: Prostate-Specific Antigen (PSA), Gleason Grade, and Clinical Stage (Tumour Stage). This stratification reflects the clinicians' belief that patients with the same risk have a similar clinical outcome and may follow a similar trajectory through the disease pathway. The National Institute for Health and Care Excellence (NICE) [19] provides guidance for the risk stratification of men with localised prostate cancer. Our dataset requires some adaptation to apply this guidance, and advice on this was obtained from the data creators. PSA is recorded as a TS. Namely, what we have done is to consider the value at diagnosis, and if there is nothing recorded at $time = 0$, then the closest value before any type of treatments. Gleason score is divided into two values; primary and secondary, thus we use the sum of both scores. The clinical stage is reported using numbers. We considered the following: clinical stage <2 as low, clinical stage $= 2$ as medium and clinical stage >2 as high risk.

– **Gleason score risk classification system:** Another well-known risk classification can be obtained by using Gleason grade alone to classify patients diagnosed with prostate cancer. Gleason grade shows the level of differentiation of the cancer cells under the microscope. High differentiation is associated with worst prognosis which indicates more aggressive tumors [20]. Gleason grade is computed as a sum of two or sometimes three scores: primary, secondary and tertiary (if applicable). Primary is the most commonly seen level of differentiation under the microscope, secondary is second most common and so on. The level of differentiation for these three scores is given from 1 to 5 and then summed together. The totals of Gleason scores in our dataset are all >5 as all the cases are definite cancer patients. We have defined two ways of groupings patients according to their Gleason score: Gleason-score-1 and Gleason-score-2. The first way of grouping, Gleason-score-1, has 3 groups: low, medium and high risk with 70, 1142 and 386 patients, respectively. Gleason-score-2, classifies patients into 4 groups: low, medium-1, medium-2 and high risk where they hold 70, 594, 548 and 386 patients respectively. The difference between the two groupings is in the medium risk group. In Gleason-score-2 the medium group is divided into two subgroups depending on the exact values of the primary and secondary scores and not only their sum (e.g. $3 + 4$ may be considered different to $4 + 3$).

– **Mortality grouping:** This labeling procedure classifies patients according to the outcome at the end of the study period, rather than looking at the potential risk of patients at diagnosis. For this grouping we used death indicators after conducting some changes on the values of the corresponding attribute. The four mortality categories and the number of patients in each category after we have made our changes on the death indicator as described in Sect. 5.1 are: died due to prostate cancer (185), died because of other reasons (28), survived for 3 years (902) and insufficient follow up to establish 3 year outcome, i.e. undetermined (510).

From now on we refer to the previous grouping systems as: NICE, GS-1, GS-2 and MG respectively. For each experiment we selected one of the four systems, then we applied k-means algorithm on individual DM and on the FM(s). We divided our experimental work into three main sets of experiments:

1. Apply cluster analysis by using individual DMs. We use here DMs that compute distances between SD elements and also between TSs elements. We then report external validation measures for the clustering generated by the SD alone, and the best and worse TS, along with the average performance of all TSs.

2. Apply cluster analysis to cluster fused distance calculations using the initial FM-1 and weighted versions of the fused matrix and compare it to the previous set of experiments. We have examined different weightings in this experiments. We report on a weighting scheme that uses clustering performance of each individual element to specify the elements' weights. In some senses, this would be the best performance we could expect, if we knew apriori which elements are more important, i.e. it represents some optimal weighting configuration for comparison. It is reported as FM-2. However, it is important to note that it will generally not be possible to establish the degree of relevance of each element in advance of clustering and hence weighted fusion may not be an option.

3. A repetition of the previous experiment using only certain distances (i.e. we filter FM-1 using UFM-1 and DFM-1). We use a filtering approach to remove objects that are related to uncertain distances and then apply k-means to the remaining objects. We set out thresholds for UFM-1 and DFM-1 in order to filter out objects. We then eliminate objects that exceed these thresholds when they are compared to half or more of the other objects. For example, we may remove an object if it holds uncertain fused distances between the object itself and 800 or more of the other 1598 objects. We have used three filters: in filter 1 we used UFM-1 and DFM-1 expressions together, whereas in filters 2 and 3 we used UFM-1 and DFM-1 individually. As a second step we can use the clustering results of the certain objects to cluster the uncertain ones. In other word, we use the centroids that were generated with the filtering approach to assign the residual objects that were removed from the analysis to the produced clusters.

In all these experiments, we aim to examine if fused DMs are more informative to the k-means algorithm than using individual DMs. To evaluate the results we calculate 3 different external validation tests: Jaccard coefficients [21], rand statistic [22] and Dice's index [23]. We choose these three methods as they are defined differently. With regards to the choice of an internal validation method, we use Dunn index [24] as it measures both compactness and clusters separation. Large values indicate the presence of compact and well-separated clusters. Finally, in Sect. 5.4 we demonstrate the significance of SMF performance using statistical testing.

5.3   Clustering Results

The number of clusters, k, was determined depending on the selected grouping system, i.e. we set $k=3$ for NICE and GS-1 experiments and $k=4$ in GS-2 and MG experiments. The presentation of clustering results are divided below into the three main sets of experiments that we have discussed earlier:

1. Applying k-means using the individual DMs for each element: Fig. 1 shows the performance of clustering evaluated according to the four possible groupings, NICE, GS-1, GS-2 and MC. The figure includes a summary of the statistics

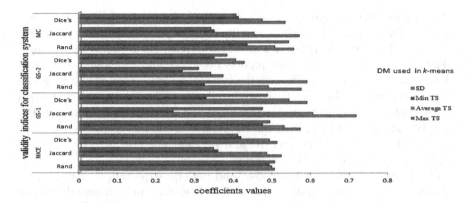

Fig. 1. Summary of the performance of k-means clustering obtained using the individual DMs for the prostate cancer dataset

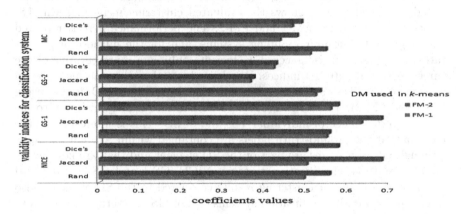

Fig. 2. Summary of the performance of k-means clustering obtained on individual DMs and fusion matrices for prostate cancer dataset

calculated using the 3 external validation indices: Jaccard, rand and Dice's coefficients. The indices evaluate the clustering configurations produced by k-means using SD and TSs DMs. In the figure we report SD and the best TS performer (Max TS), the worst(Min TS) and the average of all 22 TSs (average TS).

It can be observed from the figure that the SD element alone performs better than the worst TS in both GS-1 and GS-2 with regards to all coefficients. This may be expected since the SD element contains the information that defines the GS-1 and GS-2 groupings. This is not the case in the other classification systems according to Jaccard and Dice's methods, which report very similar results. However, SD generally performs similarly or worse than Max TS hence the TSs appear to have good information for the clustering process. In addition, Rand index, behaves different to the other two indices.

The TS on average appear to be better than the SD for both the Jaccard and the Dice's methods, whereas the Rand Index puts the performance of SD slighltly ahead of the average for all TSs in NICE, GS-2 and MC groupings.

2. The second set of experiments reports on FM-1 and the weighted version FM-2. In FM-2, for each classification system, we have selected and weighted higher the top 5 elements that produced the highest averaged evaluation coefficients. The elements that we have selected for NICE system are B1, B3, B8, B20 and B27 and for GS-1, B2, B3, B20, B22 and B27 elements. For GS-2 classification, we use P, B2, B4, B7 and B9 elements. For MC groupings, we chose P, B2, B3, B9 and B20 elements.

By looking at Fig. 2, it is obvious, and perhaps expected, that FM-2 produces better clustering results than FM-1. Therefore, prior knowledge about the most important contributors could be exploited positively. However, we need to remember that such information may not be available.

Figure 3 compares clustering performance when using individual DMs, FM-1 and FM-2. The figure shows the evaluated clustering using SD, Min TS, Average TS, Max TS, FM-1 and FM-2. The performance here is reported with regards to the four grouping systems, and for the purpose of presentation simplification, by Jaccard coefficients only as a representative of the three calculated external indices. We chose Jaccard as there seems to be an agreement with Dice's. The results show that by combining DMs (i.e. using FM-1 and FM-2), we obtained stable performance equivalent or better to the average performance of individual TSs and of SD. This is interesting in the context that the ground-truth labels are contained within SD. For example, although Gleason scores are part of the SD element and they are key factors of the first three classification systems, NICE, GS-1 and GS-2, FM-1 and FM-2 produced better clustering results compared to the SD element. Also interestingly, for all classification systems except GS-2, Average TS seems to be better than the SD element alone. Hence the information contained in the blood test time series aids in defining our groupings more accurately than the information contained in SD. The SMF approach combines data from the different type of elements to give an accurate configurations and since it may not be possible a priori to know which TS is the best performer it is reassuring that the combination approach is within some acceptable deviation from the best TSs.

3. The third set of experiments is concerned with applying k-means to cluster certain data only. As described in Sect. 5.2, we filter all the uncertain data to perform the clustering and then assign the uncertain data as a separate task. In order to screen the objects, we set three filters. In filter 1, we remove records and columns from FM-1 that correspond to 145 patients, as those patients have UFM-1 values ≥ 0.4 and DFM-1 ≥ 0.2. In filter 2, we set a threshold only for UFM-1 to remove objects that miss $\geq 30\%$ of the calculations in FM-1. In filter 3 we used DFM-1 only by setting a threshold of 0.1. As a result, by using filter 2 we removed 405 patients while by using filter 3 we excluded 383 patients. Table 1 shows a summary of the experiments compared to the results of clustering using all information (from experiment 2). We report

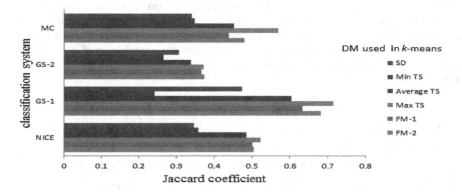

Fig. 3. Summary of the performance of k-means clustering obtained on both elements' DMs and fusion matrices for prostate cancer dataset

Jaccard coefficient as a representative for the clustering validity for all the approaches and the calculations presented evaluate the results of applying filter 1 and clustering only the certain objects (column 3) or all the objects (column 4).

The results indicate that the k-means algorithm applied to the filtered data does not produce better clustering performance. The accuracy of our model has decreased by experimenting with the filtering approach compared to the results previously reported for using the full FM-1. A similar deterioration occurs when we used the centroids of the generated clusters to assign the residual objects: 145, 405 and 383 patients removed by the 3 filters respectively. Filtering is therefore not a suitable approach. Note that Rand and Dice's indices concurred with the same conclusion. Moreover, the results of trying the other two filters came up with the same interpretation. We can concluded here that uncertain fused calculations that are related to incomplete objects and/or objects with high degree of disagreement between their elements in the similarity assessments seem to have information that aids the clustering process. Thus, we can say that they need to be included in the analysis, however, we may use the uncertainty information in a different manner. Our plan is to produce a modified version of k-means algorithm in a way that will use the uncertainty information to inform the clustering process.

Table 1. The Performance of clustering for the prostate cancer dataset filtering uncertain data

Classification system	FM-1	Filtering: certain objects only	Filtering: all objects
NICE	50.13 %	47.35 %	45.81 %
GS-1	65.83 %	51.62 %	50.19 %
GS-2	36.48 %	31.45 %	30.35 %
MC	70.22 %	39.90 %	36.67 %

Table 2. Statistical analysis of SMF performance on the prostate cancer dataset. The table reports statistics for Jaccard coefficient of k-means clustering obtained with regards to four possible groupings of the prostate cancer dataset. The first set of columns provides the statistical test that calculates the significance of SMF performance compared to using SD and the second set of columns compares to the average performance of all TSs

Groupings system	SD	FM-1	FM-2	TSs	FM-1	FM-2
NICE						
Jaccard	34.61	50.13	50.44	48.40	50.13	50.44
z score	–	±8.878	±9.052	–	±0.976	±1.153
p value	–	$< 1.0E - 5$	$< 1.0E - 5$	–	0.164532	0.124455
GS-1						
Jaccard	47.31	63.27	68.15	60.37	63.27	68.15
z score	–	±9.072	±11.924	–	±1.688	±4.59
p value	–	$< 1.0E - 5$	$< 1.0E - 5$	–	0.045706	$< 1.0E - 5$
GS-2						
Jaccard	30.73	36.48	37.36	33.87	36.48	37.36
z score	–	±3.445	±3.957	–	±1.548	±2.061
p value	–	0.000286	$< 3.8E - 5$	–	0.060811	0.019652
MC						
Jaccard	33.92	43.68	47.81	45.11	43.68	47.81
z score	–	±5.663	±7.988	–	±0.812	±1.532
p value	–	$< 1.0E - 5$	$< 1.0E - 5$	–	0.208396	0.062761

5.4   Statistical Testing

We applied a z-test to establish if the differences in performance between the various versions were statistically significant. We compare the difference in performance of data fusion and the SD element alone and also to the average of the TSs performances. Table 2 reports for each experimented classification system the Jaccard coefficient as a representative of the 3 external validation measures, as well as statistics for the test of significance of difference between the performance of fusion matrices compared to the individual DMs.

All p values that compare the performance of SD and FMs are <0.05 which indicates significant difference between accuracy percentages. With regards to p values that compare the performance of Average TSs and FMs, the statistics report them as not significant except for FM-2 in the GS-1 and in GS-2 groups. In general, these statistics show that the SMF approach produces better result than the SD element alone and comparable results to the TSs average.

Furthermore, we have evaluated the difference between FM-1 and FM-2 for all the grouping systems and the statistical tests concluded that the difference in NICE and GS-2 is is not significant with p values of 0.429754 and 0.303976 respectively. In the other two classification systems, GS-2 and MC, the difference between performances of FM-1 and FM-2 was significant with 0.001825

Table 3. The Dunn index values from the results of clustering the prostate cancer dataset: the statistics are reported for SD, FMs and the top TS for each grouping. We use − to represent the case when the DM is not the top informative TS.

DM	NICE	GS-1	GS-2	MC
SD	0.029452	0.029452	0.023402	0.030645
B26	0.055524	0.055524	−	−
B21	−	−	0.043134	0.043134
FM-1	0.017076	0.017076	0.020862	0.017076
FM-2	0.016762	0.014417	0.012754	0.00332

and 0.009565 p values. It is also interesting to note which TSs are particularly good performers, i.e. those that have the highest values according to the previous set of experiments as they are blood tests that may either be indicative of risk or mortality. They are Alkaline Phosphatase (B1, 50.59 %), Creatinine (B2, 62.16 %), Creatinine (B2, 42.98 %) and Haemoglobin (B20, 45.84 %) for NICE, GS-1, GS-2 and MC, respectively, where the figures in the brackets are the averaged external evaluation coefficients of the blood tests.

We also computed the Dunn internal index for every single DM and FMs using the results of applying clustering in relation to the four groupings. Table 3 shows the coefficients that have evaluated SD element, FM-1, FM-2 and the best individual DM for each classification system.

In general, Dunn index considers B21, B26, SD, FM-1 and FM-2 as the top DMs that have potential to produce good quality clustering results. In ascending order the top 5 DMs according to Dunn evaluations for NICE system are: B26, B21, SD, FM-1 and FM-2; for GS-1: B26, B21, SD, FM-1 and FM-2; for GS-2: B21, SD, B8, FM-1 and FM-2; and for MC: B21, B8, SD, FM-1 and FM-2. For all the experiments that we have conducted, the Dunn index ranks FM-1 as one of the top four DMs and FM-2 as one of the top five DMs in terms of quality of the clustering obtained. Accordingly, Dunn evaluates FM-1 as more informative than FM-2. Moreover, it does not rate DMs that were considered as the best performers according to external validation indices as one of the top informative DM with potential to produce good clustering configuration. More interestingly, external and internal validation metrics do not agree on the same conclusion, except for their judgment of FM-1 and FM-2 for all groupings. Thus, the fused matrices seem to have reasonable performance according to both internal and external validation. Hence the SFM approach aids the clustering towards producing good clusters according to both internal and external criteria.

6   Conclusion and Future Work

The task of managing and mining big data is considered to be a very challenging problem. We tackle the program of heterogeneous data. Our proposed approach, SMF, follows an intermediate data fusion technique where the integration process is carried out at the distance level, before a clustering algorithm is applied. Explicitly, the fusion takes place at the level of calculating similarity by combining multiple distance matrices in a weighted average manner. Then partitions can be obtained by any standard clustering algorithms. We examined SMF on a the real prostate cancer dataset. However, we are aware that we need to find other heterogeneous datasets to further validate the approach.

In the prostate cancer dataset, the main finding is that by using intermediary fusion we gain significant advantage on clustering performance in comparison to using only structured data, even though the cluster groupings we are trying to discover could be defined by using SD data alone. Also, we are able to identify which elements behave similarly with respect to distance between objects. We can also identify the elements that may produce the best clustering results. This knowledge may have some clinical relevance. For example, knowing that the progression of certain blood results over time enable us to group patients according to risk or mortality more accurately may be of interest to clinicians.

The flexibility of our approach gives it the ability to work with objects that are defined by any other combinations of data types. In bio-informatics, for example, similarity of proteins can be measured in several ways: comparing the sequences directly or comparing gene profiles.

There are many issues to be addressed in our approach in future research. For example, the identification of the optimal number of clusters for applications when no external knowledge of the clustering is available and also the development of appropriate weighting schemes. In the future we are planning to use uncertainty information differently by developing a modified implementation of the K-means clustering algorithm. Another plan is to study an alternative solution by examining late fusion approaches. For example, ensemble clustering which combines multiple partitions into a consolidate partition by consensus function.

Acknowledgments. We acknowledge support from grant number ES/L011859/1, from The Business and Local Government Data Research Centre, funded by the Economic and Social Research Council to provide economic, scientific and social researchers and business analysts with secure data services.

References

1. Wu, X., Zhu, X., Wu, G.Q., Ding, W.: Data mining with big data. IEEE Trans. Knowl. Data Eng. **26**(1), 97–107 (2014)
2. Laney, D.: 3D data management: controlling data volume, velocity, and variety. Technical report, META Group. February 2001
3. Mojahed, A., De La Iglesia, B.: A fusion approach to computing distance for heterogeneous data. In: Proceedings of the Sixth International Conference on Knowledge Discover and Information Retrieval (KDIR 2014), pp. 269–276. SCITEPRESS, Rome, Italy (2014)
4. Steinbach, M., Ertz, L., Kumar, V.: The challenges of clustering high dimensional data. In: Wille, L. (ed.) New Directions in Statistical Physics, pp. 273–309. Springer, Heidelberg (2004)
5. Johnson, R.A., Wichern, D.W. (eds.): Applied Multivariate Statistical Analysis. Prentice-Hall Inc, NJ (1988)

6. Skillicorn, D.B.: Understanding Complex Datasets: Data Mining with Matrix Decompositions. Chapman and Hall/CRC, Taylor and Francis Group, Boca Raton (2007)
7. Hall, D., Llinas, J.: An introduction to multisensor data fusion. Proc. IEEE **85**(1), 6–23 (1997)
8. Khaleghi, B., Khamis, A., Karray, F.O., Razavi, S.N.: Multisensor data fusion: a review of the state-of-the-art. Inf. Fusion **14**(1), 28–44 (2013)
9. Abidi, M.A., Gonzalez, R.C.: Data Fusion in Robotics and Machine Intelligence. Academic Press Professional Inc, San Diego (1992)
10. Faouzi, N.E.E., Leung, H., Kurian, A.: Data fusion in intelligent transportation systems: progress and challenges a survey. Inf. Fusion **12**(1), 4–10 (2011). Special Issue on Intelligent Transportation Systems
11. Dasarathy, B.V.: Information fusion, data mining, and knowledge discovery. Inf. Fusion **4**(1), 1 (2003)
12. Maragos, P., Gros, P., Katsamanis, A., Papandreou, G.: Cross-modal integration for performance improving in multimedia: a review. In: Maragos, P., Potamianos, A., Gros, P. (eds.) Multimodal Processing and Interaction. Multimedia Systems and Applications, vol. 33, pp. 1–46. Springer, New York (2008)
13. Lanckriet, G.R.G., Cristianini, N., Bartlett, P., Ghaoui, L.E., Jordan, M.I.: Learning the kernel matrix with semidefinite programming. J. Mach. Learn. Res. **5**, 27–72 (2004)
14. Bie, T.D., Tranchevent, L.C., van Oeffelen, L.M.M., Moreau, Y.: Kernel-based data fusion for gene prioritization. ISMB/ECCB (Suppl. Bioinform.) **23**(13), 125–132 (2007)
15. Shi, Y., Falck, T., Daemen, A., Tranchevent, L.C., Suykens, J.A.K., De Moor, B., Moreau, Y.: L2-norm multiple kernel learning and its application to biomedical data fusion. BMC Bioinform. **11**, 309–332 (2010)
16. Yu, S., Moor, B., Moreau, Y.: Clustering by heterogeneous data fusion: framework and applications. In: NIPS workshop (2009)
17. MacQueen, J.: Some methods for classification and analysis of multivariate observations. In: Proceedings of the Fifth Berkeley Symposium on Mathematical Statistics and Probability, vol. 1, pp. 281–297: Statistics, Berkeley. University of California Press, California (1967)
18. Bettencourt-Silva, J.H., Iglesia, B.D.L., Donell, S., Rayward-Smith, V.: On creating a patient-centric database from multiple hospital information systems in a national health service secondary care setting. Methods of Information in Medicine, 6730–6737 (2012)
19. NICE: Prostate cancer: diagnosis and treatment. NICE clinical guideline, vol. 175, pp. 1–48 (2014)
20. Chan, T.Y., Partin, A.W., Walsh, P.C., Epstein, J.I.: Prognostic significance of gleason score 3+4 versus gleason score 4+3 tumor at radical prostatectomy. Urology **56**(5), 823–827 (2000)
21. Jaccard, S.: Nouvelles researches sur la distribution florale. Bull. Soc. Vaud. Sci. Nat **44**, 223–270 (1908)
22. Rand, W.M.: Objective criteria foe the evaluation of clustering methods. J. Am. Stat. Assoc. **66**(336), 846–850 (1958)
23. Dice, L.R.: Measures of the amount of ecologic association between species. Ecology **26**, 297–302 (1945)
24. Dunn, J.C.: A fuzzy relative of the isodata process and its use in detecting compact well-separated clusters. J. Cybern. **3**(1), 32–57 (1973)

On Bicluster Aggregation and its Benefits for Enumerative Solutions

Saullo Oliveira$^{(\boxtimes)}$, Rosana Veroneze, and Fernando J. Von Zuben

School of Electrical and Computer Engineering,
Unicamp, Campinas, São Paulo, Brazil
{shgo,veroneze,vonzuben}@dca.fee.unicamp.br
http://www.fee.unicamp.br

Abstract. Biclustering involves the simultaneous clustering of objects and their attributes, thus defining local two-way clustering models. Recently, efficient algorithms were conceived to enumerate all biclusters in real-valued datasets. In this case, the solution composes a complete set of maximal and non-redundant biclusters. However, the ability to enumerate biclusters revealed a challenging scenario: in noisy datasets, each true bicluster may become highly fragmented and with a high degree of overlapping. It prevents a direct analysis of the obtained results. Aiming at reverting the fragmentation, we propose here two approaches for properly aggregating the whole set of enumerated biclusters: one based on single linkage and the other directly exploring the rate of overlapping. Both proposals were compared with each other and with the actual state-of-the-art in several experiments, and they not only significantly reduced the number of biclusters but also consistently increased the quality of the solution.

Keywords: Biclustering · Bicluster enumeration · Bicluster aggregation · Outlier removal · Metrics for biclusters

1 Introduction

Biclustering techniques aim to simultaneously cluster objects and attributes of a dataset. Each bicluster is represented as a tuple containing a subset of the rows, and a subset of the columns, as long as they exhibit some kind of coherence pattern. There are several kinds of coherence which can be found in a bicluster, and they directly interfere on the mechanism of bicluster identification. As finding all biclusters in a dataset is an NP-hard problem, several heuristics were proposed, such as CC [1] and FLOC [2]. Such heuristics may miss important biclusters, and may also return non-maximal biclusters (biclusters that can be further augmented).

In the case of binary datasets, there are a plenty of algorithms for enumerating all maximal biclusters. Some examples are Makino & Uno [3], LCM [4] and In-Close2 [5]. The enumeration of all maximal biclusters in an integer or

© Springer International Publishing Switzerland 2015
P. Perner (Ed.): MLDM 2015, LNAI 9166, pp. 266–280, 2015.
DOI: 10.1007/978-3-319-21024-7_18

real-valued dataset is a much more challenging scenario, but we already have some proposals, such as RIn-Close [6] and RAP [7].

The drawback of enumerative algorithms, particularly in the context of noisy datasets, is the existence of a large number of biclusters, due to fragmentation of a much smaller number of true biclusters. This is exemplified in one of our experiments, where we take artificial datasets, gradually increment the variance of a Gaussian noise, and get the enumerative result. As shown in Fig. 2, with enough noise, the enumerative results exhibit an strong increase on the quantity of biclusters. This fragmentation leads to a challenging scenario for the analysis of the results, which can become impractical even in small datasets. In fact, the noise is responsible for fragmenting each true bicluster into many with high overlapping, so that the aggregation of these biclusters is recommended [8,9].

We propose a way of aggregating biclusters from a biclustering result that shows a high overlapping among its components, as it is the case when enumerating biclusters in noisy datasets. For this reason, in this paper we will focus on enumerative results, but our proposal can be applied to the result of any algorithm that returns biclusters with high overlapping among them. The formulation is based on the fact that the high overlapping among biclusters may indicate that they are fragments of a true bicluster that should be reconstructed. We propose two different techniques to perform the aggregation, followed by a step that removes elements that should not be part of a bicluster. We performed experiments with three artificial datasets posing different challenges, and two real datasets from distinct backgrounds. We compared our proposals with a bicluster ensemble algorithm, and the merging/deleting steps of MicroCluster [9]. The experimental results show that the aggregation not only severely reduces the quantity of biclusters, but also tends to increase the quality of the solution.

The paper is organized as follows. In Sect. 2, we give the main definitions and discuss the related works in the literature. Section 3 outlines our proposals. The metrics used to evaluate our proposals will be presented in Sect. 4. In Sect. 5, we present the experimental procedure and the obtained results of the experiments. Concluding remarks and future work are outlined in Sect. 6.

2 Definitions and Related Work

Consider a dataset $\mathbf{A} \in \mathbb{R}^{n \times m}$, with rows $X = \{x_1, x_2, \ldots, x_n\}$ and columns $Y = \{y_1, y_2, \ldots, y_m\}$. We define a bicluster $B = (B^r, B^c)$, where $B^r \subseteq X$ and $B^c \subseteq Y$, such that the elements in the bicluster show a coherence pattern. A bicluster solution is a set of biclusters represented by $\bar{B} = \{B_i\}_{i=1}^{q}$, containing q biclusters. A bicluster is maximal if and only if we can not include any other object/attribute without violating the coherence threshold. If a solution contains non-maximal biclusters, the result is redundant because there will be biclusters which are part of larger ones.

Madeira & Oliveira [10] categorized the types of biclusters according to their similarity patterns. They also categorized the biclusters structure in a dataset based on their disposition and level of overlapping. We highlight that biclusters with constant values, constant values on rows, or constant values on columns are special cases of biclusters with coherent values, and we will focus our attention on the latter, due to its generality. For a comprehensive survey of biclustering algorithms, the reader may refer to [10,11].

The overlapping between two biclusters B and C is an important concept in this work, and is defined as:

$$ov(B,C) = \frac{|B^r \cap C^r \times B^c \cap C^c|}{min(|B^r \times B^c|, |C^r \times C^c|)}. \tag{1}$$

Now we shall proceed to the aggregation proposals in the literature. It is important to highlight that, when aggregating two maximal biclusters, the coherence threshold will be violated. Otherwise, the biclusters would not be maximal.

2.1   MicroCluster Aggregation

MicroCluster [9] is an enumerative proposal that has two additional steps after the enumeration. These steps have the task of deleting or merging biclusters which are not covering an area much different from other biclusters. The first is the deleting step. If we find a bicluster such that the ratio of its area that is not covered by any other bicluster, by its total area, is less than a threshold η, it can be removed. The second step is the merging one. Let us consider two biclusters and generate a third one with the union of rows and columns of the previous two. If the ratio of the area of the third bicluster that is not covered by any of the previous two, by its total area, is less than a threshold γ, we can aggregate the two biclusters into this third one. In this method of aggregation, non-maximal biclusters will be removed in the deleting step, thus not interfering in the final result. For more details, please refer to Zhao & Zaki [9].

2.2   Aggregation Using Triclustering

Triclustering was proposed by Haczar & Nadif [13] as a biclustering ensemble algorithm. First, they transform each bicluster into a binary matrix. After that, they propose a triclustering algorithm to find the k most relevant biclusters. As they were able to improve the biological relevance of biclustering for microarray data [14], we will use this method as a contender in this paper. One major point in ensemble is that we want to combine the results reinforcing the biclusters that seem to be important for several components, and discarding the ones that may come from noise. Due to the way the triclustering algorithm handles the optimization step, non-maximal biclusters can interfere in the final results.

Bicluster aggregation is slightly different from bicluster ensemble. While on ensemble tasks we discard biclusters that seem unimportant and combine the ones that contribute the most for the solution, in bicluster aggregation we never discard any bicluster. Given this characteristic, the bicluster ensemble solution is expected to show a high *Precision* with an impacted *Recall* (see Sect. 4), as it eliminates biclusters.

2.3   Other Aggregation Methods

Gao & Akoglu [12] used the principle of Minimum Description Length to propose CoClusLSH, an algorithm that returns a hierarchical set of biclusters. The hierarchical part can be seen as an aggregation step. This step is done based on the LSH technique as a hash function. Candidates hashed to the same bucket are then aggregated until no merging improves the final solution. Their work is focused in finding biclusters in a checkerboard structure, that does not allow overlapping, thus being not suitable for the kind of problem we are dealing with.

Liu *et al.* [8] proposed OPC-Tree, a deterministic algorithm to mine Order Preserving Clusters (OP-Clusters), a general case of Order Preserving Sub Matrices (OPSM) type of biclusters. They also have an additional step for creating a hierarchical aggregation of the OP-Clusters. The Kendall coefficient is used to determine which clusters should be merged and in which order the objects should participate in the resultant OP-Cluster. The highest the Rank Correlation using the Kendall coefficient, the highest the similarity between two OP-Clusters. The merging is allowed according to a threshold that is reduced in a level-wise way. OPC-Tree considers the order of the rows in the bicluster. In this work, we are dealing with biclusters of coherent values. In this case, a perfect coherent values bicluster keeps the order of its rows and the hierarchical step of OPC-Tree would be able to be used in this case as well. But we are considering noisy datasets, in which this assumption probably will not hold, thus the hierarchical step of OPC-Tree is not suitable for the problem we are dealing with.

3   New Proposals for Aggregation

3.1   Aggregation with Single Linkage

Our first proposal receives as input a biclustering solution $\bar{B}$, from enumeration or from a result presenting high overlapping among its components. With this solution, we transform each bicluster into a binary vector representation as follows: Given the dimensions of the dataset $\mathbf{A} \in \mathbb{R}^{n \times m}$, each bicluster will be a binary vector $\mathbf{x}$ of length $n + m$. For a bicluster B transformed into the binary vector $\mathbf{x}$, the first n positions represent the rows of the dataset $\mathbf{A}$ and if the bicluster contains the ith row, $\mathbf{x}_i = 1$, otherwise $\mathbf{x}_i = 0$. The last m positions represent the columns of the dataset $\mathbf{A}$ and if the bicluster contains the ith column, $\mathbf{x}_{n+i} = 1$, otherwise $\mathbf{x}_{n+i} = 0$. After this transformation, we use the Hamming distance to apply the single linkage clustering on the existing biclusters. Notice that the Hamming distance on this transformation will just count how many rows and columns are different among the two biclusters. In this case, a non-maximal bicluster may be distant from the bicluster that covers its maximal area, thus impacting the quality of the results of this method of aggregation. In this case, it is necessary that this proposal receives a biclustering solution $\bar{B}$ containing only maximal biclusters.

After choosing a cut on the dendrogram, we aggregate all biclusters that belong to a junction using the function *aggreg*, defined as:

$$aggreg(B, C) = (B^r \cup C^r, B^c \cup C^c), \tag{2}$$

that is simply the union of rows/columns of the biclusters. It is important to note that the *aggreg* function is associative, since it is based on the union operation. Moreover, we want to highlight that the direct union of rows/columns may include elements that should not be part of a bicluster. In Sect. 3.3 we will present a way to remove rows/columns that may be interpreted as outliers.

3.2   Aggregation by Overlapping

It seems intuitive to aggregate the biclusters with an overlapping rate above a defined threshold. This proposal is based on the aggregation by pairs: while having two biclusters with an overlapping rate higher than a pre-determined threshold th, we remove them from the set of biclusters, and include the result of the function *aggreg*, defined on Eq. 2, taking these two biclusters as the arguments.

Let B, C, D, and E be biclusters. Note that for $D = aggreg(B, C)$, $ov(D, E) \geq ov(B, E)$ and $ov(D, E) \geq ov(C, E)$. So, for all biclusters E where $ov(B, E) \geq th$ or $ov(C, E) \geq th$, we have $ov(D, E) \geq th$. For this reason, the order of the aggregation does not interfere on the final result. It is also important to note that the new bicluster D can have $ov(D, E) \geq th$, for some bicluster E where $ov(B, E) < th$ and $ov(C, E) < th$. In this aggregation proposal, maximal biclusters will properly merge with non-maximal biclusters.

3.3   Outlier Removal

After aggregating the results, we need to process each final bicluster to look for objects and/or attributes that may be interpreted as outliers. In this work, this step will always be executed after the aggregation using any of our two proposals.

Let $B = (B^r, B^c)$ be an aggregated bicluster, with $|B^r| = o, |B^c| = p$. We define a participation matrix $\mathbf{P} \in \mathbb{Z}^{o \times p}$, where each element p_{ij} indicates the quantity of biclusters in which this element takes part in B. For example, if an element is part of 15 biclusters that compose B, then its value on the $\mathbf{P}$ matrix will be 15.

So, we will explain the process of outlier removal with the help of Fig. 1. We have two steps of outlier removal: one for the objects, the other for the attributes. To remove possibly outlier objects, we take the mean and the standard deviation of all columns on the participation matrix $\mathbf{P}$. The left side of Fig. 1 illustrates this step. After that, we check the values of each element of the columns. If the value is less than the mean minus one standard deviation, then we check this element as a potential outlier. In Fig. 1, we can see that the entire first row was checked as potential outlier because $1 < 7.75 - 4$. If we mark the entire row as a potential outlier, it is removed from the bicluster. In our example, that is the case.

We execute the same process for the columns, calculating the mean, standard deviation and checking for potential outliers on the rows. We remove the column if it is entirely marked as a potential outlier.

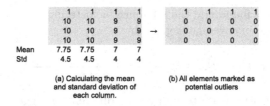

(a) Calculating the mean and standard deviation of each column.

(b) All elements marked as potential outliers

Fig. 1. Example of outlier removal.

4   Metrics for Biclustering

In this paper we will use only external metrics, except for the Gene Ontology Enrichment Analysis (GOEA). External metrics compare a given solution with a reference one. For an extensive comparison of external metrics for biclustering solutions, the reader may refer to [15].

The Gene Ontology Project[1] (GO) is an initiative to develop a computational representation of the knowledge of how genes encode biological functions at the molecular, cellular and tissue system levels. The GOEA compares a set of genes with known information. For example, given a set of genes that are up-regulated under certain conditions, an enrichment analysis will find which GO terms are over-represented (or under-represented) using annotations for that gene set[2]. This method is commonly used to analyze results from biclustering techniques on microarray gene expression datasets.

Precision, Recall and *F-score* are often used on information retrieval for measuring binary classification [16]. If we take pairs of elements, we can extend these metrics to evaluate clustering/biclustering solutions with overlapping. The pairwise definition of *Precision* and *Recall* can be found in [17]. It is important to highlight that these metrics do not consider the quantity of biclusters. *Pairwise Precision*, or just *Precision* for simplicity, is the fraction of retrieved pairs that are relevant; while *Pairwise Recall*, or just *Recall* for simplicity, is the fraction of relevant pairs that are retrieved. The *F-score* is the harmonic mean of *Precision* and *Recall*.

Clustering Error (*CE*) is an external metric that considers the quantity of biclusters in its evaluation. This metric severely penalizes a solution with more biclusters than the reference, thus not being recommended for evaluating enumerative results. The definition and more details can be found in [15].

We propose the difference in coverage, that measures what the reference biclustering solution covers and the found biclustering solution does not cover, and vice versa. Although very similar, when compared with the pairwise definitions of *Precision* and *Recall*, this metric gives a more intuitive idea of how two solutions cover distinct areas of the dataset. It also can be computed much faster. Let $\cup_{\bar{B}} = \bigcup B_i^r \times B_i^c$ be the usual union set of a biclustering solution $\bar{B}$.

---

[1] http://geneontology.org.

[2] http://geneontology.org/page/about Acessed on 2015, January, 16.

Let $\bar{B}$ and $\bar{C}$ be the found and the reference biclustering solutions, respectively. Then the difference in coverage is given by:

$$dif\_cov(\bar{B},\bar{C}) = \frac{|\cup_{\bar{B}} - \cup_{\bar{C}}| + |\cup_{\bar{C}} - \cup_{\bar{B}}|}{m \times n}. \tag{3}$$

We will use this measure to verify how different an aggregated solution is from the enumerative one.

5   Experiments

In our experiments, we employed three artificial datasets: *art1*, *art2*, and *art3*; and two real datasets: GDS2587 and *FOOD*. We designed the artificial datasets to present different scenarios with increasing difficulty. They have 1000 objects and 15 attributes. Each entry is a random integer, drawn from a discrete uniform distribution on the set $\{1, 2, ..., 100\}$. Then we inserted: 5 bicluster arbitrarily positioned and without overlapping on *art1*; 5 bicluster arbitrarily positioned and with a similar degree of overlapping on *art2*; and 15 bicluster arbitrarily positioned and with different degrees of overlapping on *art3*.

For each bicluster, the quantity of objects was randomly drawn from the set $\{50, ..., 60\}$, and the quantity of attributes was randomly drawn from the set $\{4, 5, 6, 7\}$. To insert a bicluster, we fixed the value of the first attribute and obtained the values of the other attributes by adding a constant value to the first column. This characterizes biclusters of coherent values. This constant value was randomly drawn from the set $\{-10, -9, ..., -1, 1, ..., 9, 10\}$.

GDS2587[3] is a microarray gene expression dataset, with 2792 genes and 7 samples, collected from the organism *E. coli*. We removed every gene with missing data in any sample, and the data was normalized by mean centralization, as usual in gene expression data analysis [18]. In this dataset we aim to validate our contribution when devoted to microarray gene expression data analysis, as it is considered a relevant application of biclustering methods.

FOOD[4] is a dataset with 961 objects, which represent different foods, and 7 attributes, which represent nutritional information. As the values of each attribute are in different ranges, we used the same pre-processing as Veroneze *et al.* [6]. In this dataset our goal is to illustrate the usefulness of bicluster aggregation in a different scenario and to verify if the aggregation leaves uncovered areas that the enumeration has covered at first.

5.1   Experiments on Artificial Datasets

Our goal is to verify the impact of noise in the enumeration of biclusters, and how the aggregation can improve the quality of the final results. To this end, we will add a Gaussian noise with $\mu = 0$ and $\sigma \in \{0, 0.01, ..., 1\}$, to each dataset,

---

[3] http://www.ncbi.nlm.nih.gov/sites/GDSbrowser?acc=GDS2587.
[4] http://www.ntwrks.com/chart1a.htm.

and then run the RIn-Close algorithm. This procedure will be repeated for 30 times and all reported values will be the average of this 30 executions. We will set RIn-Close to mine coherent values biclusters, with at least 50 rows and 4 columns. Also, we will use crescent values for ϵ due to the importance of the parameter. If ϵ is too small, we may miss important biclusters expressing more internal variance. If ϵ is too high, the biclusters may include unexpected objects or attributes.

As we know the biclusters, we will use *Precision, Recall* and *F-score* to assess the quality of the results after the enumeration. After that, we will perform the aggregation on the results with the value of ϵ that led to an initial *Precision* closest to 0.85. This value was chosen because if the *Precision* is too low, it means that the ϵ value is allowing too many undesired objects or attributes in the enumerated biclusters. In this case, the aggregation may not improve the quality of the final results because their input is not of good quality. If the *Precision* is too high, we will only be able to see improvements in the reduced quantity of biclusters, but the aggregation may increase the *Precision* too.

We will consider the following algorithms as contenders:

Triclustering [13]. We set k to the true number of biclusters. The authors supplied the code for this algorithm.

Merging and Deleting steps of MicroCluster [9]. To parameterize this algorithm, we ran a grid search with the values in the set $0.15, 0.1, 0.05$, getting 9 results for each run. Also, as the aggregation step of the algorithm is composed of two steps, merging and deleting, we ran each experiment twice: with the merging step first (MD) and with the deleting step first (DM). Unless we want to draw attention to some particular fact, we will report only the best result. The authors supplied the code for this algorithm[5].

Single Linkage (see Sect. 3.1). We cut the dendrogram with the proper quantity of biclusters: for *art1* and *art2*, 5 biclusters; for *art3*, 15 biclusters.

Aggregation by Overlapping (see Sect. 3.2). We tested several values for the rate of overlapping.

After getting the results for all executions of the listed algorithms, we will choose the best result from each one and compare them using the *CE* metric.

Figure 2 shows the quantity of enumerated biclusters on the artificial datasets, for several values of ϵ. In all datasets, for every value of ϵ, the behavior is the same: as the noise increases the quantity of enumerated biclusters starts to increase. In Fig. 2a and b, we know that the real quantity of biclusters is 5, but when the noise increases, the enumerated quantity reaches approximately 800 biclusters, depending on the value of ϵ. In Fig. 2c, we can see that the quantity of biclusters reaches high values too. At some level of noise, the number of biclusters starts to decrease to a point that the algorithm is not able to find any bicluster.

In Fig. 3, we can see the quality of the enumeration without considering the quantity of biclusters.

---

[5] http://www.cs.rpi.edu/~zaki/www-new/pmwiki.php/Software/Software.

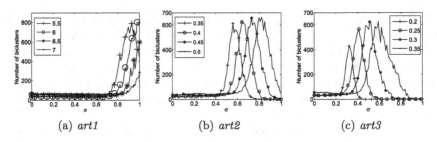

Fig. 2. Quantity of enumerated biclusters by the variance of the Gaussian noise in the artificial datasets. Each curve is parameterized by ϵ.

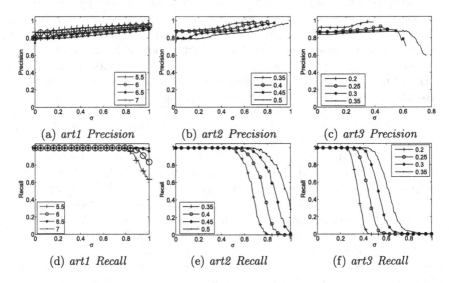

Fig. 3. *Precision* and *Recall* for the solutions of RIn-Close, with several values of ϵ, by the variance of the Gaussian noise in the artificial datasets.

As we can see in Fig. 3d, the noise has almost no interference in the recall for *art1*. It means that this dataset has biclusters very well defined, that even with some noise they are not missed. On the other hand, when the variance of the noise is too low, Fig. 3a shows that the found biclusters contains more elements than expected. It is happening because the parameter ϵ is high, allowing some elements to be part of the biclusters even without being part of the original solution. As the noise increases, less of these intruder elements are going to satisfy the ϵ restriction to be thus included in some bicluster. In this dataset, the effect of the noise were not so severe on the quality, given that the recall started to decrease only when the variance of the noise was close to 1.

In dataset *art2* the effect of noise can be better observed. Figure 3e shows that the noise starts to affect the solutions very early. When $\epsilon = 3$, the recall starts to decrease very soon, with $\sigma \approx 0.5$. However, for more relaxed values of ϵ we can still see the decrease on the recall. Being the most difficult, dataset

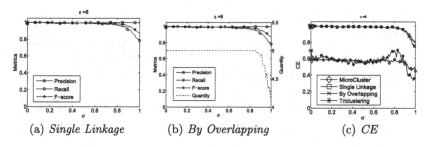

Fig. 4. Solutions of aggregation as a function of the variance of the noise in dataset *art1*. The scale on the right refers to quantity.

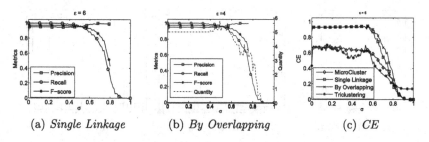

Fig. 5. Solutions of aggregation as a function of the variance of the noise in dataset *art2*. The scale on the right refers to quantity.

art3 is the most affected by noise. Independently of the value of ϵ, the RIn-Close was not able to find any biclusters after some levels of variance in the noise. For example, when $\epsilon = 2$, after $\sigma \approx 0.4$ the *Precision* gets undefined. This happens because the metric is not defined when the quantity of biclusters is zero. In Fig. 3f, we can see that the decline of the recall starts when $\sigma \approx 0.3$ for $\epsilon = 2$.

Now we will discuss the results of the aggregation with the previously listed algorithms. As stated earlier, we will use the results from a value of ϵ that led to an initial *Precision* close to 0.85. In this case, we have $\epsilon = 6, 4, 3$ for *art1*, *art2* and *art3*, respectively.

Figure 4a shows the quality of the aggregation with single linkage for dataset *art1*. We can see that, with the proper number of biclusters, the aggregation was able to get an almost perfect result. The same thing happened with the aggregation by overlapping, reported in Fig. 4b. Figure 4c shows the CE metric for all solutions of aggregation. We can see that our proposals were capable of producing the best performance on this dataset.

Figure 5a shows the quality of the aggregation with single linkage for the dataset *art2*. This time, the solution was close to the maximum achievable performance, but not so close as it was in *art1*. Figure 5b shows the quality of the aggregation by overlapping for the same dataset. The quality of this solution is very similar to the one obtained with single linkage. Figure 5c shows the CE metrics obtained by all the methods of aggregation. Again, our proposals outperformed the other two algorithms.

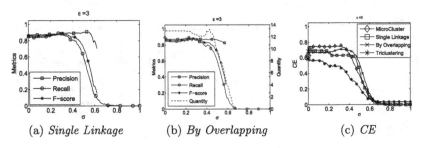

Fig. 6. Solutions of aggregation as a function of the variance of the noise in dataset *art3*. The scale on the right refers to quantity.

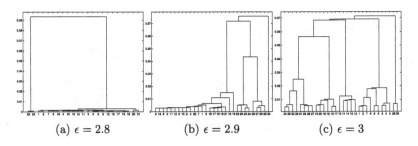

Fig. 7. Dendrograms of the aggregation with single linkage on *GDS2587* dataset.

Figure 6a and b show the quality of aggregation with single linkage and by overlapping, respectively. We can see that this dataset is more challenging than the previous ones. However, the aggregation was able to significantly reduce the quantity of biclusters, while keeping a good quality. Figure 6c shows the CE metric for all aggregation methods. Initially MicroCluster had a better performance, but our proposals were more robust to noise, getting a better result when $\sigma \gtrsim 0.4$.

The aggregation was not only able to reduce the quantity of biclusters of the enumeration, but also improve the quality of the final result. Now we are going to verify the behavior of the aggregation in real datasets.

5.2 Experiments on Real Datasets

We will start with the *GDS2587* dataset by running RIn-Close to enumerate its coherent values biclusters. We set $minRow = 50, minCol = 4$. When $\epsilon < 2.8$ no biclusters were found, and when $\epsilon = 3.0$ the quantity of biclusters was already huge. We found 23, 2.825 and 19.649 biclusters when $\epsilon = 2.8, 2.9$, and 3.0, respectively.

Proceeding to the aggregation, Fig. 7 shows the dendrograms of the aggregation with single linkage. In this case, the cuts are straightforward, having 2, 4, and 5 clusters respectively. The aggregation by overlapping with a rate of 75 % reached the same quantity of biclusters. We used these quantities to parameterize the

Table 1. Enrichment analysis of one bicluster from the aggregation by overlapping with rate of 70 %, on *GDS2587* dataset.

GO Term	p-val	counts	definition
GO:0044464	0.00000000	39 / 774	Any constituent part of a cell, the basic structural and functional unit of all organisms...
GO:0044444	0.00000011	19 / 608	Any constituent part of the cytoplasm, all of the contents of a cell excluding the plasma membrane...
GO:0044424	0.00000350	19 / 578	Any constituent part of the living contents of a cell; the matter contained within (but not including) the plasma membrane

triclustering algorithm. The results of the aggregation with MicroCluster were very similar, and they depended only on the γ parameter. We got 7, 8 and 11 biclusters when $\gamma = 0.15, 0.1$, and 0.05, respectively. We will now compare the results with the *gene ontology enrichment analysis*. A bicluster is called 'enriched' when any ontology term gets a p-value less than 0.01.

When $\epsilon = 2.8$, except for triclustering (only the first bicluster was enriched), all the algorithms returned only enriched biclusters. In fact, the four main enriched terms were always the same, sometimes on different orders but with very close p-values.

When $\epsilon = 2.9$, all algorithms returned only enriched biclusters, including triclustering. When $\epsilon = 3$, all algorithms except for triclustering returned only enriched biclusters. Triclustering returned 4 from 5 enriched biclusters.

Table 1 shows the main enriched terms of one bicluster from the aggregation by overlapping after outlier removal, when $\epsilon = 2.8$. In this case, the expert should choose which solution fits better the goal of the data analysis.

We will now proceed to the analysis of the *FOOD* dataset. We are going to verify how the aggregation changes the coverage of the dataset when compared to the enumeration. As the aggregation will severely reduce the quantity of final biclusters, it is important to see if it will leave uncovered areas that were previously covered.

We replicated the experiment from Veroneze *et al.* [6] on this dataset and we will use $\epsilon = 1.25$ as recommended on that work. With $minRow = 48, minCol = 2$ and looking for coherent values biclusters, the quantity of enumerated biclusters for $\epsilon = 1.25$ is 8.676.

Figure 8 shows the dendrogram of the aggregation with single linkage. We can see that the cuts between 2 and 7 are acceptable. In fact, cutting in two groups seems the best option, but it may be considered a small quantity of biclusters. As from 4 to 5 the height is more pronounced, for the comparison it seems acceptable to cut the dendrogram on 4 objects. The aggregation by overlapping with a rate of 70 % was also able to recover 4 aggregated biclusters.

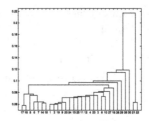

Fig. 8. Dendrogram for the aggregation with single linkage when $\epsilon = 1.25$ on *FOOD* dataset.

MicroCluster with the deleting operation first was not able to properly aggregate the biclusters, keeping more than 800 biclusters when $\eta = 0.15$. This behavior is the opposite of what happened with the artificial datasets. There, when the deleting operation came first the results were more effective. Here when the merging operation came first, the aggregation was able to reach 13 to 27 biclusters, depending on the γ parameter. As on the artificial datasets the best parameters were $\eta = \gamma = 0.15$, for the comparison we will use this parameterization with the merging operation occurring first, that gives us 13 biclusters. For the triclustering algorithm we set $k = 4$, using insider information from the aggregation by overlapping. Table 2 shows the comparison of difference in coverage (see Eq. 3) between the aggregated solutions with the enumerated solution from RIn-Close. We can see that the triclustering algorithm produces the most distinct solution when compared with the enumerated solution obtained with RIn-Close, exhibiting $\approx 61.33\%$ of difference in coverage. The solutions from the aggregation by overlapping and with single linkage are relatively close to each other, as on the artificial datasets, showing a difference in coverage of $\approx 12.50\%$. At the end, the closest solution to the RIn-Close results was the aggregation by overlapping, with a difference in coverage of 9.1%. If we consider that this solution reduced the quantity of biclusters from 8.676 to 4 biclusters, the difference in coverage of only 9.1% seems very promising.

Table 2. Difference in coverage of the solutions with the enumeration on *FOOD* dataset.

	Single Linkage	MicroCluster	Triclustering	RIn-Close
By Ov.	12.50%	35.50%	70.31%	9.1%
Single Linkage	-	46.60%	81.51%	20.17%
MicroCluster	-	-	45.73%	27.38%
Triclustering	-	-	-	61.33%

6   Considering Remarks and Future Work

We have compared the performance of our proposals against the most similar proposal in the literature, using artificial and real datasets. The artificial datasets were characterized by a controlled structure of biclusters and were useful to show that the aggregation can severely reduce the quantity of biclusters, while increasing the quality of the final solution. Our proposals outperformed the compared algorithms on the first two artificial datasets, and showed to be more robust to noise on the third artificial dataset.

We also verified if the aggregation could get enriched biclusters in the case of a gene expression dataset. For different values of ϵ on the RIn-Close algorithm, we could see that the different methods of aggregation reached very similar results. The main challenge of the aggregation with single linkage is to decide where to cut the dendrogram, but as we could see, this task was straightforward on the tested datasets. Except for the triclustering, all aggregations returned only enriched biclusters. And finally, we applied the aggregation methods to the *FOOD* dataset and analyzed how the aggregation changed the coverage area when compared to the enumeration without aggregation. Triclustering led to the most distinct result, and the aggregation by overlapping covered an area very similar to the area covered by the enumeration.

We can conclude that the aggregation is strongly recommended when enumerating all biclusters from a dataset. The aggregation will not only significantly reduce the quantity of biclusters, but will also reduce the fragmentation and increase the quality of the final result. A post-processing step for outlier removal brings additional robustness to the methodology. As a further step of the research, we can adapt our proposals to work on an ensemble configuration. We can also extend this work to deal with time series biclusters, which require contiguous attributes.

Acknowledgment. The authors would like to thank CAPES and CNPq for the financial support.

References

1. Cheng, Y., Church, G.M.: Biclustering of expression data. In: Proceedings of International Conference on Intelligent Systems for Molecular Biology; ISMB, vol. 8, pp. 93–103 (2000)
2. Jiong, Y., Wang, H., Wang, W., Yu, P.: Enhanced biclustering on expression data. In: Proceedings of Third IEEE Symposium on Bioinformatics and Bioengineering, 2003, pp. 321–327 (2003)
3. Makino, K., Uno, T.: New algorithms for enumerating all maximal cliques. In: Hagerup, T., Katajainen, J. (eds.) SWAT 2004. LNCS, vol. 3111, pp. 260–272. Springer, Heidelberg (2004)
4. Uno, T., Kiyomi, M., Arimura, H.: Lcm ver. 2: efficient mining algorithms for frequent/closed/maximal itemsets in FIMI, vol. 126 (2004)

5. Andrews, S.: In-close, a fast algorithm for computing formal concepts. In: Seventeenth International Conference on Conceptual Structures (2009)
6. Veroneze, R., Banerjee, A., Zuben, F.J.V.: Enumerating all maximal biclusters in real-valued datasets (2014). arXiv:1403.3562v3, vol. abs/1403.3562
7. Pandey, G., Atluri, G., Steinbach, M., Myers, C.L., Kumar, V.: An association analysis approach to biclustering. In: Proceedings of the 15th ACM SIGKDD International Conference on Knowledge Discovery and Data Mining, KDD 2009, pp. 677–686 (2009)
8. Liu, J., Wang, J., Wang, W.: Biclustering in gene expression data by tendency. In: IEEE Computer Society on CSB, pp. 182–193 (2004)
9. Zhao, L., Zaki, M.J.: Microcluster: efficient deterministic biclustering of microarray data. IEEE Intell. Syst. 20(6), 40–49 (2005)
10. Madeira, S.C., Oliveira, A.L.: Biclustering algorithms for biological data analysis: a survey. IEEE/ACM Trans. Comput. Biol. Bioinformatics 1, 24–45 (2004)
11. Tanay, A., Sharan, R., Shamir, R.: Biclustering algorithms: a survey. In: Aluru, S. (ed.) Handbook of Computational Molecular Biology. Chapman & Hall/CRC Computer and Information Science Series, London (2005)
12. Gao, T., Akoglu, L.: Fast information-theoretic agglomerative co-clustering. In: Wang, H., Sharaf, M.A. (eds.) ADC 2014. LNCS, vol. 8506, pp. 147–159. Springer, Heidelberg (2014)
13. Hanczar, B., Nadif, M.: Ensemble methods for biclustering tasks. Pattern Recogn. 45(11), 3938–3949 (2012)
14. Hanczar, B., Nadif, M.: Improving the biological relevance of biclustering for microarray data in using ensemble methods. In: 2011 22nd International Workshop on Database and Expert Systems Applications, pp. 413–417. August 2011
15. Horta, D., Campello, R.J.G.B.: Similarity measures for comparing biclusterings. IEEE/ACM Trans. Comput. Biol. Bioinf. 11, 942–954 (2014)
16. Salton, G.: Evaluation parameters. In: Salton, G. (ed.) The SMART Retrieval System, Experiments in Automatic Document Processing, pp. 55–112. Prentice-Hall, Englewood Cliffs (1971)
17. Menestrina, D., Whang, S.E., Garcia-Molina, H.: Evaluating entity resolution results (extended version), Technical report, Stanford University (2009)
18. Prelić, A., Bleuler, S., Zimmermann, P., Wille, A., Bühlmann, P., Gruissem, W., Hennig, L., Thiele, L., Zitzler, E.: A systematic comparison and evaluation of biclustering methods for gene expression data. Bioinformatics 22, 1122–1129 (2006)

Semi-Supervised Stream Clustering Using Labeled Data Points

Kritsana Treechalong, Thanawin Rakthanmanon, and Kitsana Waiyamai[✉]

Department of Computer Engineering, Kasetsart University, Bangkok, Thailand
{g5514552527,thanawin.r,fengknw}@ku.ac.th

Abstract. Semi-supervised stream clustering performs cluster analysis of data streams by exploiting background or domain expert knowledge. Almost of existing semi-supervised stream clustering techniques exploit background knowledge as constraints such as must-link and cannot-link constraints. The use of constraints is not appropriate with respect to the dynamic nature of data streams. In this paper, we proposed a new semi-supervised stream clustering algorithm, SSE-Stream. SSE-Stream exploits background knowledge in the form of single labeled data points to monitor and detect change of the clustering structure evolution. Exploiting background knowledge as single labeled data points is more appropriate for data streams. They can be immediately utilised for determining the class of clusters, and effectively support the changing behavior of data streams. SSE-Stream defines new cluster representation to include labeled data points, and uses it to extend the clustering operations such as merge and split for detecting change of the clustering structure evolution. Experimental results on real-world stream datasets show that SSE-Stream is able to improve the output clustering quality, especially for highly complex and drift datasets.

Keywords: Semi-supervised learning · Semi-supervised clustering · Stream clustering · Background knowledge · Domain expert knowledge · Constraint

1 Introduction

In recent years, semi-supervised stream clustering has become a research topic of growing interest. Semi-supervised stream clustering performs cluster analysis of data streams by exploiting background or domain expert knowledge. Besides a higher clustering result, background knowledge also has potential to enhance the clustering execution-time and can be used to reduce the possibility of empty clusters generation [1]. The use of background knowledge can be found as a potential improvement in many applications such as Intrusion Detection System (IDS) where background knowledge is exploited in terms of rules of known intrusion types.

Recently, many stream clustering algorithms have been proposed such as HPStream [2], CluStream [3], E-Stream [4], Denstream [5], and SED-Stream [6].

© Springer International Publishing Switzerland 2015
P. Perner (Ed.): MLDM 2015, LNAI 9166, pp. 281–295, 2015.
DOI: 10.1007/978-3-319-21024-7_19

However, few of them have been involved in a semi-supervised stream learning. Ruiz et al. [10] proposed a conceptual model for using constraints over data streams and extended the constraint-based K-means [9]. C-Denstream [11], the extension of Denstream, used constraints during the clustering process. CE-Stream [12] extends E-Stream, an evolution-based stream clustering method to support constraints. Constraint is a pair of data points, which can be must-link (two points must be in the same cluster) or cannot-link (two points cannot be in the same cluster). Exploiting background knowledge as constraints is not appropriate with respect to the dynamic nature of data streams. Within a time period, a pair of constraint points must be verified before each constraint is activated, thus small number of constraints is utilised. Moreover, all the previous techniques do not take into account the dynamic nature of constraints which can be changed over time. Exploiting background knowledge as single data points (not pair of points) is more appropriate for data streams.

In this paper, we propose a new semi-supervised stream clustering using labeled data points called SSE-Stream. SSE-Stream exploits background knowledge as single labeled data points to detect change of the clustering structure evolution. Compared to constraint-based approach, labeled data points can be immediately utilised for determining the class of clusters, and effectively support the changing behaviour of data streams. SSE-Stream is an extension of SED-stream which is an efficient evolution-based stream clustering technique. In order to manage the labeled data, a new cluster representation is proposed. Existing evolution-based clustering operations in SED-stream such as merge and split are redefined and extended. Experimental results on three stream datasets (Network Intrusion, Forrest Cover Type and Electricity) show that the output clustering quality of SSE-Stream is higher than SED-Stream, especially for highly complex and drift dataset.

The remaining of the paper is organized as follows. Section 2 presents basic definitions and concepts of stream clustering. Section 3 presents our proposed semi-supervised stream clustering technique called SSE-Stream. Section 4 compares performance of SSE-Stream with SED-Stream on three real-world stream datasets. Section 5 concludes the paper.

2   Basic Definitions and Concepts of Stream Clustering

In this section, basic concepts and techniques related to the stream clustering are given. Assume that **data streams** consists of a set of multidimensional records $X_1 \ldots X_k$ arriving at time stamps $T_1 \ldots T_k \ldots$. Each **data point** X_i is a multidimensional record containing d dimensions, denoted as $X_i = (x_i^1 \ldots x_i^d)$.

2.1   Cluster Representation Using Fading Cluster Structure with Histogram

Due to the limited size of memory, it is not possible to store a large amount of data points arriving during the progression of data streams. Instead, a *Fading*

Cluster structure (FCS) has been proposed in [13] to use as cluster representation. Extension of *FCS* named *FCH* (*Fading Cluster Structure with
Histogram*) was introduced in E-Stream. An α-bin histogram was added to
FCH in order to detect change of the clustering structure. SED-Stream
extends *FCH* to support dimension selection and is defined as $FCH = (FC1(t), FC2(t), W(t), BS(t), H(t)$. Following is the description of *FCH*.

Let N be the total number of data points of such cluster, T_i be the time
when data point x_i is retrieved, and t be the current time. The *fading weight* of
data point x_i is defined as $f(t - T_i)$ where $f(t) = 2^{-\alpha t}$ and α is the user-defined
decay rate.

FC1(t) is a vector of weighted summation of each dimension at time t. The
j^{th} dimension is

$$FC1^j(t) = \sum_{i=1}^{N} f(t - T_i) \cdot (x_i^j). \tag{1}$$

FC2(t) is a vector of weighted sum of square of each dimension at time t.
The jth dimension is

$$FC2^j(t) = \sum_{i=1}^{N} f(t - T_i) \cdot (x_i^j)^2. \tag{2}$$

W(t) is a sum of all weights of data points in the cluster at time t, i.e.,

$$W(t) = \sum_{i=1}^{N} f(t - T_i). \tag{3}$$

BS(t) is a bit vector of projected dimensions at time t. For the j-th dimension is

$$BS^j(t) = \begin{cases} 1 & \text{if dimension } j \text{ is within a set of} \\ & \text{relavant cluster dimensions} \\ 0 & \text{Otherwise.} \end{cases} \tag{4}$$

H(t) is a α-bin histogram of data values with α equal width intervals. For
the l-th bin histogram of j-th dimension at time t, the elements of H^j are

$$H_l^j(t) = \sum_{i=1}^{N} f(t - T_i) \cdot (x_i^j) \cdot (y_{i,l}^j) \tag{5}$$

where

$$y_{i,l}^j = \begin{cases} 1 \text{ if } l \cdot r + min(x^j) \leq x_i^j \leq (l+1) \cdot r + min(x^j); \\ r = \frac{max(x^j) - min(x^j)}{\alpha} \\ 0 \text{ Otherwise.} \end{cases} \tag{6}$$

Clusters can be categorized into 2 types: **active** and **inactive** cluster regarding to their weight. Active cluster is a cluster having its weight greater than

user-specified threshold *active_cluster_weigth*. Only active clusters can assemble incoming data point located nearby. Meanwhile, such inactive clusters with lower weight than *active_cluster_weigth*, can be merged together with other inactive clusters or active clusters to produce an active cluster.

2.2   Distance Functions

To deal with projected dimensions, distance functions are modified to take into account only the relevant cluster dimensions. Suppose that $BS(t)$ is a bit vector represented cluster dimensions at timestamp t. The distances functions can be defined as follows.

Cluster-Point Distance is a distance from a data point to a center of active cluster. For each dimension, the distance is normalized by the radius (standard deviation) of the cluster data. This function is used to find the closet active cluster for an incoming data point. The Cluster-Point distance function $dist(C, X_i)$ of cluster C and incoming data point X_i at timestamp t can be formulated as:

$$dist(C, X_i) = \frac{1}{n} \cdot \sum_{j \in BS(t)} |\frac{center_C^j - x_i^j}{radius_C^j}| \tag{7}$$

where n is the number of relevant dimensions of cluster C in the bit vector $BS(t)$.

Cluster-Cluster Distance is a distance between two cluster centers. It is used to determine whether pair of clusters that can be merged together. The cluster-cluster distance function $dist(C_a, C_b)$ of cluster C_a and C_b at timestamp t can be formulated as:

$$dist(C_a, C_b) = \frac{1}{n} \cdot \sum_{j \in bs(t)} |center_{C_a}^j - center_{C_b}^j| \tag{8}$$

where n is the total number of relevant dimensions and $bs(t)$ represents the bit vector of n relevant dimensions. Notice that the number of relevant dimensions and dimensions of two clusters may differ. $bs(t)$ is the union set of the relevant dimensions of these two clusters.

2.3   Evolution-Based Stream Clustering

Behavior of data streams can evolve over time. The evolution of the clustering structure can be classified into five categories: appearance, disappearance, self-evolution, merge, and split.

Appearance: A new cluster can appear if there is a sufficiently dense group of data points in one area.

Disappearance: Existing clusters can disappear because the existence of data is diminished over time.

Self-evolution: Data can change their behaviors, which cause size or position of a cluster to evolve.

Merge: A pair of clusters can be merged if their characteristics are very similar. Merged clusters must cover the behavior of the pair.

Split: A cluster can be split into two smaller clusters if the behavior inside the cluster is obviously separated.

3   SSE-Stream: Semi-supervised Stream Clustering Using Labeled Data Points

In this section, we first give an overview of the proposed semi-supervised stream clustering technique called SSE-Stream. SSE-Stream exploits the background knowledge in the form of labeled data points, and uses them to guide execution of the clustering structure evolution operations. Then, a new cluster representation is defined. Finally, extension of the clustering structure evolution operations with the new cluster representation is explained.

3.1   Overview of Semi-supervised Stream Clustering Using Labeled Data Points

In SSE-Stream, we have 2 types of data streams. (1) Un-labeled data points are multi-dimensional records containing d dimensions, denoted as $X_i = (x_i^1 \ldots x_i^d)$. (2) Labeled data points are a multi-dimensional records containing d dimensions and y_i is class label, denoted as $X_i = (x_i^1 \ldots x_i^d, yi)$. To include labeled data points, a new cluster representation, named Class of cluster (Coc), is proposed. When a labeled data point enters into the clustering process, it will be stored in a buffer. Based on the new cluster representation, Coc, SSE-Stream tries to determine the cluster label by using class label y, denoted as

$$Coc = majority(y) \; in \; cluster \tag{9}$$

In Fig. 1, a cluster has five labeled data points, four points is of red class and one point of blue class. Thus, majority class of this cluster is red, then Coc = red.

SSE-Stream is an evolution-based algorithm that supports the monitoring and change detection of clustering structure that can evolve over time. SSE-Stream has Coc to improve merge and split operation. The result of all data processing compare with stream clustering to show in Fig. 2.

Figure 2, compares result of any stream clustering technique with SSE-Stream. With SSE-Stream, labeled data points can be used to identify correct class of cluster, thus nearby clusters will not be merged even if they are close each other. However, without knowledge about class, all data points are grouped into one cluster. This will result in a low output clustering quality.

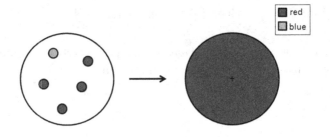

Fig. 1. Determination of class of cluster (Coc) (Color figure online)

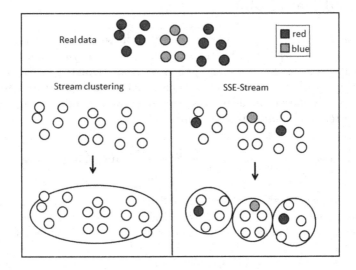

Fig. 2. Comparison of the output clustering between any stream clustering technique and SSE-Stream (Color figure online)

3.2 Extension of the Clustering Structure Evolution Operations with New Cluster Representation

(1) ForceSplit Operation: SSE-Stream is able to detect the number of classes in a cluster by means of labeled data points. When a cluster has more than one class, SSE-stream will check split condition by calling the ForceSplit operation. If the split condition is true, it will result in an improvement of the quality of the output clusters.

When a labeled data point enters into the stream clustering process, SSE-Stream assigns its closest cluster. If the cluster has more than one class, Force-Split operation tries to identify the split dimension which best separates two classes, together with its best split position (from the lowest value in histogram). Then, labeled data points are assigned to their closest clusters. Finally, Coc of the new clusters is determined.

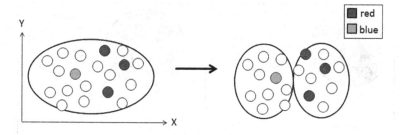

Fig. 3. Example of ForceSplit operation execution (Color figure online)

Figure 3 shows execution of the ForceSplit operation. First, ForceSplit operation selects the best dimension for split. In this case, X is the selected split dimension. After a cluster is splitted into two smaller clusters, ForceSplit divides existing labeled data points from the old cluster, assign them new closest clusters, and set Coc. Here, the red cluster was split into blue and red clusters.

(2) Merge Operation: Stream clustering can merge cluster when clusters are adjacent. However, in SSE-Stream, with knowledge about cluster class from Coc, nearby clusters will not be merged because those clusters are of different classes. Merging two clusters of different classes will result in decreasing the quality of the output cluster. Merge operation of SSE-Stream will perform either clusters are of same Coc or their Coc is undefined. In Fig. 4, red and blue are adjacent clusters, but they will not be merged in SSE-Stream.

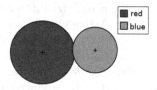

Fig. 4. Example of Merge operation execution (Color figure online)

(3) Find Closest Cluster: When a labeled data point enters into the clustering process, SSE-Stream will assign the most appropriate cluster by calling the FindClosestCluster operation. Two conditions are verified by the FindClosestCluster operation. (1) If distance between of the cluster and the labeled data point is less than the cluster radius, then it will be assigned to that cluster. (2) If distance between the labeled data point and all clusters is more than their radius, then it will be assigned to the closest cluster having same or undefined Coc.

Figure 5 shows the two FindClosestCluster conditions. First condition, the blue data point is within the red cluster. SSE-Stream will put the blue data point into the red cluster. Second condition, the blue data point is outside all clusters. SSE-Stream will choose the closest cluster having same or undefined Coc.

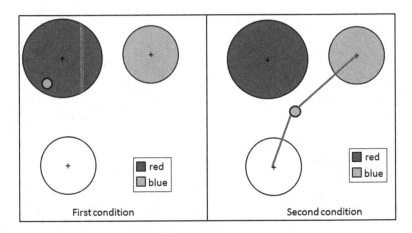

Fig. 5. Example of FindClosestCluster operation execution (Color figure online)

3.3   SSE: Algorithm for Semi-supervised Stream Clustering Using Labeled Data Points

Figure 6, describes SSE-Stream main algorithm. SSE-Stream is an evolution-based algorithm that supports the monitoring and the change detection of clustering structure that can evolve over time. It is designed for Semi-supervised stream clustering. In SSE-Stream, data streams are of 2 types which are data point and labeled data point. Various types of clustering structure evolution are supported which are appearance, disappearance, self-evolution, merge and split. The clusters in SSE-Stream can be categorized into 2 types: active and inactive clusters regarding to their weight. Active cluster is a cluster having its weight greater than user-specified threshold active cluster weight. Only active clusters can assemble incoming data point located nearby. Meanwhile, such inactive clusters with lower weight than active cluster weight, can be merged together with other inactive clusters or active clusters to produce an active cluster.

In line 1, the algorithm starts by retrieving a new data point. In line 2, it fades all clusters and deletes those have weight less than user-specified threshold. In line 3–5, if incoming data point is labeled data point and storage of labeled data point is full, it fading labeled data point and set all Coc. In line 6, it splits a cluster when behavior inside the cluster is obviously separated. In line 7, it checks Coc for overlapping clusters and merges them. In line 8, when the number of cluster count exceeds the limit, it checks Coc and find the closest pair of clusters and merges them until the number of cluster count does not exceed limit maximum number of clusters. In line 9, it checks all clusters whether their statuses are active. In line 10, it finds the closest cluster to the incoming data point. In line 11–22, if the distance between closest cluster and incoming data point is less than $radius_factor$ (as an input parameter) then the point is assigned to the cluster. If incoming data point is labeled data point, it stores labeled data point and sets Coc of closest cluster, then if the cluster has more than one

```
Algorithm : SSE-Stream
1 : Retrieve new data $X_i$
2 : FadingAll
3 : if( $X_i$ is label data and |All_label| > MaximumLabel )
4 :          FadingLabelData
5 :          SetAllCoc
6 : CheckSplit
7 : MergeOverlapCluster
8 : LimitMaximumCluster
9 : FlagActiveCluster
10: (minDistance,index) ← FindClosetCluster
11: if(minDistance < radius_factor)
12:          Add $X_i$ to $FCH_{index}$
13:          if($X_i$ is label data)
14:                    StoreLabel(index)
15:                    SetCoc(index)
16:                    if(this cluster has more than one class)
17:                              ForceSplit(index)
18: else
19:          Create new $FCH$ from $X_i$
20 :         if($X_i$ is label data)
21 :                   StoreLabel(index of new FCH)
22 :                   SetCoc(index of new FCH)
23: Waiting for new data
```

Fig. 6. SSE-Stream main algorithm

class, call ForceSplit in closest cluster. If the distance between closest cluster and incoming data point is more than $radius_factor$ then creates isolated data point. If incoming data point is labeled data point, it stores labeled data point and set Coc. Finally, the flow of control returns to the top of algorithm and waits for a new data point.

Following is the explanation of each sub-algorithm that has been extended from SED-Stream. Details of each sub-algorithm is given in Fig. 7.

FadingAll: SSE-Stream performs fading of all clusters and deletes clusters with weight less than the input parameter.

CheckSplit: SSE-Stream finds the split point in dimensions. If a splitting point is found in active cluster, it is split and stored with index pairs of split in S. Then, index of the closest splitted cluster is assigned to the labeled data point in All_label and Coc of the splitted cluster is determined.

MergeOverlapCluster: SSE-Stream finds the pairs of active clusters that are overlapped and same Coc or one of Coc is undefined. For each pair of active clusters, cluster-cluster distance is calculated. If the distance is less than the $merge_threshold$ (an input parameter) and the merged pair is not already in S then the two clusters are merged. Then, index of the cluster is assigned to the labeled data point in All_label and Coc of cluster is determined.

Fig. 7. SSE-Stream sub-algorithms

LimitMaximumCluster: SSE-Stream checks weather the total number of clusters reaches its maximum $maximum_c luster$ (an input parameter). If it exceeds the maximum, then the closest pair of clusters as same Coc or one of Coc is undefined is merged until the number of the remaining clusters is less than or equal to the $maximum_c luster$.

FlagActiveCluster: SSE-Stream checks current cluster. If the weight of any cluster is greater or equal to $active_w eight_t hreshold$ then it is flagged as an active cluster. Otherwise, the flag is cleared.

FindClosestCluster: SSE-Stream calculates cluster-point distance. Then, determines the closest active cluster to contain an incoming data point. But if the incoming data point is a labeled data point, SSE-Stream will consider

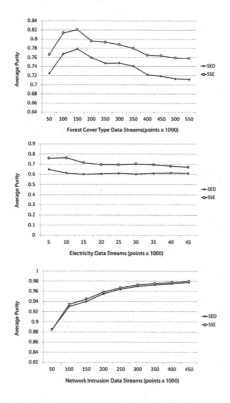

Fig. 8. Average purity of the clustering algorithms on the three standard streams datasets

two conditions. (1) cluster-data point distance is less than the cluster radius. (2) Class label of the labeled data point is of same or undefined Coc.

StoreLabel: SSE-Stream stores the labeled data point into a All_label buffer for other operations called. Index of the closest cluster is assigned to the labeled data point before being stored.

SetCoc and SetAllCoc: SetCoc and SetAllCoc SSE-Stream calls All_label to set Coc of the cluster. Coc is determined based on majority class cluster.

FadingLabelData: SSE-Stream deletes first point of All_label if All_label is full. A labeled data point is faded by using a first-in first-out technique.

ForceSplit: SSE-stream will check split condition. If the split condition is true, a cluster is split. Then, index of the closest splitted cluster is assigned to the labeled data point in All_label and Coc of the splitted cluster is determined.

4 Evaluation and Experimental Results

SSE-Stream is semi-supervised stream clustering that is extended from SED-Stream. Thus, in this section, we compare its clustering quality with SED-Stream in terms of purity measurement.

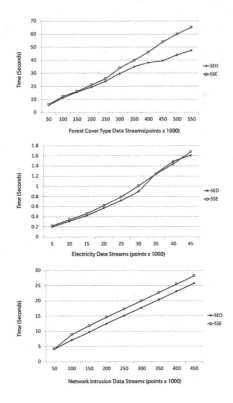

Fig. 9. Execution time of clustering algorithms on the three standard streams datasets

4.1 Experimental Setup

In our experiments, three standard datasets are used: Network Intrusion and Forest Cover Type datasets from UCI repository and Electricity dataset. Only numerical attributes are selected and used in all the experiments. For the Network Intrusion dataset, it contains 494020 records with 22 classes. The second dataset, Forest Cover Type datasets, contains 581012 records with 7 classes. The last dataset, Electricity dataset, contains 45312 records with 2 classes. Notice that from all the data points, only 1 % of them are the labeled data points. SSE-Stream uses a 100-size buffer to store labeled data points.

4.2 Clustering Quality in Terms of Purity

The clustering quality is evaluated using average purity measured for every 50,000 incoming data points. Figure 8 shows the results of SSE-Stream and SED-Stream on the three datasets. The purity of SSE-Stream is increased from SED-Stream by 0.5 %., 4.2 % and 10 % on Network Intrusion, Forest cover type and Electricity datasets, respectively.

Network Intrusion is a very simple dataset, where its distribution is clear and gradual. Thus, every stream clustering algorithm generates high clustering quality. For a more complex stream dataset such as Forest cover type, SSE-stream is able to generate higher clustering quality than SED-Stream. For a very complex and drift dataset such as Electricity dataset, which is extremely difficult for clustering, SSE-Stream still generates very high clustering quality than SED-Stream. SSE-Stream is excellent for managing complex and highly drift datasets.

4.3   Execution Time

Figure 9 compares execution time of SSE-Stream and SED-Stream. It shows that the execution time of SSE-Stream is a bit slower than SED-Stream. This can be explained by the fact that SSE-Stream required more time to manage labeled data points and execute the clustering operations.

4.4   Clustering Quality with Respect to the Percentage of Labeled Data Points

In our experiments, number of labeled data points is retrieved from 1 % and 0.1 % of all the data points. Figure 10 shows the clustering quality of the algorithms

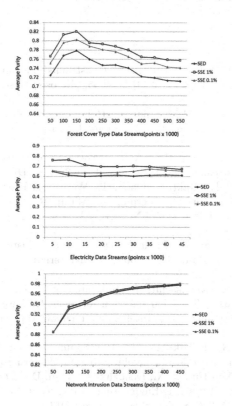

Fig. 10. Average purity in terms of percentage of labeled data

in terms of percentage of labeled data points used (0.1 % and 1 %). We notice that with very small percentage of labeled data points, the clustering quality of SSE-Stream is still increased. Thus, usage of labeled data has a very high positive effect to the output clustering quality.

5   Conclusion and Future Work

In this paper, we proposed a new evolution-based semi-supervised stream clustering algorithm, SSE-Stream. SSE-Stream exploits labeled data points as background knowledge to monitor and detect change of the clustering structure evolution. A new cluster representation that takes into account labeled data is proposed and the evolution-based clustering operations such as merge and split are extended. Experimental results show that the output clustering quality of SSE-Stream is higher than the state of the art evolution-based stream clustering such as SED-Stream, especially for highly complex and drift datasets.

In future work, we will extend SSE-stream to support other types of background knowledge such as the different types of constraints (must-link, cannot-link constraints). Our idea is transform those constraints into labeled data points, for example by assigning a label on each pair of the constraints data points.

References

1. Bradley, P.S., Bennett, K.P., Demiriz, A.: Constrained k-means clustering. Technical report. Technical report MSR-TR-2000-65, Microsoft Research, Redmond, WA (2000)
2. Milenova, B.L., Campos, M.M.: Cpar: clustering large databases with numeric and nominal valuees using orthogonal proections. In: Proceedings of the 29th VLDB Conference (2003)
3. Aggarwal, C., Han, J., Yu, P.S.: A framework for projected clustering of high dimensional data streams. In: Proceeding of the 30th VLDB Conference (2004)
4. Udommanetanakit, K., Rakthanmanon, T., Waiyamai, K.: E-stream: evolution-based technique for stream clustering. In: Alhajj, R., Gao, H., Li, X., Li, J., Zaïane, O.R. (eds.) ADMA 2007. LNCS (LNAI), vol. 4632, pp. 605–615. Springer, Heidelberg (2007)
5. Cao, F., Ester, M., Qian, W., Zhou, A.: Density-based clustering over an evolving data stream with noise. In: SIAM 2006: SIAM International Conference on Data Mining (2006)
6. Chairukwattana, R., Kangkachit, T., Rakthanmanon, T., Waiyamai, K.: SED-stream: discriminative dimension selection for evolution-based clustering of high dimensional data streams. Int. J. Intell. Syst. Technol. Appl. Arch. **13**(3), 187–201 (2014)
7. Chairukwattana, R., Kangkachit, T., Rakthanmanon, T., Waiyamai, K.: SE-stream: dimension projection for evolution-based clustering of high dimensional data streams. In: Van Huynh, N., Denoeux, T., Tran, D.H., Le, A.C., Pham, S.B. (eds.) Knowledge and Systems Engineering. Advances in Intelligent Systems and Computing, pp. 365–376. Springer, Heidelberg (2013)

8. Meesuksabai, W., Kangkachit, T., Waiyamai, K.: HUE-stream: evolution-based clustering technique for heterogeneous data streams with uncertainty. In: Tang, J., King, I., Chen, L., Wang, J. (eds.) ADMA 2011, Part II. LNCS, vol. 7121, pp. 27–40. Springer, Heidelberg (2011)
9. Wagstaff, K., Cardie, C., Rogers, S., Schroedl, S.: Constrained K-means clustering with background knowledge. In: ICML 2001: Proceedings of 18th International Conference on Machine Learning, pp. 577–584 (2001)
10. Ruiz, C., Spiliopoulou, M., Menasalvas, E.: User constraints over data streams. In: IWKDDS (2006)
11. Ruiz, C., Menasalvas, E., Spiliopoulou, M.: C-Denstream: using domain knowledge on a data stream. In: Gama, J., Costa, V.S., Jorge, A.M., Brazdil, P.B. (eds.) DS 2009. LNCS, vol. 5808, pp. 287–301. Springer, Heidelberg (2009)
12. Sirampuj, T., Kangkachit, T., Waiyamai, K.: CE-stream: evaluation-based techniquefor stream clustering with constraints. In: The 10th International Joint Conference on Computer Science and Software Engineering (JCSSE 2013) (2013)
13. Han, J., Wang, J., Philip, S.Y.: A framework for projected clustering of high dimensional data streams. In: Proceedings of the Thirtieth International Conference on Very Large Data Bases, VLDB 2004, pp. 852–863 (2004)
14. Zhu, X.: semi-supervised learning literature survey

Avalanche: A Hierarchical, Divisive Clustering Algorithm

Paul K. Amalaman[(✉)] and Christoph F. Eick

Department of Computer Science, University of Houston, Houston
TX 77204-3010, USA
{pkamalam, ceick}@uh.edu

Abstract. Hierarchical clustering has been successfully used in many applications, such as bioinformatics and social sciences. In this paper, we introduce *Avalanche,* a new top-down hierarchical clustering approach that takes a dissimilarity matrix as its input. Such a tool can be used for applications where the dataset is partitioned based on pairwise distances among the examples, such as taxonomy generation tools and molecular biology applications in which dissimilarity among gene sequences are used as inputs — as opposed to flat file attribute/value pair datasets. The proposed algorithm uses local as well as global information to recursively split data associated with a tree node into two sub-nodes until some predefined termination condition is met. To split a node, initially the example that is furthest away from the other examples — the *anti-medoid* — is assigned to right sub-node and then additional examples are progressively assigned to this node which are nearest neighbors of the previously added example as long as a given objective function improves. Experimental evaluations done with artificial and real world datasets show that the new approach has improved speed, and obtained comparable clustering results as the well-known UPGMA algorithm on all datasets used in the experiment.

Keywords: Divisive clustering · Hierarchical clustering · Agglomerative clustering · Quality of clusters

1 Introduction

This paper proposes a divisive approach to hierarchical clustering for applications that use as input a dissimilarity matrix. To the best of our knowledge such a method has not yet been proposed in literature. Clustering is the unsupervised grouping of examples into clusters. The clustering problem has received a lot of attention by researchers in many disciplines. Hierarchical clustering algorithms aim at organizing datasets hierarchically as dendrograms based on the distances of examples and clusters. Hierarchical clustering approaches are either agglomerative or divisive. In the general case, agglomerative (bottom-up) approaches start by merging the most similar examples/clusters and continue until the last two clusters have been merged. Divisive approaches — which are less popular —, on the other hand, start with a cluster containing all the examples of the dataset, and recursively split the dataset until a termination condition has been met. Both agglomerative and divisive clustering approaches are greedy algorithms, making decisions based on local patterns or based on global objective functions. The ability of the

© Springer International Publishing Switzerland 2015
P. Perner (Ed.): MLDM 2015, LNAI 9166, pp. 296–310, 2015.
DOI: 10.1007/978-3-319-21024-7_20

top-down approaches to use global information about the dataset to select splits is often viewed as a potential advantage [16]. On the other hand, the same publication claims that bottom-up approaches are perceived to produce better clustering result than the top-down methods but usually run significantly slower. The ability of the top-down methods to stop growing the tree when a predefined termination condition is met has made them very popular in applications such as document searching/indexing, web query. However, thus far, most of the divisive approaches involve the computation of a centroid (K-mean like or principal direction divisive partitioning (PDDP) based approaches) which restricts their use to flat file datasets with numerical attributes. In applications that use dissimilarity matrices as inputs, centroids cannot be computed due to the lack of numerical attributes; consequently, novel divisive methods are needed for such datasets and introducing such methods is the main focus of this paper.

This paper proposes Avalanche, a novel top-down hierarchical clustering algorithm which uses as input a dissimilarity matrix. The problem is to come up with a test that splits a set into two subsets, maximizing an underlying objective function. During the top-down process a node is split into two sub-nodes such that the distance between the two sub-node clusters is maximized, and the sum of the "intra-cluster distances" of both clusters is minimized. To split a node, initially the example that is furthest away from the other examples — the *anti-medoid* — is assigned to the right sub-node and then — using the one nearest neighbor chain approach (1-NNC) — additional examples are progressively assigned to this node which are nearest neighbors of the previously added example as long as the objective function improves. Given a set of objects, the *medoid* is the centermost object (the one with the shortest total distance to the other objects in the set); the *anti-medoid* on the other hand, is the outermost object (the object with the largest total distance to the other objects in the set).The heuristic used by Avalanche to find a solution for the set split is a two-step method based on the one nearest neighbor chain approach (1-NNC). First, it selects the anti-medoid to start the node split and assigns it to the right node. Next, it follows the 1-NN chain that originates from the anti-medoid to add further examples to the right node until doing so degrades the objective function.

The general idea of the Avalanche algorithm is illustrated in Fig. 1. Figure 1(a) displays an initial dataset where s1 is the anti-medoid since s1 contribution to total intra-cluster distance is 24 (largest value). Next, the nearest neighbor of s1, s2, is selected as candidate for assignment. The objective function which will be explained in detail in Sect. 3 is computed to determine if adding s2 to the right node improves the inter-cluster distance between the two sets. The answer is "yes" and s2 is added. At this point the right node dataset is {s1, s2} and the left node dataset is {s3, s4, s5}. Figure 1 (b) shows the first node split. Since 1-NN(s2) = s3 the algorithm attempts to add s3 to the right node. However, adding s3 to set {s1, s2} will decrease total inter-cluster distance and increase total intra-cluster distance. Therefore the node split stops and the obtained clusters for this step are: {s1, s2} and {s3, s4, s5}. The tree construction continues in Fig. 1(c) where the sub-node {s3, s4, s5} is further split.

In Fig. 1(c) s5 is the anti-medoid. 1-NN(s5) = s4 however, adding s4 to right node would increase total intra-cluster distance and decrease inter-cluster distance, which would decrease the objective function. Therefore the split stops and we obtain {s3, s4}

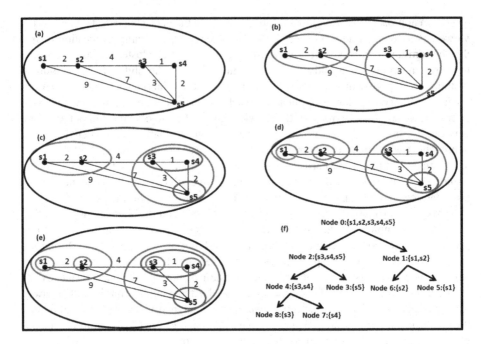

Fig. 1. Avalanche splitting process starts from the outermost example toward the center of the cluster

and {s5}. Figure 1(d) and 1(c) show that the two-object clusters are further split. The final tree is shown in Fig. 1(f). More details of this approach are provided in Sect. 3.
The main contributions of this paper include:

1. The introduction of a new unsupervised approach to divisive clustering which uses both global and local information in its splitting decision. Most divisive approaches make splitting decision based on global statistics information about the dataset (variance, sum of square errors). Likewise, most agglomerative approaches make merging decision based on proximity of pair of examples which is perceived to be local information. Avalanche splitting decision is based on both concepts since it computes the anti-medoid from the dataset (global) then uses 1-NNC approach to merge neighboring examples in the proximity of the anti-medoid (local).
2. A new unsupervised approach to top-down hierarchical clustering which takes as input a dissimilarity matrix. Divisive algorithms that have been proposed in the literature so far cannot be applied to datasets were examples do not have any attributes.
3. The introduction of a novel objective function which considers the inter-cluster distance in addition to the intra-cluster distance when evaluating different clustering solutions.
4. Incremental methods are proposed which save time in the objective functions computations.
5. The algorithm runs faster than its main competitor while maintaining comparable or better clustering results!

The rest of the paper is organized as follows. In Sect. 2 we present related work. Section 3 details the proposed algorithm. Finally, in Sect. 4 we discuss results of experimental evaluation.

2 Background and Motivation

Hierarchical clustering approaches are either agglomerative or divisive. In order to decide which clusters should be combined/divided they require some evaluation measure. Divisive approaches rely heavily on Ward's minimum variance method [19] which states that the cluster with the largest sum of square error should be chosen for the split. The agglomerative approaches, on the other hand, rely on dissimilarity among the clusters/examples to make their merging decision. An important choice required in agglomerative hierarchical clustering is how to measure the distance between the clusters. Commonly used distances — often known as "linkages" — include complete link, single link, average link, centroid link [2, 5, 8, 9, 17]. Given two clusters C_i, and C_j, and x_{ip}, x_{jq} examples in cluster C_i, and C_j respectively; the different distance measures are defined as follows:

- Single link: $d(C_i, C_j) = \min\{d(x_{ip}, x_{jq})\}$ which is the smallest distance between an element in one cluster and an element in the other
- Complete link: $d(C_i, C_j) = \max\{d(x_{ip}, x_{jq})\}$ which is the largest distance between an element in one cluster and an element in another cluster
- Average link: $d(C_i, C_j) = \text{avg}\{d(x_{ip}, x_{jq})\}$ which is the average distance between elements in one cluster and elements in the other
- Centroids/medoids link: $\text{cen}\{d(x_{ip}, x_{jq})\}$ or $\text{med}\{d(x_{ip}, x_{jq})\}$ which is the distance between two centroids/medoids of the two clusters. A centroid is the central point of a cluster (mean point), while a medoid is an actual example which has the smallest largest distance to any point in the cluster.

The most popular agglomerative approach is the average link approach also known as the **Un**-weighted **P**airs **G**roup **M**ethod with **A**rithmetic **M**ean (UPGMA) algorithm. It takes as input a distance matrix. When UPGMA is used the distance between two clusters C_1, and C_2 is defined as

$$d(C_1, C_2) = \frac{1}{|C_1||C_2|} \sum_{x \in C_1} \sum_{y \in C_2} d(x, y)$$

where $|C_1|$ and $|C_2|$ denote the number of examples in cluster C_1, and C_2 respectively. First, it locates the two closest examples/clusters in the matrix, and merges them into a cluster, say C_k; that is $C_k = C_1 \cup C_2$. The rows and columns occupied by the examples/clusters thus merged are removed from the matrix and a new row and column for C_k are entered. For any cluster C_i the distance between C_i and C_k is computed as

$$d(C_i, C_k) = \frac{|C_1|}{|C_1| + |C_2|} d(C_1, C_i) + \frac{|C_2|}{|C_1| + |C_2|} d(C_2, C_i)$$

Divisive hierarchical clustering is a top-down approach which starts with the root node having all the data associated with it, and the approach recursively splits it into two node/set pairs until sets containing one example are obtained or it terminates earlier if a predefined termination condition is met. The most popular approach is the "Bisecting K-mean" approach [6, 7, 10, 11, 14, 16]. To split a node, it first computes the centroid of the dataset. Next, it iteratively computes the centroids K_1, and K_2 of two regions of the dataset using K-mean (K = 2). Then it bisects the data with a hyper-plane passing through the centroid and perpendicular to segment K_1K_2. The process continues until a termination condition is reached. Another divisive approach is the principal direction divisive partitioning (PDDP) algorithm [3]. It computes the singular value decompositions (SVD) and derives the principal direction then divides the dataset with a hyper-plane passing through the origin and perpendicular to the principal direction vector. The principal direction of the data is its direction of maximum variance [15]. Recently a hybrid method has been proposed [12]. It uses both agglomerative and divisive hierarchical clustering algorithms to get the best of both approaches. In a first step, the algorithm uses the bisect K-means approach to cluster the dataset in K′ clusters and then uses UPGMA on the centroids of the computed K′ clusters. If two centroids end up in same cluster then all of their examples are merged in one cluster. This approach is different from our proposed method in that it applies sequentially the top-down method then the bottom-up method, whereas our approach incorporates bottom-up merging criteria into its top-down splitting decision. One of the main drawbacks of both the bisect K-mean and PDDP is that they are sensitive to outliers.

Divisive algorithms proposed thus far are unsuitable for clustering datasets where the only input is a distance matrix because they rely on centroids computation which is not possible with distance matrix. For those applications we propose Avalanche, a top-down hierarchical algorithm that does not use centroid in its splitting method. The detail of the Avalanche algorithm is provided in Sect. 3.

3 The Avalanche Approach

The general idea of the Avalanche approach is that at each intermediate node the algorithm tries to split the node into two child-nodes (binary split) such that the distance between the clusters associated with the child nodes is maximum and the sum of the "within cluster" distances is minimum. Table 1 summarizes the notations that are used throughout the paper.

3.1 Problem Definition

Let T_P be a node, and T_L, and T_R its associated sub-nodes (to be computed). We let I (T_L,T_R) be the inter-cluster distance between the sub-nodes and $U(T_L)$ and $U(T_R)$ be respectively the intra-cluster distance of node T_L and T_R. The problem is to split T_P into T_L, and T_R in such a way that inter-cluster distance between T_L and T_R, $I(T_L,T_R)$, is maximized and total intra-cluster distance, $U(T_L) + U(T_R)$, is minimized.

This splitting problem can formally be defined as follows.

Table 1. Notations and symbols used in this paper

Notation	Description
D	Input matrix
$D(x_i, x_j)$	Distance between object x_i, and x_j
T_p, T_L, T_R	Respectively parent node and its associated left, and right nodes
$I(T_L, T_R)$ or I	Total inter-cluster distance between node T_L, and T_R
$U(T_P)$ $U(T_L), U(T_R)$	Sum of all distances between objects within a given set; respectively in parent node T_P, left node T_L, and right node T_R
$U(x_i)$ or $U(x_i, T)$	Sum of distances from object x_i to every object in a given set T.
$H(T_L, T_R)$ or H	The objective function
1-NN(x_i)	The nearest neighbor object to x_i

Given a set T_P of objects $\{x_1,..., x_n\}$ find two sub-sets T_L and T_R, with $T_P = T_R \cup T_L$ and $T_R \cap T_L = \varnothing$ such that:

$$H(T_L, T_R) = (1 - \alpha) I(T_L, T_R) - \alpha [U(T_R) + U(T_L)] \tag{1}$$

is maximized.

α is a parameter between 0 and 1 which balances the importance of each objective.

When $\alpha = 1$ one favors to minimize the intra-cluster distance over maximizing the inter-cluster distance. Similarly when $\alpha = 0$ one favors to maximize the inter-cluster distance over the intra cluster distance. We note that intra-cluster distance of the parent node is equal to the sum of intra-cluster distances of both sub-nodes plus inter-cluster distance between both sub-nodes.

$$U(T_P) = I(T_L, T_R) + [U(T_R) + U(T_L)] \tag{2}$$

We also remark that $U(T_P)$ is a constant value because assigning an example to left or right node, does not change total intra-cluster distance of the parent node. Substituting $I(T_L, T_R) = U(T_P) - [U(T_R) + U(T_L)]$ into Eq. (1) we obtain $H(T_L, T_R) = - [U(T_R) + U(T_L)] + U(T_P)(1 - \alpha)$. Since $U(T_P)(1 - \alpha)$ is a constant value, maximizing H is equivalent to minimizing $[U(T_R) + U(T_L)]$. Likewise replacing $[U(T_R) + U(T_L)] = U(T_P) - I(T_L, T_R)$ into Eq. (1) we obtain $H(T_L, T_R) = I(T_L, T_R) - \alpha U(T_P)$. Since $\alpha U(T_P)$ is a constant value, maximizing H is equivalent to maximizing $I(T_L, T_R)$. Therefore the objective function expressed in Eq. (1) is equivalent to the objective function given by Eq. (3) or equivalently by Eq. (4).

$$\text{Maximize } H_1(T_L, T_R) = I(T_L, T_R) \tag{3}$$

$$\text{Minimize } H_2(T_L, T_R) = [U(T_L) + U(T_R)] \tag{4}$$

We note that maximizing/minimizing the objective function is independent of α. The total inter-cluster distance, $I(T_L, T_R)$, is given by Eq. (5) and total intra-cluster distances for node T_L is by given by Eq. (6).

$$I(T_L, T_R) = \sum_{\substack{x \in TL \\ y \in TR}} D(x, y) \tag{5}$$

$$U(T_L) = \sum_{\substack{x, y \in Tl \\ x \neq y}} D(x, y) \tag{6}$$

Finding the optimum solution for H_1 or H_2 is computationally complex. A somewhat similar problem is the well-known "the balanced number partitioning problem" [20] which is known to be NP-hard. Computing the optimum solution for our objective function may require that all possible assignments to either set be tried out. Avalanche proposes a heuristics to split the dataset which is based on identifying first, the anti-medoid and then assigning it to a cluster (previously empty set). Next, neighboring examples to the anti-medoid are added one at a time to grow the cluster so long as the objective function improves. Therefore unlike traditional bottom up hierarchical approaches that consider only proximity of clusters as sole criterion for merging (local information), or top-down hierarchical clustering which use variance reduction as splitting strategy (global information), the proposed algorithm uses both local as well as global information about the data to recursively split it with the aim of obtaining improved clustering result. Our approach incorporates global information about the dataset because it computes the anti-medoid from of the dataset, and local information is considered by identifying examples in the neighborhood of the anti-medoid. The detail of this method is provided in Sect. 3.2. Since both Eqs. (3) and (4) are equivalent, for the rest of the paper we consider Eq. (3). Avalanche computes an approximate value for the objective function which we describe in Sect. 3.2.

3.2 Node Evaluation

An approximated solution to H_1 is provided in this section. We use T_L, T_R, and T_P to not only designate the parent node and both child-nodes but also to refer to the datasets in the respective nodes. Given T_P a parent node, the algorithm starts with one of the sub-node being an empty set, (say $T_R = \emptyset$) and the other sub-node filled with all the data of the parent node, (say $T_L = T_P$). Next, it attempts to move examples from T_L to T_R one example at a time, maximizing H_1. One important challenge is how to determine the examples that need be removed from T_L to T_R. A brute force approach would be to try all possible splits then select the one that maximizes the objective function. Such an approach would be impractical. We use the nearest neighbor chain approach (1-NNC) for this purpose. Firstly, it is cost effective to compute 1-NNC from an input distance matrix. Secondly, by adding the examples one at a time incremental optimization of the objective function can be achieved. With this approach, first the anti-medoid is computed and assigned to T_R. Then its nearest neighbor, S_{last}, is computed and tentatively assigned to T_R. If H_1 improves when S_{last} is assigned to T_R then the assignment holds and the nearest neighbor of S_{last} residing in T_L is computed and tentatively assigned to T_R; the chain of assignment continues until H_1 does not improve. To compute the anti-medoid (the outermost object) Avalanche first computes the contribution of each example to total intra-cluster distance of T_L then selects the object with the largest value.

This process is detailed in Algorithm 1 which gives the pseudo code of the Avalanche algorithm.

Algorithm 1: Node Evaluation for the Top-Down Tree

1. Input: T_P ## Parent node
  2. Outputs: T_L, T_R ## Sub-nodes
  3.
  4. $T_L \leftarrow T_P$; $T_R \leftarrow \emptyset$
  5. IF size of $T_L = 1$ THEN
  6.   Return;
  7.
  8. $S_{last} \leftarrow$ Anti-Medoid(T_L)
  9. current_H$\leftarrow$ Compute $H(T_L, T_R)$
10. WHILE (TRUE)
11.   previous_ H$\leftarrow$ current_H
12. $T_R \leftarrow T_R \cup \{S_{last}\}$
13. $T_L \leftarrow T_L - \{ S_{last}\}$
14.   current_H$\leftarrow$ Compute $H(T_L, T_R)$
15.   IF ((current_H - previous_H) > 0) THEN
16. $S_{last} \leftarrow$ 1-NN(S_{last})
17.   ELSE
18.   exit
19. END WHILE

If T_L has only one example, there is no need for a split (line 5–6); otherwise the anti-medoid is computed. The algorithm then enters the loop and iteratively uses 1-NN approach to move objects from T_L to T_R as long as H_1 improves (line 10–18). In line 12 to 14 both sets are updated (tentatively) and a new H_1 computed. If there is no improvement in H_1 the algorithm exits the loop (line 18); otherwise the assignment is confirmed and the algorithm fetches from T_L the nearest object to the object last assigned to T_R (1-NN(S_{last})), and assigns it to S_{last} (line 16).

3.3   Implementation

The anti-medoid (line 8) is computed using Eq. (7). Equation (7) computes each object x_i's ($x_i \in T_L$) contribution to total intra-cluster distance of T_L.

$$U(x_i, T_L) = \sum_{k=1}^{N} D(x_i, x_k) \tag{7}$$

$U(x_i, T_L)$ can be stored as a vector say, vU. Matrix D is augmented with a row representing vU (last row of D).When object x_k located at row k of matrix D is moved to T_R a new row is entered in matrix D as new_vU $\leftarrow$ old_vU $-$ row(k,D) where row(k, D) means "row k of matrix D". Doing so updates both the intra-cluster distances for T_L and the inter-cluster distances at the same time in vector vU. Thus to compute $I(T_L, T_R)$ we only need to sum the values in the cells of vU corresponding to the objects already moved to T_R; and to compute $U(T_L)$ we sum the values in the cells corresponding to objects still remaining in T_L.

3.4   Runtime Complexity

Given a node dataset of size t, it cost $O(t^2)$ to compute the anti-medoid (to construct vector vU and to select the largest value from it). It cost O(t) to compute H_1 from vector vU; therefore the cost to split a node is $O(t^2) = O(t^2) + m*O(t)$ where m (m < t) is the number of objects assigned to T_R. When only one example is assigned per node split, H is computed only 1 time (per node split); therefore it cost $O(n^2)$ for splitting the root node, then O $((n-1)^2)$ to split the two nodes at depth 1, etc. and hence $O(n^2 + (n-1)^2 + \cdots + 2^1) = O$ (n^3) to build the tree. In case half of the examples are assigned per node split, it cost $O(n^2) = O(n^2) + n/2*O(n)$ to split the root node, $O([(n/2)^2 + (n/2)^2] = O(2*(n/2)^2)$ to split both nodes at depth 1, etc. Therefore the cost to build the tree in this scenario is $O(n^2 + 2^1*$ $(n/2^1)^2 + 2^2*(n/2^2)^2 + \cdots + 2^p*(n/2^p)^2]$ where p = log(n). This value can further be simplified as $O(n^2(1 + (1/2^1)^2 + \cdots + (1/2^{p-1})^2) = O(n^2 + n^2 s_n)$ with $s_n =$ $(1/2^1)^2 + \cdots + (1/2^{p-1})^2 < 1$. Therefore total cost is $O(n^2)$. That it cost $O(n^2)$ for the best case scenario and $O(n^3)$ for the worst case scenario to build the tree.

3.5   Illustrating How Avalanche Splits Nodes

Figure 2 demonstrates one implementation of algorithm 1. Initially, we assume that $T_L \leftarrow T_P$ and $T_R = \emptyset$. The highlighted columns contain data moved to T_R and the white columns represent data still in T_L. We use H_1 as objective function for this illustration. Column H hosts the values of the objective function and the last row of the table contains the current vector vU.

Figure 2(a) hosts the input matrix of dataset shown in Fig. 1, augmented with the initial vector vU in the last row. We note row(U_0) to mean "vector vU located in row U_0 of the table". At this early stage no example has yet been assigned to T_R. Next, the algorithm computes Umax, the largest value in row(U_0) and found that U (s1) = Umax = 24. Hence the anti-medoid is s1 and s1 is assigned it to T_R.

In Fig. 2(b) row(s1) and column(s1) are highlighted to signify that T_L = {s2, s3, s4, s5}, T_R = {s1}. Next, a new vU is computed. To do so, a new row is entered at the bottom of the table and updated as row(U_1) $\leftarrow$ row(U_0) $-$ row(s1). This is so because after s1 has been assigned to T_R, we must subtract its contribution to T_L's intra-cluster distances $(U(T_L) \leftarrow U(T_L) - U(s1))$ leading to U(s2) = 16, U(s3) = 8, U(s4) = 8, and U (s5) = 12. Hence the new $U(T_L)$ = U(s2) + U(s3) + U(s4) + U(s5) = 44. $U(S_1)$ contains the contribution of s1 to total inter-cluster distance. Since T_R only contains s1, it comes that H = $I(T_R, T_L)$ = I({s1}, {s2, s3, s4,s5}) = U(s1) = 24.

In Fig. 2(c) the algorithm computes the nearest neighbor of s1; 1-NN(s1) = s2 (circled value); s2 is tentatively assigned to T_R. This is shown in Fig. 2(d) where column(s1) and column(s2) are highlighted. A new row is entered and its values updated as row(U_2) ← row(U_1) − row(s2). H = I({s1, s2},{s3, s4, s5}) = U(s1) + U (s2) = 22 + 16 = 38. (We remark that the intra-cluster distance of T_L is now $U(T_L)$ = U (s3) + U(s2) + U(5) = 4 + 3+5 = 12).

Next, in Fig. 2(e) 1-NN(s2) = s3. A new row, (row(U_3), is entered and updated in the same manner as previously done for row(U_2) and H = 16 + 12 + 4 = 32 < 38. Therefore s3 candidacy is rejected and the node split stops. The result of the node split is shown in Fig. 2(f).

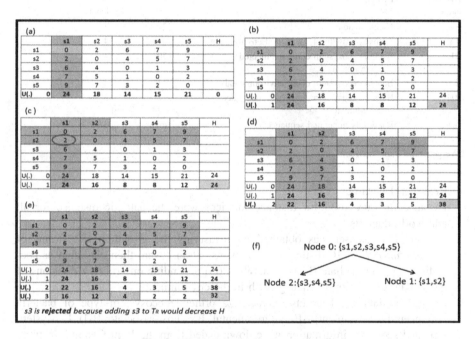

Fig. 2. Illustrating Avalanche node split

4 Experimental Evaluation

We compare the results of our proposed method to those obtained with the widely used UPGMA algorithm. This is because a direct comparison to divisive approaches that have been proposed in the literature is not possible as our approach uses dissimilarity matrices whereas current divisive approaches use the pair (attribute/attribute values). The current approach was evaluated using two types of performance criteria: (1) runtime complexity, (2) intra-cluster distances of the generated clusters. One artificial dataset and five real world datasets were used to evaluate the algorithms. All experiments reported in this paper were performed on a 64-bit PC i7-2630 CPU at 2 Ghz running Windows 7. With respect to the runtime evaluation, we generated various size datasets (artificial dataset). We implemented Avalanche and UPGMA in the same machine. UPGMA was implemented in its canonical form without optimization. We ran both algorithms and compared the average speeds (Sect. 4.2). We used total average intra-cluster distances to measure the quality of the clustering result (Sect. 4.3).

4.1 Datasets Description

We used the datasets summarized in Table 2 throughout the experiments.

Table 2. Datasets

Dataset name	Description	Size	Number of class labels
E.coli	Niche breadth	82	3
AV	Archaea growth rate in natural environment	70	4
BE	Bacteria ecosystem class: engineered environment	120	4
BV	Bacteria ecosystem class: environmental	311	4
BH	Bacteria ecosystem class: host-associated environment	571	3
Art	Randomly generated sequences	*many sizes	3

*We generated various sizes of this dataset

- Real-world Datasets

  Distance measure used for distance matrices were the patristic distance for all real-world datasets.

  E. coli: This dataset was obtained by measuring the growth of 82 strains of *Escherichia coli* in 10 distinct environments. Strains were then classified as specialists (S), intermediate (I), or generalists (G) depending on arbitrary divisions of the standard deviation of their growth in the environments.

  Ecosystem datasets: Datasets characterize principle ecosystem type of bacteria (engineered environment, BE; environmental, BV; host-associate, BH). Ecosystem type and sequence information were downloaded from the Joint Genome Institute website [21].

  In addition, dataset for growth rate of various types of archaea in natural environment was used in this experiment (AV).

- Artificial Datasets

  Art dataset: Artificial dataset was generated using a random sequence generator and have a sequence length of 200. The publicly available software MEG6 (Molecular Evolutionary Genetics Analysis version 6) [13] was used to compute the distance matrix using p-value distance.

4.2 Runtime

We used the artificial dataset with various sizes to evaluate the runtime speed of the approaches. The trees were fully grown until each leaf-node contains one example. Figure 3 summarizes the obtained results.

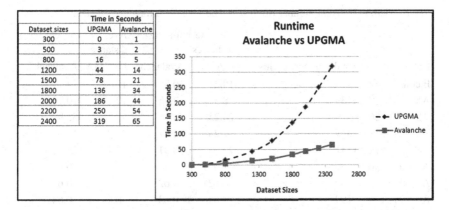

	Time in Seconds	
Dataset sizes	UPGMA	Avalanche
300	0	1
500	3	2
800	16	5
1200	44	14
1500	78	21
1800	136	34
2000	186	44
2200	250	54
2400	319	65

Fig. 3. Avalanche runtime complexity

As expected, the result in Fig. 3 confirms that Avalanche is a lot faster than the UPGMA algorithm (not optimized form of the UPMGA).

4.3 Intra-cluster Distance

To measure the compactness of clusters created by both algorithms, we use the average intra-cluster distances criterion. The weighted average method is used to compute overall average "within cluster" distances

$$\text{Av\_Intra}(X) = \sum_{C_i \in X} = (\frac{|C_i|}{N} * \text{intraclusterDistance}(C_i))$$

where N is the sum of the cardinality of all clusters C_i. $N = \sum_{C_i \in X}(|C_i|)$, X is the clustering result (set of clusters generated by the algorithm), $|C_i|$ cardinality of cluster C_i. We made the assumption that the lower the average intra-cluster distance, the better the algorithm with respect to generating compact clusters. Table 3 summarizes the obtained results. The result shows that Avalanche produces smaller size clusters on four datasets, tie on three datasets with UPGMA and lost on one dataset (slightly under perform on the E.coli dataset). The obtained result therefore supports the assumption made that incorporating local and global information into the splitting decision leads to improved clustering result.

Dataset Art20, Art50, and Art100 are three different sizes of the Art dataset. Using the AV dataset, we illustrate in Fig. 4 the number of clusters generated by Avalanche, the clusters' sizes, the inter cluster distance that was generated after the cluster was created, and the resulting average intra-cluster distance. The table in Fig. 4(c) contains in its first row sizes of clusters generated by Avalanche (on the AV dataset). The second row contains the number of clusters per size. For example, there are 70 clusters of size 1 (first column) and one cluster of size 70 (last column). It can be observed from Fig. 4 that with the AV dataset, the small size clusters do not generate large inter-cluster distances (in general). However, cluster of size 8 and cluster of size 19

Table 3. Total number of clusters and average intra-cluster distance

Dataset	Size	Total number of clusters	Average intra cluster distances		Average inter cluster distances	
			UPGMA	Avalanche	UPGMA	Avalanche
E.coli	82	163	**10.34**	10.5	15.32	14
AV	70	139	8.32	**7.98**	12.1	10.13
BE	120	239	26.25	**24.7**	43.96	52.9
BV	311	621	17.28	**14.67**	31.34	17.09
BH	571	1141	29.05	**24.31**	49.1	29.47
Art 20	20	39	0.6	0.59	0.61	0.6
Art 50	50	99	0.62	0.62	0.63	0.63
Art 100	100	199	0.63	0.63	0.64	0.64

generated after split, a large inter cluster distance of 30 (Fig. 4(a)). This indicates that the two clusters are well separated clusters. Likewise, cluster of size 27 and 43 generated another spike. The actual clusters (size 8 and 19) and (size 43 and size 27) are shown in Fig. 4 (b). As shown in Fig. 4(b) the first split generated clusters of size 27 and 43. Then the second split divided the cluster of size 27 into cluster of size 8 and cluster of size 19. By analyzing the inter-cluster distances and the size of the generated clusters the structure in the dataset can be understood.

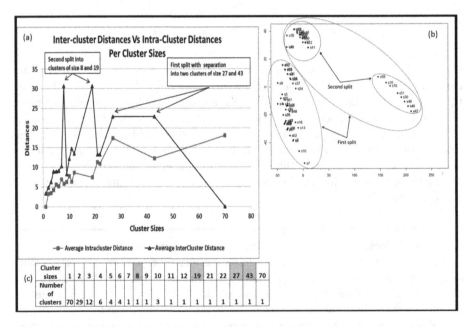

Fig. 4. Illustrating Avalanche ability to generate clusters that are both compact and far apart

5   Conclusion

This paper introduces a novel hierarchical clustering algorithm called Avalanche. Avalanche forms clusters by splitting node datasets using a nearest neighbor chaining approach that originates from the anti-medoid of the dataset to be split and the chaining approach is controlled by a novel objective function that can consider both intra-cluster and inter-cluster distances. An incremental method to save time in the objective functions computations is proposed. The approach takes as input a dissimilarity matrix and therefore can be a useful tool for applications where the dataset is formed by pairwise distances among the examples; taxonomy generation tools and molecular biology application need such capabilities. Divisive clustering algorithms that have been proposed in the literature cannot be used for such datasets because they rely on centroid computation which is not feasible when the input is a distance matrix. Unlike other approaches, Avalanche incorporates in its splitting decision local as well as global information which provides the algorithm with the capability of generating better clustering results. Experimental evaluations confirm that the new approach generates comparable or better clustering results than the well-known UPGMA algorithm.

References

1. Ao, S.I., Yip, K., Ng, M., Cheung, D., Fong, P.-Y., Melhado, I., Sham, P.C.: Clustag: hierarchical clustering and graph methods for selecting tag SNPs. Bioinformatics **21**(8), 1735–1736 (2005)
2. Bien, J., Tibshirani, R.: Hierarchical clustering with prototypes via minimax linkage. J. Am. Stat. Assoc. **106**, 1075–1084 (2011)
3. Boley, D.L.: Principal direction divisive partitioning. Data Min. Knowl. Disc. **2**(4), 325–344 (1998)
4. Chitta, R., Narasimha Murty, M.: Two-level k-means clustering algorithm for $k-\psi\psi$ relationship establishment and linear-time classification. Pattern Recogn. **43**(3), 796–804 (2010)
5. Defays, D.: An efficient algorithm for a complete link method. Comput. J. Br. Comput. Soc. **20**(4), 364–366 (1977)
6. Forgy, E.: Cluster analysis of multivariate data: efficiency versus interpretability of classification. Biometrics **21**, 768–780 (1965)
7. Gose, E., Johnsonbaugh, R., Jost, S.: Pattern Recognition & Image Analysis. Prentice-Hall, New York (1996)
8. Hastie, T., Tibshirani, R., Friedman, J.: The Elements of Statistical Learning; Data Mining, Inference and Prediction, 2nd edn. Springer, New York (2009)
9. Everitt, B., Landau, S., Leese, M.: Cluster Analysis, 4th edn. Arnold, London (2001)
10. Jain, A.K., Dubes, R.C.: Algorithms for Clustering Data. Prentice-Hall advance reference series. Prentice-Hall, Upper Saddle River (1988)
11. Jain, A.K., Murty, M.N., Flynn, P.J.: Data clustering: a review. ACM Comput. Surv. **31**(3), 264–323 (1999)
12. Murugesan, K., Zhang, J.: Hybrid bisect K-means clustering algorithm. In: 2011 Second International Conference on Business Computing and Global Informatization, pp. 216–219

13. Tamura, K., Stecher, G., Peterson, D., Filipski, A., Kumar, S.: MEGA6: molecular evolutionary genetics analysis version 6.0. Mol. Biol. Evol. **30**, 2725–2729 (2013)
14. Selim, S.Z., Ismail, M.A.: K-means-type algorithms: a generalized convergence theorem and characterization of local optimality. IEEE Trans. Pattern Anal. Mach. Intell. **6**(1), 81–86 (1984)
15. Savaresi, S.M., Boley, D.L., Bittanti, S., Gazzaniga, G.: Choosing the cluster to split in bisecting divisive clustering algorithms. In: SIAM International Conference on Data Mining (2002)
16. Steinbach, M., Karypis, G., Kumar, V. A comparison of document clustering techniques. In: Proceedings of World Text Mining Conference, KDD 2000, Boston (2000)
17. Sibson, R.: SLINK: an optimally efficient algorithm for the single-link cluster method. Comput. J. Br. Comput. Soc. **16**(1), 30–34 (1973)
18. Tan, P.-N., Steinbach, M., Kumar, V.: Introduction to Data Mining, 1st edn. Addison-Wesley, Boston (2005)
19. Ward Jr, J.H.: Hierarchical grouping to optimize an objective function. J. Am. Stat. Assoc. **58**, 236–244 (1963)
20. Mertens, S.: Computational the easiest hard problem. In: Percus, A., Istrate, G., Moore, C. (eds.) Complexity and Statistical Physics. Oxford University Press, Oxford (2006)
21. The Joint Genome Institute: https://img.jgi.doe.gov/cgi-bin/w/main.cgi (2015)

Text Mining

Author Attribution of Email Messages Using Parse-Tree Features

Jagadeesh Patchala, Raj Bhatnagar[✉], and Sridharan Gopalakrishnan

Department of Electrical Engineering and Computing Systems,
University of Cincinnati, Cincinnati, OH, USA
{patchajh,gopalasa}@mail.uc.edu
raj.bhatnagar@uc.edu

Abstract. Most existing research on authorship attribution uses various types of lexical, syntactic, and structural features for classification. Some of these features are not meaningful for small texts such as email messages. In this paper we demonstrate a very effective use of a syntactic feature of an author's writing - text's parse tree characteristics - for authorship analysis of email messages. We define author templates consisting of context free grammar (CFG) production frequencies occurring in an author's training set of email messages. We then use similar frequencies extracted from a new email message to match against various authors' templates to identify the best match. We evaluate our approach on Enron email dataset and show that CFG production frequencies work very well and are robust in attributing the authorship of email messages.

Keywords: Stylometry · Emails · Author attribution · Grammar productions

1 Introduction

Authorship attribution has been pivotal in identifying authors of texts and solving various authorship controversies in works of literature [1], in micro-blogging messages [2], and for identifying original author of emails [3]. The attribution of authorship is accomplished by extracting various types of lexical, syntactic and semantic information from the texts and forming necessary features to help discriminate among the authors.

When dealing with a closed set of authors we need to design a classifier to identify one author from among a fixed set of known authors. When dealing with a situation in which the real author may not be among the known set of authors, the classifier needs to include the possibility of *"none-of-these"* in addition to the known set of authors.

There are numerous authorship attribution studies that deal with larger sized texts such as books, newspaper articles, and journal papers. In this paper, we discuss the problem of authorship attribution for much smaller bodies of text such as email messages. Usually email messages are short and certain lexical

© Springer International Publishing Switzerland 2015
P. Perner (Ed.): MLDM 2015, LNAI 9166, pp. 313–327, 2015.
DOI: 10.1007/978-3-319-21024-7_21

features such as lengths of paragraphs, and the average number of words in a sentence are not meaningful. The authors in [3] have included content specific features (greeting or farewell messages) for attributing authorship of email messages. Some authors include such content specific features in their emails but every message does not contain them. It is easy to deceive the authorship attribution system that depends primarily on content specific features by just copying some features used by a different person. Syntactic features represent characteristics of messages in terms of writing styles and language proficiencies of authors and are therefore more reliable for classification. These features are innate to authors' language capabilities and get embedded in the writings without authors' conscious attention. An author's grammatical style does not change with the topic, context, and length of the message, and therefore we can use the grammar features to build a relatively topic and context independent authorship attribution system. We have presented in this paper the results of the use of relative frequencies of the grammar's production rules as the feature set for author attribution. The CFG grammar productions can be seen in the parse trees of sentences and the relative frequencies of various CFG productions for a text can be easily determined. These relative frequencies, we believe, represent an innate aspect of the language-writing style of an author and are expected to be unique to an author, especially when compared within a small set of authors. We show by the results presented in this paper that the grammar production frequencies are indeed very informative, effective, and robust features for author attribution.

Figure 1 shows a few productions in which the non-terminal S represents a sentence, NP represents a noun phrase, VP represents a verb phrase, etc. We are proposing to count the number of times each production rule is used while parsing the text written by an author. Typically, for a natural language such as English, popular parsers depend on at least a few thousand potential productions. We seek to use from among these only a few hundred most frequent and informative productions.

Production rule frequencies, by themselves, have not been tested as features for author attribution in any of the approaches in the existing literature. The work presented in [4] mentions a group of syntactic features for large text messages and that group includes n-grams, function words, and high level grammar rewrite rule frequencies. These features as a group are used to train an SVM clas-

S -> NP VP

NP -> DT NN

NP -> PRP

NP -> NN

VP -> VB NP

VP -> VB PP

PP -> PRP NP

Fig. 1. Sample CFG grammar productions.

sifier to separate messages written by a very small number of (three) authors. The work presented in [5] uses production frequencies but seeks to discriminate only between two authors' large texts by building a PCA based discriminant in the space of production frequencies. Here we examine the effectiveness of CFG production frequencies, by themselves, and show that they are very effective in author attributions of small texts in groups as large as thirty authors.

Each author template and each new message's signature consist of relative frequencies of grammar productions and these can be viewed as probabilities of grammar productions in an author's writings. We normalize the relative frequencies of grammar productions included in templates to sum to 1.0. We then use divergence measures to determine the differences between the probability distributions represented by message signatures and author templates.

2   Related Work

Traditional feature sets that have been used for relatively large texts such as long articles and literary works are not suitable for email messages due to small sizes and very different overall composition [3] of the email messages. For example, the n-grams are a popular set of features for use with large texts but for short email messages the n-gram signatures vary too much from one message to the other for the same author. Results presented in [6] have used various lexical and syntactic features for email authorship analysis. They have used ratio of short words (word size <3) to the total number of vocabulary words, ratio of words used once (Hapax legomenon) to the total number of vocabulary words, and function-words attributes as features. Authors in [7] have reported that the frequencies of such lexical features do not discriminate among the authors of email messages. They did observe some significant between-author differences particularly with the function word related features. However, they observed a significant drop in author attribution accuracy with the increase in the number of function words in the message texts. Authors of [8] have used the lexical and structure-specific items such as the greeting and farewell messages, and presence/absence of the sender signature and his/her contact information. They used Support Vector Machines (SVMs) to evaluate their approach and showed that combining SVM with other machine learning algorithms increased the accuracy of author attribution. Authors in [9] have used content free features such as vocabulary richness and frequency of function words, and also structural features and content specific features such as frequency of particular keywords and special characters for attribution of authors to messages. With an author set size of three, they achieved an accuracy of 75 % with content free features alone and the combination of content free words and structural features increased the performance to 81 %. Authors in [10] have presented a clustering based approach for email authorship analysis and have evaluated their approach on the Enron email messages dataset. They used lexical features and function words' frequencies as features. They reported an accuracy of 90 % for 5 authors, 80 % for 10 authors. With 20 authors, they achieved an accuracy of 74 % [11]. For our proposed feature set and template-based classification approach we show in this paper significantly better accuracy and also robustness in author attribution.

Most of the previous research on authorship analysis has followed two types of approaches. The first are the nearest-neighbor type of methods that keep multiple training data samples for every author and develop a k-nearest neighbor type of classification model [4,6]. In the second type of approach [12,13] the combined corpus of all training messages for an author is used to extract the features and characterize an author's writing style. Thus, every author has a profile built for him or her in terms of the shared features across all the messages. The authorship of the test email message is assigned to the author whose features' profile best matches the signature of the test message.

3   Methodology

After concatenating all the training messages of an author in a single text file we generated parse trees for all the sentences. We used the publicly available Stanford statistical parser [14] to separately parse the message texts for each author. We then went through the parse trees and counted the frequencies of various grammar productions. A template is then built for each author from these production frequencies. We have fifteen authors in our pool and for each author we selected the most frequent 250 grammar productions to be included in the authors' style template. It is possible that some productions are encountered in the writings of only a few authors and not in the writings of all the authors. We performed a union of the sets of the top 250 productions for each author and this gave us a set of 265 grammar productions, implying that there were 15 productions that were missing from the most frequent set of 250 productions for some authors. The main steps followed to create the author profiles are summarized in Fig. 2.

The frequencies in a template represent the writing style of an author and we generate a template for each author in our selected set of fifteen authors. We

---

1. Combine all the training email messages written by an author into a single text file.
2. Parse the sentences in this text file and generate the parse trees for sentences.
3. Count the occurrence of each grammar production in the parse trees of the author text.
4. Take the frequent 250 productions (in terms of their counts of occurrence) for each author.
5. Construct a union set of the frequent 250 productions taken from each author. These productions constitute the common template structure. Each template for an author contains the relative frequencies for these productions found in the author's training text.
6. Normalize the frequencies in a template so that they all add up to 1.0

Fig. 2. Steps to generate the author templates

"The project coordinators believe they can reach a solution with Accenture in which Enron pays no additional fees but the project is completed in a slightly scaled back form. This allows Accenture to have a completed project and no more cash goes out the door."

Fig. 3. Sample message from training emails

```
(ROOT
  (S
    (S
      (NP (DT The) (NN project) (NNS coordinators))
      (VP (VBP believe)
        (SBAR
          (S
            (NP (PRP they))
            (VP (MD can)
              (VP (VB reach)
                (NP (DT a) (NN solution))
                (PP (IN with)
                  (NP
                    (NP (NNP Accenture))
                    (SBAR
                      (WHPP (IN in)
                        (WHNP (WDT which)))
                      (S
                        (NP (NNP Enron))
                        (VP (VBZ pays)
                          (NP (DT no) (JJ additional) (NNS fees)))))))))))))
    (CC but)
    (S
      (NP (DT the) (NN project))
      (VP (VBZ is)
        (VP (VBN completed)
          (PP (IN in)
            (NP (DT a)
              (ADJP (RB slightly) (JJ scaled))
              (JJ back) (NN form))))))
    (. .)))
(ROOT
  (S
    (NP (DT This))
    (VP (VBZ allows)
      (S
        (NP (NNP Accenture))
        (VP (TO to)
          (VP (VB have)
            (UCP
              (NP (DT a) (VBN completed) (NN project))
              (CC and)
              (S
                (NP (DT no) (JJR more) (NN cash))
                (VP (VBZ goes)
                  (PRT (RP out))
                  (NP (DT the) (NN door)))))))))
    (. .)))
```

Fig. 4. Parse tree generated by Stanford parser for email text in Fig. 2.

have chosen most frequent 250 productions of each author for inclusion in the templates. This choice is on a very high side for the number of features needed to identify an author. Our results with including only 150 frequent productions for each author were only very slightly worse than those with the frequent 250 productions. We experimented with different techniques in selecting the features and found out that the most frequent features have high discrimination potential. If we increase the number of productions beyond 250 in author template, the accuracy decreases as the newly added features are adding a lot of noise. Figures 3 and 4 show a sample email text and its corresponding parse tree generated by the parser. Figure 5 shows a typical production frequency template for an author.

To determine the author of an email message in the test sets, we parse the message text by the same parser and create a production frequency profile for the message similar to the author templates, containing the relative frequencies of grammar productions. Then, we use the divergence value as a metric to calculate the dissimilarity between the test message profile and each of the known author templates.

Divergence is a measure of the difference between two probability distributions. A low divergence value between two probability distributions indicates that the two distributions are similar to each other and a high divergence value

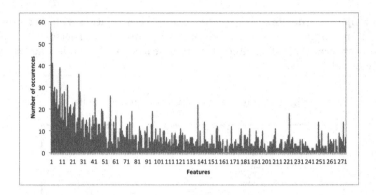

Fig. 5. Frequencies of productions in an author template.

means that the distributions are very dissimilar. We have used three different divergence metrics, namely, Kullback-Leibler (KL) divergence [15], Jeffrey's
Kullback-Leibler (J) divergence [16], and Arithmetic and Geometric mean (AGM)
divergence [17] for computing the dissimilarity between a message profile and
each of the authors' templates.

The three different measures of divergence have some semantic differences and
measure the differences between two distributions from different perspectives.
In this paper our main purpose in including these three different divergence
measures is to show the robustness of the grammar production features even
when different divergence measures are used. But the effect of using D_{KL} vs the
other two has also been highlighted in the results section.

There may be few productions in authors' templates and in message profiles that have zero probabilities assigned to them. The productions with zero
probabilities provides a large amount of information about the two distributions
being different. To maintain the spirit of getting large information contribution
while also preserving the computability of divergence value, we replace each zero
production frequency with a very small value. We have considered three cases
for substitution for zero frequencies (before normalization to make them sum to
1.0), namely, substituting each zero with 0.1, 0.25, or 0.5 in all the templates
and signatures uniformly. The lower the substituted value for the zeroes, the
more it enhances the influence on divergence value from a missing production in
an author's texts. The divergence values reported in this paper are the averages
for the three cases of the zeroes in the distributions being replaced by the three
small values.

A decision rule needs to be devised for assigning an author to an email
message's text. We consider two situations for creating such decision rules. The
first case is of a closed author set, when the author of the text message being
examined is contained in a set of known authors. In the second case it is assumed
that the author may or may not be in the set of known authors. Almost all
existing approaches work with the case of the closed author set and assign the
most likely author to the message. For the second case it may not be good

enough to assign an author to a message just because the divergence between the author's template and message profile is smallest among the divergence values of the message from all the authors' templates. It may happen that a message's divergence values from all the authors' templates are about the same in value and do not differ significantly from each other. It is expected that a message text's profile when compared to the template of its actual author will have much smaller divergence compared to its divergence from the templates of the other authors. We would like to place a larger confidence in our attribution of an author to an email message, and we achieve it as follows. We compute the divergence value of the test message with each of authors' templates. We determine the smallest of these divergence values and determine the z-score of this smallest divergence value in the context of the fifteen divergence values. That is, we find out the number of standard deviation values by which the smallest value is below the mean value of all the divergence values. Only when the smallest divergence value is significantly smaller than all the other divergence values, we get a high magnitude (and a negative sign) for the z-score of the smallest divergence value. We call this z-score as the *MinSep* (minimum's separation) value and compute it as follows:

$$MinSep(D) = [Min(D) - \mu(D)]/\sigma(D) \tag{1}$$

Here, D is the set of divergence values of a message against all the author templates. We have designed the decision rule for authorship attribution as follows. An author is attributed to an email message when the divergence value for the message profile and the author's template is the smallest among all the profile-template divergence values and the z-score of this smallest divergence value is less than or equal to -2.0. That is, the message profile is significantly closer to the selected author's template when compared to the templates of all the other authors. This second decision rule is certainly desirable for the case of non-closed set of possible authors, and may also be used for the case of closed author set to gain greater confidence in author attribution. For the non-closed set of authors it will help us detect the situations when no single author stands out as potential author and we can then say that the author is outside the set of known authors.

4   Results and Discussion

We performed the author attributions on two datasets created from the Enron email dataset. We have selected 2 distinct sets of 15 authors and have extracted a random set of 120 email messages sent by each author. We selected 60 email messages for each author as part of the training set and remaining 60 as test set. We considered only those emails that have at least five sentences.

We computed the author attribution accuracy for each of the three selected divergence measures - *KL*-divergence, *J*-divergence and *AGM*-divergence. The accuracy values indicate the overall prediction performance and are defined as follows:

Accuracy = Number of emails whose author is correctly identified/Total number of email messages tested for the authors. There are 15 authors in both the

Table 1. Accuracy values for dataset 1

Accuracy	KL divergence	J divergence	AGM divergence	Naïve Bayes	SVM
Test run1	91.0	79.09	84.32	81.73	89.03
Test run 2	82.73	74.56	78.64	78.29	84.31
Test run 3	84.60	78.23	76.87	75.62	83.49
Overall	86.11	77.29	79.94	78.54	85.61

Table 2. Accuracy values for dataset 2

Accuracy	KL divergence	J divergence	AGM divergence	Naïve Bayes	SVM
Test run1	83.28	72.31	54.32	63.18	85.27
Test run 2	77.35	66.72	57.64	69.24	70.11
Test run 3	81.23	71.36	49.87	74.74	78.42
Overall	80.62	70.13	53.94	69.05	77.93

datasets and we have three test runs, where we have 12 emails for each author in the test run 1 (4 sets with 3 emails in each run for each author), 24 emails for each author in test run 2 (4 sets with 6 emails in each run for each author), and 40 emails for each author in test run 3 (4 sets with 10 emails in each run for each author). Given a test email message, we parse it, construct its production frequencies' profile, and compute its divergence values from each of the 15 authors' templates.

Tables 1 and 2 show the accuracy results when we simply attribute the authorship of a message to the author whose template results in the minimum divergence value. We see that the performance results for our production rule based templates, for each of the three test runs, are better than the other instance based approaches for both the datasets. We used the Matlabs naïve Bayes classifier with multinomial distribution and LIBSVM [18] one-vs-one classifier for multiclass SVM based classification. The results in Tables 1 and 2 show that the frequencies of CFG productions in an author's text contain enough information to effectively discriminate the real author from the rest of the authors.

For both the datasets, we see that the performance using the KL divergence is superior to the performance obtained by the other two divergence measures. Explanation for this behavior is as follows. An authors template is generated using a large number of messages written by him/her and it is likely that most of the grammar productions will be observed in one or the other message by the author. When we create a production frequency profile for a single message it is likely that fewer productions will be observed in the parse tree of test message, mainly because it is a much smaller body of text. Therefore, typically a message profile has more productions having zero probability values than an author's template. As discussed above, we replace the zero frequencies in profiles and templates by a very small value (0.1, 0.25 or 0.5). We now look at how these productions with zero probability values affect the divergence value. Each term in the KL divergence summation is of the form: $P_i* \log(P_i/Q_i)$. The probability P_i comes from one distribution and the Q_i from the other distribution. If the message profile's distribution is used as the "Q"distribution and author's template is used as the "P"distribution then more of the exceptionally small values will appear in the denominators, making the values of the terms larger. And when message profiles are used as "P"distributions and author templates are used as the "Q"distributions then the exceptionally small probability values will be in the numerators resulting in the terms to have very small values. In our test runs we have used message profiles as "Q"distributions and therefore, there is a pronounced effect on increasing the divergence values whenever zero probability values are encountered in the message profile. When the message profile and author template have more of such zero-frequency-mismatches in probability values, the divergence value quickly become very large. It is expected that such zero-frequency-mismatches between a message profile and its true author will be fewer and therefore the divergence value will not get too large. This effect of the KL divergence provides an improved discrimination between the true author and the other authors in the pool. The other two divergence measures are symmetric in the sense that they consider the distributions P and Q in both roles, once in numerator and once in denominator, and then average the values of the two cases to give the final divergence values. Therefore, the effect of zero probability productions is not as pronounced for the J and AGM divergence measures.

Table 3. Accuracy values for different author sizes

Author set size	KL divergence	J divergence	AGM divergence
5	93.8 ± 1.4	89.7 ± 2.2	91.0 ± 2.3
10	90.3 ± 2.2	88.2 ± 1.7	88.6 ± 1.8
15	88.7 ± 1.1	82.5 ± 1.3	83.2 ± 2.7
20	82.9 ± 2.8	76.9 ± 2.2	75.3 ± 3.2
25	78.2 ± 3.2	68.1 ± 1.3	65.8 ± 1.8
30	74.6	65.8	60.1

We randomly selected author sets of 5, 10, 15, 20, 25 authors and calculated attribution accuracy using various divergence measures. We repeated the process 10 times and the accuracy values are reported in Table 3.

Figure 6 shows the effect of the feature set size for various feature types for 30 authors. We consider the frequencies of character trigrams obtained from the authors' texts as the features and treat these features as a baseline. We use the frequencies of function words (most frequent words) as other baseline to compare our production features. From Fig. 6, we can observe that grammar productions attained high accuracy compared to character trigrams and function words with few numbers of features. The reason for this could be that the emails are short and the character trigrams and function words vary a lot from one message to

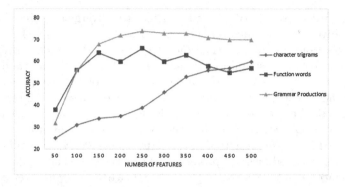

Fig. 6. Accuracy values vs number of features for various feature types.

the other with in an author whereas the grammatical style mostly remains stable from one email to the other for an author and this is well captured by the stylistic features. The accuracy of the productions decrease slightly after 250 features and the reason for this could be that beyond the first 250, the following grammar productions are too infrequent and cause more noise than adding information. From this, we can say that the performance of author attribution by our methodology is better than the results reported in the literature using other syntactic and lexical features of the messages.

We now consider the effect of using the stricter decision rule that announces an author as belonging to a text message only if the divergence of his/her template is at least two standard deviations below the mean of divergence values for all the author's templates. The accuracy values for the three test runs with the three divergence measures after applying the *"Min separation $<= -2.0$"* inference rule are shown in Tables 4 and 5. The results we got without applying this inference rule are presented in Tables 1 and 2. Even though the accuracy values decrease when we apply this inference rule, it provides a high level of confidence in the author being attributed to the message text. The *KL*-divergence still performs better than the other two divergence measures.

We analyzed the effectiveness of our inference rule by selecting different cut off values for z-score and accuracy, precision and recall values are plotted in Fig. 7. For this analysis, we considered 60 emails for creating the author profile,

Table 4. Accuracy values with z-score inference rule for dataset 1

Accuracy	KL divergence	J divergence	AGM divergence
Test run1	86.67	71.33	77.78
Test run 2	75.56	57.78	62.25
Test run 3	77.63	63.89	68.89
Overall	79.95	64.33	69.64

Table 5. Accuracy values with z-score inference rule for dataset 2

Accuracy	KL divergence	J divergence	AGM divergence
Test run1	74.67	65.17	47.31
Test run 2	63.92	52.65	50.72
Test run 3	72.35	61.94	43.74
Overall	70.31	59.92	47.25

and the remaining 60 emails as the testing data for each author. We observed that there is no significant increase in the precision values for z-score of 0, −0.5, and −1 but precision improves significantly for larger magnitudes of the z-score. The reason for this is that most of the misclassified messages do not have a significantly low divergence value when compared to the rest of the authors, and accuracy and recall suffer because our standard for attributing an author to a text has become stricter.

4.1 Non-Closed Author Set

In order to test and validate the effectiveness of our z-score based decision rule we performed the following test. We created a text message containing at least one sentence taken from each of the fifteen authors and named it *"jumbled"*. We also considered three sets of randomly selected texts from newspaper articles, namely *"random1"*, *"random2"*, and *"random3"*, and added them to the set of test messages. The reason to include them is to test whether our decision rule refuses to attribute authors to these jumbled and random texts. As shown in Fig. 8, the z-scores for the divergence values (*MinSep* values) of the closest templates for the jumbled and random texts come out to be between −1.4 and −1.7 in test run 1. Similar z-score values, between −1.3 and −1.5 for test run 2, and between −1.1 and −1.6 for test run 3 are encountered. As these *MinSep* values are not less than −2.0, our decision rule refuses to assign any of the known authors in the pool to these texts. Figure 8 plots the *MinSep* values for the fifteen email messages and four random texts. This stricter and more conservative decision rule also refuses to assign authors to 7.5 % of texts whose true authors are in the

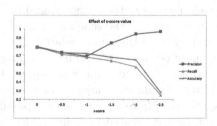

Fig. 7. Precision, recall, and accuracy values for different z-score values.

pool of authors, and those authors may have been picked up if the divergence value happened to be the smallest among all divergence values. The remaining 92.5 % of the texts in the test set were assigned the same authors that would have been assigned by the decision rule that went by selecting author with the minimum divergence value. It can be seen that the random texts always stay very high in terms of their *MinSep* scores and no author is attributed to them. This shows the effectiveness of our z-score based decision rule in case of a non-closed set of authors.

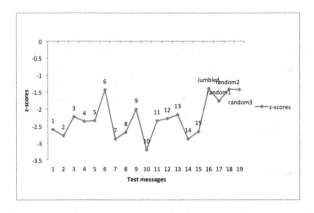

Fig. 8. Plot of min separation values.

We also selected 100 emails from authors who are not in our training author profiles and analyzed the effectiveness of our inference rule. With z-score threshold of −2.0, our inference rule rejected 79 emails and assigned an author for 21 emails. The average z-score of the 21 emails that we assigned some author is −2.29. That is, given a set of 30 authors profiles, 79 % of messages from non-members of this group were not assigned an author.

4.2   Robustness of Methodology

To analyze the robustness of our methodology, we performed tests to determine how much the results are affected when some noise is added to the text messages being tested. We created five different texts from an authors message as follows: T1 is the original email from the author containing eight sentences. T2 is obtained from T1 by removing half the sentences from T1. T3 is created from T2 by adding one random sentence from an unrelated essay to T2. T4 is created from T3 by adding one more random sentence from an English essay to T3. T5 is created by adding a random sentence from a newspaper to T4.

These five texts always contain the majority of the text from the original message's author. The average *MinSep* values after repeating our methodology five times with different random sentences being added are shown in Figs. 9 and 10 (for two different author's messages). In Fig. 9 the first author is the true author and the *MinSep* score is close to −3.0 before any alterations are

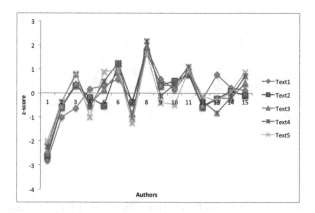

Fig. 9. z-scores of KL divergence for T1, T2, T3, T4, T5. T1 belongs to author one.

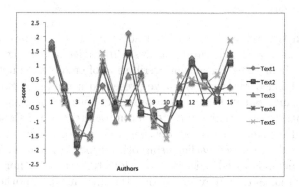

Fig. 10. z-scores of KL divergence for T1, T2, T3, T4, T5. T1 belongs to author three.

made to the text. For all the other authors in the pool (along the x-axis) the *MinSep* score is much higher than the -3.0 value. As the message is distorted slowly we see that the *MinSep* values degrade slowly. This test shows that while retaining at least the half of the original message and adding some noise text to the message the *MinSep* scores degrade slowly and not in some catastrophic manner to significantly alter the authorship attribution decision. The case shown in Fig. 10 shows a very similar behavior.

We created 4 different texts from an author's email as follows: T1 is the original email from author 9 containing 12 sentences. T2 is obtained by removing 5 sentences from T1.T3 is obtained by adding 2 sentences from author 3 email message to T2. T4 is created by adding an email (with 7 sentences) from author 3 to T2. The texts T1 and T2 are pure text written by author 9, where as T3 contains text from author 3 and 9 with majority text from author 9. T4 contains text from both the authors with same proportions. The results from Fig. 11 shows that our approach is robust to small changes in text from the same domain. However, for text T4 we got a low z-score for author 3 and we reject to assign an author for T4 as it has z-score less than -2.0.

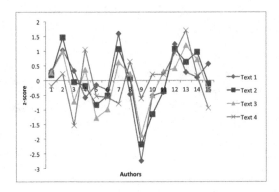

Fig. 11. z-score plot of KL divergence for texts T1,T2, T3, and T4

5 Conclusion

Most of the existing research in email authorship attribution has used the content specific, lexical and structural features. Properties of parsing trees of texts as features, by themselves, for author attribution of emails have not been examined in any published research. We have presented a template matching based methodology using relative frequencies of productions occurring in parse trees as component features of author templates. From our results we showed that this feature set and a divergence based decision rule performs very well, and better than the results reported earlier using other types of features. We have also presented and demonstrated a decision rule for attributing authors to messages when the potential author may be outside the known set of authors. We have also shown the robustness of our author attribution method in the presence of noise in text. In future, we plan to use the author stylistic information to capture the bogus reviews written by paid reviewers on websites.

References

1. Hope, W., Holston, K.: The Shakespeare Controversy: An Analysis of the Authorship Theories. McFarland, Jefferson (2009)
2. Sousa Silva, R., Laboreiro, G., Sarmento, L., Grant, T., Oliveira, E., Maia, B.: 'twazn me!!!;(' Automatic authorship analysis of micro-blogging messages. In: Muñoz, R., Montoyo, A., Métais, E. (eds.) NLDB 2011. LNCS, vol. 6716, pp. 161–168. Springer, Heidelberg (2011)
3. De Vel, O., Anderson, A., Corney, M., Mohay, G.: Mining e-mail content for author identification forensics. ACM Sigmod Rec. **30**, 55–64 (2001)
4. Gamon, M.: Linguistic correlates of style: authorship classification with deep linguistic analysis features. In: Proceedings of the 20th International Conference on Computational Linguistics, p. 611. Association for Computational Linguistics, Stroudsburg (2004)
5. Baayen, R., Van Halteren, H., Tweedie, F.: Outside the cave of shadows: using syntactic annotation to enhance authorship attribution. Literary Linguist. Comput. **11**, 121–131 (1996)

6. Teng, G.F., Lai, M.S., Ma, J.B., Li, Y. :E-mail authorship mining based on SVM for computer forensic. In: Proceedings of 2004 International Conference on Machine Learning and Cybernetics, pp. 1204–1207. IEEE Press, New York (2004)

7. De Vel, O.: Mining e-mail authorship. In: Proceedings of Workshop on Text Mining, ACM 6th International Conference on Knowledge Discovery and Data Mining (2000)

8. Nizamani, S., Memon, N.: CEAI: CCM-based e-mail authorship identification model. Egypt. Inf. J. **14**, 239–249 (2013)

9. Zheng, R., Qin, Y., Huang, Z., Chen, H.: Authorship analysis in cybercrime investigation. In: Chen, H., Miranda, R., Zeng, D.D., Demchak, C.C., Schroeder, J., Madhusudan, T. (eds.) ISI 2003. LNCS, vol. 2665, pp. 59–73. Springer, Heidelberg (2003)

10. Iqbal, F., Binsalleeh, H., Fung, B., Debbabi, M.: Mining writeprints from anonymous e-mails for forensic investigation. Digital Invest. **7**, 56–64 (2010)

11. Iqbal, F., Binsalleeh, H., Fung, B., Debbabi, M.: A unified data mining solution for authorship analysis in anonymous textual communications. Inf. Sci. **231**, 98–112 (2013)

12. Peng, F., Schuurmans, D., Wang, S., Keselj, V.: Language independent authorship attribution using character level language models. In: Proceedings of the 10th Conference on European Chapter of the Association for Computational Linguistics, pp. 267–274. Association for Computational Linguistics, Stroudsburg (2003)

13. Mosteller, F., Wallace, D.L.: Applied Bayesian and Classical Inference. Springer Series in Statistics. Springer, New York (1984)

14. Klein, D., Manning, C.D.: Accurate unlexicalized parsing. In: Proceedings of the 41st Annual Meeting on Association for Computational Linguistics, pp. 423–430. Association for Computational Linguistics, Stroudsburg (2003)

15. Leibler, R.A., Kullback, S.: On information and sufficiency. Ann. Math. Stat. **22**, 79–86 (1951)

16. Jeffreys, H.: An invariant form for the prior probability in estimation problems. Proc. Roy. Soc. Lon. **186**, 453–461 (1946)

17. Inder, J.E.T.A.: New developments in generalized information measures. In: Hawkes, P.W. (ed.) Advances in Imaging and Electron Physics, vol. 91, pp. 37–135. Academic Press, New York (2006)

18. Chang, C.C., Lin, C.J.: LIBSVM: a library for support vector machines. ACM Trans. Intell. Syst. Technol. **2**, 27:1–27:27 (2011)

Query Click and Text Similarity Graph
for Query Suggestions

D. Sejal[1]([✉]), K.G. Shailesh[1], V. Tejaswi[2], Dinesh Anvekar[3], K.R. Venugopal[1],
S.S. Iyengar[4], and L.M. Patnaik[5]

[1] Department of Computer Science and Engineering, University Visvesvaraya College
of Engineering, Bangalore University, Bangalore-1, India
sej_nim@yahoo.co.in
[2] National Institute of Technology, Surathkal, Karnataka, India
[3] Nitte Meenakshi Institute of Technology, Bangalore, India
[4] Florida International University, Miami, USA
[5] Indian Institute of Science, Bangalore, India

Abstract. Query suggestion is an important feature of the search engine
with the explosive and diverse growth of web contents. Different kind of
suggestions like query, image, movies, music and book etc. are used every
day. Various types of data sources are used for the suggestions. If we
model the data into various kinds of graphs then we can build a general
method for any suggestions. In this paper, we have proposed a general
method for query suggestion by combining two graphs: (1) query click
graph which captures the relationship between queries frequently clicked
on common URLs and (2) query text similarity graph which finds the
similarity between two queries using Jaccard similarity. The proposed
method provides literally as well as semantically relevant queries for
users' need. Simulation results show that the proposed algorithm outperforms heat diffusion method by providing more number of relevant
queries. It can be used for recommendation tasks like query, image, and
product suggestion.

Keywords: Image suggestion · Query suggestion · Query relevance ·
Recommendation

1 Introduction

Exponential growth of information on the Web is a challenging task for the
search engines to meet the need of the users. How organising and utilizing the
Web information effectively and efficiently has become more and more critical.
To get any information from web, the user issues queries, follows some links in
web snippets, clicks on advertisement, and spends some time on pages. If the
user is not satisfied with the information which he has received from the clicked
page, he reformulates his query. In order to enhance the user experience, it is a
common practice in a search engine to provide some types of query suggestion.

© Springer International Publishing Switzerland 2015
P. Perner (Ed.): MLDM 2015, LNAI 9166, pp. 328–341, 2015.
DOI: 10.1007/978-3-319-21024-7_22

On the Web, query suggestion is a technique to recommend a list of relevant queries to users' input by mining correlated queries from the previous knowledge. The simple way to suggest a query is spelling correction. In this paper, our interest is to suggest more elaborate forms of queries. For example, if the user submits a query *Java*, then user may be prompted to other queries like *Java for windows, Java 32 bit* or *Java download*, and also a related concept like *sun micro system*.

Basically, query suggestions on commercial sites like flipkart.com, Myntra.com etc., suggests products based on collaborative filtering [1,2]. Collaborative filtering is a method for automatic predictions about the interests of a user by collecting rating information or preferences from many users. Therefore collaborative filtering algorithm needs to build product-user matrix which determines relationship between user preferences to that product. The constraint with this approach is that in most of the cases, rating data are not available. However on the web, search log is always available to us. This is used to retrieve information about how people search information on the web and how they rephrase their query.

Motivation: There are several challenges to design suggestion framework on the web. First, most of the time users tend to submit short queries with only one to three terms. Short queries are more likely to be ambiguous. We observe that 9.82 % of web queries contain one term, 27.31 % of web queries contain two terms, and 26.99 % of web queries contain three terms. Second, in most of the cases, the users do not have enough knowledge about the topic they are searching for, and they are not able to clearly phrase the query words. Then, users have to rephrase the query words and rephrase their queries frequently. So, it is necessary to solve the above mentioned problems for query suggestion to satisfy users' information need and to increase search engine usability. Different types of data sources are used for suggestion on the web. In most cases, these data sources can be converted to graphs. We can solve many suggestion problems by designing a general graph suggestion approach.

Contribution: In this paper, we propose a generic method for the query suggestions on the web by using query relevance directed graph generated from a search log. We have constructed query click graph by capturing the relationship between queries frequently clicked on common URLs. Then, we have constructed query text similarity graph using Jaccard similarity between queries. Finally, we have combined query click graph and query text similarity graph to construct query relevance graph. This method has several advantages. It is a general method which can be used in many suggestion tasks such as query, image, and product suggestion. It can provide semantically relevant results to meet the original users' need and is scalable to a large dataset.

Organization: This paper is organized as follows: We have reviewed various query suggestion techniques under Sect. 2. Section 3 describes the Background Work. Section 4 presents Query Relevance model and algorithm. Section 5 discusses experiment results, query suggestion results comparison, and efficiency analysis. Finally, conclusions are presented in Sect. 6.

2 Related Work

It is a great challenge for any search engine to understand users' search intention. Various techniques have been studied extensively in the past decade to improve performance and quality of query suggestions. In this section, we have reviewed several papers related to query suggestion, query expansion and query term suggestion methods.

Udo et al. [3] have presented a systematic study on different query modification methods applied to query log collected on a local Web site. They have distinguished methods that derive query suggestions from previously submitted queries using logs or from actual documents. Mohamed et al. [4] have proposed a novel location aware recommender system (LARS) that uses location based rating to produce recommendations. LARS produces recommendations using taxonomy of three types of location based ratings within a single framework: (1) Spatial ratings for non-spatial items, (2) non-spatial ratings for spatial items, and (3) spatial ratings for spatial items.

Yang et al. [5] have proposed a new user friendly patent search paradigm, which can help users to find relevant patents more easily, and improve their search experience. They have developed three techniques, error correction, topic-based query suggestions, and query expansion, to make patent search more user-friendly. Brian et al. [6] have proposed a method for improving content-based audio similarity by learning from sample of collaborative filter data. First, a method for deriving item similarity is developed from a sample of collaborative filter data. Then, the sample similarity is used to train an optimal distance metric over audio descriptors. The resulting distance metric can then be applied to previously unseen data for which collaborative filter data is unavailable.

Gao et al. [7] have presented their work to cross-lingual query suggestion, i.e., for a input query in one language, it suggests similar or relevant queries in another language. Support Vector Machine (SVM) regression algorithm is used to learn the cross-lingual term similarity function. Aris et al. [8] have developed a framework which models the querying behaviour of users for query recommendation by a query flow graph. A sequence of queries submitted by a user can be seen as a path on this graph. It is based on the use of a Markov chain random walk over the query-flow graph. As user clicked information is not considered to create a graph, this approach results in lower accuracy.

Hossein et al. [9] have presented a new query recommendation technique based on identifying orthogonal queries in an ad-hoc query answer cache. This approach requires no training, is computationally efficient, and can be easily integrated into any search engine with an answer cache. Rodrygo et al. [10] have proposed a ranking approach for producing effective query suggestion. A structured representation of candidate suggestions is generated from related query with common clicks and common sessions. This representation helps to overcome data sparsity for long-tail queries.

Term Suggestion is a method in which, as the user types in queries letter by letter, suggest the terms that are topically coherent with the query. In [11,12], authors have recommended queries based on terms of the queries. A user can modify a part of the query by adding terms after the query, deleting terms within the query or modifying terms to new terms.

In [13–18] authors have used snippet information of clicked URLs or search results returned from a query for query recommendation in different ways. These methods are not general and the extensibility is very low.

Adam [19] has described a clustering-by-directions (CBD) algorithm for inter-active query expansion. When a user executes a query, the algorithm shows potential directions in which the search can be continued. The CBD algorithm first selects different directions, and afterward, it determines how the user can move in each direction. Pawan et al. [20] have provided theoretical analysis of a parametric query vector, which is assumed to represent the information needs of the user. A global query expansion model is derived based on the lexical association between terms.

Huanhuan et al. [21] have proposed a novel contexts aware query suggestion approach, which considers immediately preceding queries in query log as context in query suggestion. Qiaozhu et al. [22] have proposed a novel query suggestion algorithm ranking queries with the hitting time on large scale bipartite graph. Every query is connected with a number of URLs, on which the users have clicked when submitting query to a search engine. The weights on the edges present the number of times, the users used this query to access this URL.

Hao et al. [23] have presented a method to suggest both semantically relevant and diverse queries to web users. The proposed approach is based on Markov random walk and hitting time analysis on query-URL bipartite graph. Guo et al. [24] have proposed a method to recommend queries in a structured way for satisfying both search and exploratory interests of users. A Query-URL-Tag tripartite graph is obtained by connecting query logs to social annotation data through URLs. A random walk on the graph and hitting time is employed to rank possible recommendations with respect to the given query. Then, the top recommendations are grouped into clusters with label and social tags. This approach satisfies users' interest and significantly enhances users' click behaviour on recommendations.

Yang et al. [25] have introduced the concept of diversifying the content of the search result from suggested queries while keeping the suggestion relevant. First, query suggestion candidate is generated by applying random walk with restart (RWR) model to query-click log. Zhu et al. [26] have proposed query expansion method based on query log mining. This method extracts correlations among queries by analysing the common documents selected by a user. Then, expansion terms are selected by analysing relation between queries and documents from the past queries. Nick et al. [27] have applied a Markov random walk model on click graph to produce a probabilistic ranking of documents for a given query from a large click log. Click graph is a bipartite graph, with two types of nodes: queries and documents. An edge connects a query and document if the user has clicked for that query-document pair.

In [21–27], authors have used query-URLs clicked information to construct query-URL bipartite graph for query suggestion in various ways. There are two major problems with query-URL based query suggestion: (1) the number of common clicks on URLs for different queries is limited. (2) Although two queries may lead to the same URLs clicking, they may still be irrelevant because they may point to totally different contents of the web document. These methods ignore the rich information between two queries relevance. Our approach in this work is to consider relevance information between two queries to enhance user suggestion.

3 Background

Hao et al. [28] have proposed query suggestions based on heat diffusion method on directed query-URL bipartite graph. An undirected bipartite graph is considered where, $B_{ql} = (\ V_{ql},\ E_{ql}\ )$; $V_{ql} = (\ Q \cup L\ )$, $Q = \{q_1,\ q_2,\dots q_n\}$ and L = $\{l_1,\ l_2,\dots l_p\}$. $E_{ql} = \{(q_i,\ l_j),$ there is an edge from q_i to $l_j\}$ is the set of all edges. The edge (q_i, l_j) exists if and only if a user u_i clicked a URL l_k after issuing a query q_j. The weight on the edges is calculated by number of times a query is clicked on a URL.

This undirected bipartite graph cannot accurately interpret the relationship between queries and URLs. Hence, they are converted into directed query-URL bipartite graph. In this converted graph, every undirected edge is converted into two directed edges. The weight on a directed query-URL edge is normalized by the number of times that the query is issued. The weight on a directed URL-query edge is normalized by the number of times that the URL is clicked. Query suggestion algorithm is applied on the converted graph.

A converted bipartite graph G = (V ∪ U , E) consists of query set V and URL set U. For given a query q in V, a sub-graph is constructed by using depth first search in G. The search stops when the number of queries is larger than a predefined number. Then, heat diffusion process is applied on the sub-graph. Top-k queries are suggested with largest heat value. This method outperforms SimRank, Forward random walk and backward random walk.

Heasoo et al. [29] has developed online query grouping method by generating graph which combines the relationship between queries frequently issued together by users, and the relationship between queries frequently leading to clicks on similar URLs. Related query clusters are generated by Monte Carlo random walk simulation method for a given query.

4 Query Relevance Model and Algorithm

4.1 Problem Definition

Given a user input query q and search log of search engine, we convert search log into a graph, where nodes represent queries and edges represent relationship between queries. The objective is to provide semantically relevant query suggestions to meet the original users' need.

4.2   Assumptions

It is assumed that the user is online and enters input query with less than six terms.

4.3   Query Relevance Model

The query relevance model captures the relevant queries from user search log. This model constructs query relevance graph by combining query click graph which captures the relationship between (i) queries frequently clicked on common URLs and (ii) query text similarity graph using Jaccard similarity between queries.

Query Click Graph. Relevant queries from the search logs can be obtained by considering those queries that stimulate the users to click on the same set of URLs. For example, the queries *solar system* and *planet* are not textually similar, but they are relevant. This information can be achieved by analyzing common clicked URLs on queries in the search log.

Consider a URL-Query undirected bipartite graph, $BG_{qu} = (V_{qu}, E_{qu})$, where $V_{qu} = Q \cup U$, $Q=\{q_1,q_2,...q_m\}$ and $U=\{u_1,u_2,...u_n\}$. E_{qu} is the set of all edges. The edge (q_i,u_j) exists if and only if user has clicked a URL u_j after issuing query q_i. Often, the user issues query and by mistake clicks on some URL, which has no relation. In order to reduce noise and outliers, those edges which have only one click between query and URL are removed.

From BG_{qu} Query Click directed Graph, QC=(V_q,E) is constructed, where V_q are queries and E is a directed edge from q_i to q_j which exists if and only if there is atleast one common URL u_k, that both q_i and q_j link in BG_{qu}.

The weight of edge (q_i,q_j) in QC, $w_c(q_i,q_j)$ is calculated by counting occurrence of the pair (q_i,q_j) in URL-Query group. A URL-Query group $UQ_i = \{Q_i \in Q\}$ is a set of queries Q_i generated by unique URL u_i clicked by user for different queries. Figure 1a shows Query click graph.

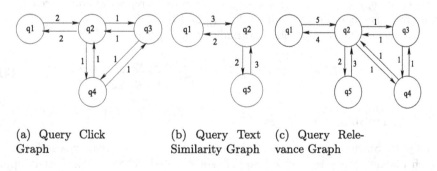

(a) Query Click
Graph

(b) Query Text
Similarity Graph

(c) Query Rele-
vance Graph

Fig. 1. (a) Query click graph; (b) Query text similarity graph; (c) Query relevance graph

Query Text Similarity Graph. It is not necessary for the user to always click the same URL for different queries. In such a case, we may not get two or more queries having common URL, but queries may be relevant since they share common words. For example, queries *cloud computing* and *cloud computing books* shares common words. To get these relevant queries, query text similarity graph is constructed.

For this graph, query text similarity by Jaccard coefficient J_c is calculated. Jaccard coefficient is defined as the fraction of common words between two queries as given in Eq. 1.

$$J_c(q_i, q_j) = \frac{words(q_i) \cap words(q_j)}{words(q_i) \cup words(q_j)} \tag{1}$$

Query Text Similarity directed graph is defined as $QG_{ts}=(V_q,\mathrm{E})$, where V_q is distinct queries in search log and E an edge between q_i to q_j exists if $J_c(q_i,q_j)>0.6$. The weight of edge (q_i,q_j) in QG_{ts}, $w_{ts}(q_i,q_j)$ is calculated by counting the occurrence of q_j in the search log with respect to q_i. Figure 1b shows the example of Query Text Similarity Graph.

In Fig. 1b, q_1, q_2 and q_5 are distinct queries in search log. The Jaccard value of q_1 and q_2, is assumed to be greater than 0.6, therefore q_1 and q_2 are included as nodes in the graph. The weight of q_1 to q_2 is 3 as occurrence of q_2 is 3 with respect to q_1 in the search log and weight of q_2 to q_1 is 2 as occurrence of q_1 is 2 with respect to q_2 in the search log. Similarly, Jaccard value of q_2 and q_5 is greater than 0.6. The weight between q_2 and q_5 is calculated using the same procedure as q_1 and q_2.

Query Relevance Graph. The query click graph QC and query text similarity graph QG_{ts} captures two important properties of relevant queries. In order to utilize both the properties, the query click graph and query text similarity graph are combined into a single graph, Query Relevance Graph $QRG = (V_q,\mathrm{E})$, where V_q is set of queries either from QC or QG_{ts} and E an edge between q_i to q_j exists either from QC or QG_{ts}. The weight of edge (q_i,q_j) in QRG is calculated as follows : $w_r(q_i,q_j) = w_c(q_i,q_j) + w_{ts}(q_i,q_j)$. Figure 1c represents query relevance graph from the combined graphs of Fig. 1a and b.

The query relevance graph is normalized by Eq. 2.

$$Normalized w_r(q_i, q_j) = \frac{w_r(q_i, q_j)}{\sum_{n=1}^{k} w_r(q_i, q_n)} \tag{2}$$

4.4 Query Suggestion Algorithm

The Query Suggestion algorithm is shown as Algorithm 1. Given a search log of search engine, query relevance graph is constructed. Then, given a user input query, depth first search algorithm is used to suggests queries to meet the original users' need.

Algorithm 1. Query Suggestion Algorithm

Input   : Input query q
Output: Top-5 Suggested Queries
begin

1   Construct query relevance graph $G=(V,E)$ using the method shown in the previous section. The directed edges E are weighted by normalization.

2   Given a query q, apply depth first search method on query relevance graph's nodes V.

3   The first Top-5 results are the suggested queries.

5 Experiments

5.1 Data Collection

Publicly available America On-line(AOL) search engine data [30] is used to construct query suggestion graph. Nearly 3813395 click through information with 1293620 unique queries and 400694 unique URLs have been considered.

The Web user activities can be obtained by click through data records. The users' interest and latent semantic relationship between users and queries as well as queries and clicked URLs can be retrieved. Each line of click through data contains information about anonymous user ID number, the query issued by the user, the time at which the query was submitted for search, the rank of the item on which they clicked and the domain portion of the URL in the clicked result. Each line in the data represents one of the two types of events: a query that was not followed by the user clicking on a result item and a click through an item in the result list returned from a query.

In this paper, we have used the relationship of queries and URLs for construction of query click graph and query to query relationship for construction of query text similarity graph. We have ignored user ID, rank and time information of click through data.

5.2 Data Cleaning

Click through data is the raw data recorded by search engine that contains a lot of noise which affects the efficiency of query suggestion algorithm. Hence, data is filtered by keeping well formatted, frequent, English queries i.e., queries which only contain characters $a, b .. z$ and space. After cleaning, data is reduced by 58.45 %. Nearly 490866 distinct queries and 334224 distinct URLs have been used in this experiments.

As discussed previously, most of the time, user tend to submit short queries with only one, two or three terms and therefore filtered data is obtained by keeping queries less than six terms. Further, data is reduced by 6.20 %, resulting in 448299 distinct queries and 318644 distinct URLs. Query suggestion results are generated by including all queries and by including only those queries which have less than six terms. We observe that both the results are similar.

5.3   Varying of Parameter-Jaccard Coefficient

As discussed in Sect. 4, Jaccard coefficient(J_c) is defined as the fraction of common words between two queries. Query suggestion results are generated by varying the value of J_c greater than 0.5 and 0.6. Finally, it is observed from the results, that when the optimal value for Jaccard coefficient is greater than 0.6, it yields more related queries.

When the query text similarity graph is constructed with J_c value greater than 0.5, queries which are only literally similar are obtained from the generated query relevance graph. For example, for the query *java*, queries *new one campaign commercial* and *one campaign commercial* are also obtained which are not relevant for the given input query. For the query *bank of america*, queries *bank of america banking centers*, *bank of america online banking* and *bank of america banking on line* are obtained which are only literally similar.

When the value of J_c is greater than 0.6, it results in those queries which are not only literally similar but also latent semantically relevant queries. For input query *java*, queries *java sun systems* and *sun java* are obtained which are latent semantically relevant queries.

If the value of J_c is equal to 0.5, then irrelevant queries are being selected. For example, consider two queries *California state university* and *California state polls* in which the value of J_c is 0.5, and both the queries are not relevant. Even when the value of J_c is less than 0.5, words matching between the two queries reduces, and hence relevant queries are not obtained.

5.4   Query Suggestion Results

We have displayed the Top-5 suggestion results of Heat Diffusion algorithm and our algorithm in Table 1. For Heat Diffusion Algorithm the numeric values shows the heat value for that query. For our algorithm, numeric value shows the normalized value of query in the query relevance graph.

5.5   Performance Analysis

From the results shown in Table 1, it is observed that our query relevance model is suggesting literally similar queries as well as latent semantically relevant queries. As discussed earlier, most of the time, the user tends to submit short queries with only one, two or three terms. Therefore, 60 test queries with one, two or three terms have been considered and different topics for input queries, such as Health, Shopping, Computer, Art have been covered in our experiments.

For example, given an input query *java*, the algorithm suggests *java download*, and *download java script*, which are literally similar queries. While a *sun microsystems* query suggests the company name of the java platform. This query is latent semantic to the input query *java*. Similarly for the input query *free music*, the query suggestion results are *free music downloads*, *music downloads*, *shareware music downloads*, *broadway midi files* and *midi files*. Midi file is the Musical Instrument Digital Interface protocol for music. Here it is observed that

Table 1. Query suggestion results comparison

Sr.No	Query	Heat diffusion algorithm	Our algorithm
1	Java	Word whomp game 0.9044	Download java 0.012
			Download java script 0.087
			Java download 0.32
			Sun microsystems 0.0455
			Sun microsystems colorado 0.0625
2	Fireworks	Phantom fireworks 0.8651	Firecrackers 0.25
			Phantom fireworks 0.25
			How to make a firework 0.25
			Phantomfireworks 0.25
3	Free music	Sad songs 0.9174	Free music downloads 0.2523
		Bored 0.9075	Music downloads 0.0588
		Humorous pictures 0.9051	Shareware music downloads 0.007
		Free music downloads 0.8962	Broadway midi files 0.0159
		Funny quotes 0.8928	Midi files 0.5
4	Wedding	Wedding channel 0.8653	Bridal shows nj 0.0714
			Wedding channel 0.0119
			Wedding dresses 0.125
			Lavender wedding dresses 0.0287
			Tea length wedding dress 1.0

the resulting queries are literally similar and latent semantically to the given input query.

We have compared the performance of our algorithm against Heat Diffusion algorithm (Hdiff) [28]. In the heat diffusion algorithm, the query-URL bi-partite graph is constructed from the search log. For a given input query, a sub graph is constructed by using depth-first search on bi-partite graph and then the diffusion process is applied on the sub graph to the ranked result. The top-5 queries based on heat value are used as suggested results. The graph construction method is the major difference between our proposed model and heat diffusion model. In heat diffusion model, query-URL bi-partite graph is used. While in proposed model URL-query relation and query text similarity relation is used to construct query-query graph i.e., query relevance graph.

To evaluate the quality of semantic relation is not easy in query suggestion as the queries taken as input is generated by users, and there are no linguistic resources available. In this paper, we have evaluated quality of semantic relation manually by three human experts and automatic evaluation based on Open Directory Project (ODP) database. We have adopted the method used in [31] to evaluate the quality of query recommendation results.

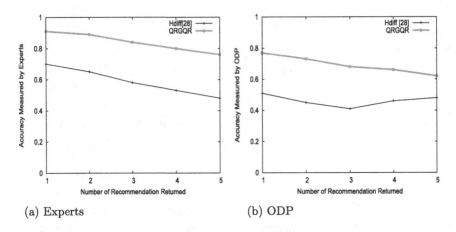

(a) Experts   (b) ODP

Fig. 2. Accuracy comparison

In manual evaluation, three research students are asked to rate the query suggestion results. We have asked them to evaluate relevance between testing queries and suggested results in the range of 0 to 1, in which 0 means totally irrelevant and 1 means totally relevant. The average value of rating results is shown in Fig. 2a. It is observed that the accuracy of our algorithm increases by 25.2 % in comparison with Hdiff.

In automatic evaluation, ODP database is used; also known as *dmoz* human edited directories of the web. When a user types a query in ODP, besides site matches, we can also find categories matches in the form of paths between directories and these categories are ordered by relevance. For example, the query *solar system* would provide the category *Science : Astronomy : Solar System*, while one of the results for *planets* would be *Science: Astronomy : Solar System : Planets*. Here, "*:*" is used to separate different categories. Hence, to measure the relation between two queries, we use a notion of similarity between the corresponding categories as provided by ODP and measure the similarity between two categories D and D_1 as the length of their longest common prefix $CP(D,D_1)$ divided by the length of the longest path between D and D_1. More precisely, denoting the length of a path with $|D|$, this similarity is defined as $sim(D,D_1) = |CP(D,D_1)|/max\{|D|,|D_1|\}$. The similarity between the two queries above mentioned is 3/4. Since they share the path *Science : Astronomy : Solar System* and the longest path is made of the four directories. We have evaluated the similarity between two queries by measuring the similarity between the most similar categories of the two queries among the top five answers provided by ODP.

Accuracy comparison measured by ODP is shown in Fig. 2b, from which it is observed that the accuracy of our algorithm increases by 23.2 % in comparison with Hdiff. From the above two evaluation process it can be concluded that our proposed query suggestions algorithm is efficient. The major difference between our proposed model and heat diffusion model is the method of graph construction. The query relevance graph can identify richer query relevance information as there are more available paths to follow in the graph. Hence our proposed algorithm outperforms the heat diffusion algorithm.

5.6 Efficiency Analysis

Experiments have been conducted on 8 GB memory and intel Xeon(R) CPU E31220 @ 3.10 GHz Quad Core processor workstation. Dataset used for both Hdiff and our method are same. Though our algorithm gives more number of relevant queries than heat diffusion method, the retrieval time is same. The computation time for the query suggestion of both Hdiff and our method is around 0.01 seconds.

6 Conclusions

In this paper, we have proposed a general method for query suggestions by combining query click graph which captures the relationship between queries frequently clicked on common URLs and query text similarity graph which finds the similarity between two queries using Jaccard similarity. The proposed method provides literally as well as semantically relevant queries for users' need. Simulations performed on America On-line (AOL) search data and as compared with Heat diffusion method, have shown that our method outperforms Heat diffusion method [28] by providing more relevant queries for a given input query with almost the same computation time as Heat diffusion method. It can be used for many recommendation tasks like query, image, and product suggestions. The query click graph has relationship between query and URLs, and hence the our algorithm can also be used to suggest URLs.

References

1. Das, A., Datar, M., Garg, A., Rajaram, S.: Google news personalization: scalable online collaborative filtering. In: WWW 2007: The Proceedings of 16th International Conference on World Wide Web, pp. 271–280 (2007)
2. Ma, H., King, I., Lyu, M.R.: Effective missing data prediction for collaborative filtering. In: SIGIR 2007: The Proceedings of 30th International ACM SIGIR Conference on Research and Development in Information Retrieval, pp. 39–46 (2007)
3. Kruschwitz, U., Lungley, D., Albakour, M.-D., Song, D.: Deriving query suggestions for site search. J. Am. Soc. Inf. Sci. Technol. **64**(10), 1975–1994 (2013)
4. Sarwat, M., Levandoski, J.J., Eldawy, A.: LARS*: an efficient and scalable location-aware recommender system. IEEE Trans. Knowl. Data Eng. **26**(6), 1384–1399 (2014)
5. Yang Cao, J., Fan, J., Li, G.: A user-friendly patent search paradigm. IEEE Trans. Knowl. Data Eng. **25**(6), 1439–1443 (2013)
6. McFee, B., Barrington, L., Lanckriet, G.: Learning content similarity for music recommendation. IEEE Trans. Audio Speech Lang. Process. **20**(8), 2207–2218 (2012)

7. Gao, W., Niu, C., Nie, J.-Y., Zhou, M., Hu, J., Wong, K.-F., Hon, H.-W.: Cross-lingual query suggestion using query logs of different languages. In: SIGIR 2007: The Proceedings of 30th Annual International ACM SIGIR Conference on Research and Development in Information Retrieval, pp. 463–470 (2007)

8. Anagnostopoulos, A., Becchetti, L., Castillo, C., Gionis, A.: An optimization framework for query recommendation. In: WSDM 2010: The Proceedings of 3rd ACM International Conference on Web search and Data Mining, pp. 161–170 (2010)

9. Vahabi, H., Ackerman, M., Baeza-Yates, D.L.R., Lopez-Ortiz, A.: Orthogonal query recommendation. In: RecSys 2013: The Proceedings of the 7th ACM Conference on Recommender System, pp. 33–40 (2013)

10. Santos, R.L.T., Macdonald, C., Ounis, I.: Learning to rank query suggestions for adhoc and diversity search. ACM J. Inf. Ret. $16(4)$, 429–451 (2013)

11. Song, Y., Zhou, D., He, L.: Query suggestion by constructing term-transition. In: WSDM 2012: The Proceedings of 5th ACM International Conference on Web Search and Data Mining, pp. 353–362 (2012)

12. Fan, J., Wu, H., Li, G., Zhou, L.: Suggesting topic based query terms as you type. In: The Proceedings of 12th International Asia-Pacific Web Conference, pp. 61–67 (2010)

13. Liu, Y., Miao, J., Zhang, M., Ma, S., Liyun, R.: How do users describe their information need : query recommendation based on snippet click model. Int. J. Expert Syst. Appl. $38(11)$, 13874–13856 (2011)

14. Sharma, S., Mangla, N.: Obtaining personalized and accurate query suggestion by using agglomerative clustering algorithm and P-QC method. Int. J. Eng. Res. Technol. $1(5)$, 1–8 (2012)

15. Narawit, W., Chantamune, S., Boonbrahm, S.: Interactive query suggestion in Thai library automation system. In: The Proceedings of 10th International IEEE Conference on Computer Science and Software Engineering, pp. 76–81 (2013)

16. Leung, K.W.-T., Ng, W., Lee, D.L.: Personalized concept-based clustering of search engine queries. IEEE Trans. Knowl. Data Eng. $20(11)$, 1505–1518 (2008)

17. Chen, Y., Zhang, Y.-Q.: A personalized query suggestion agent based on query-concept bipartite graphs and concept relation trees. Int. J. Adv. Intell. Paradigms $1(4)$, 398–417 (2009)

18. Kim, Y., Seo, J., Croft, W.B., Smith, D.A.: Automatic suggestion of phrasal-concept queries for literature search. Int. J. Inf. Process. Manage. $50(4)$, 568–583 (2014)

19. Kaczmarek, A.L.: Interactive query expansion with the use of clustering-by-directions algorithm. IEEE Trans. Ind. Electron. $58(8)$, 3168–3173 (2011)

20. Goyal, P., Behera, L., McGinnity, T.M.: Query representation through lexical association for information retrieval. IEEE Trans. Knowl. Data Eng. $24(12)$, 2260–2273 (2012)

21. Cao, H., Jiang, D., Pei, J., He, Q., Lian, Z., Chen, E., Li, H.: Context-aware query suggestion by mining click-through and session data. In: KDD 2008: The Proceedings of 14th International ACM SIGKDD Conference on Knowledge Discovery and Data Mining, pp. 875–883 (2008)

22. Mei, Q., Zhou, D., Church, K.: Query suggestion using hitting time. In: CIKM 2008: The Proceedings of 17th ACM Conference on Information and Knowledge Management, pp. 469–477 (2008)

23. Ma, H., Lyu, M.R., King, I.: Diversifying query suggestion results. In: AAAI 2010: The Proceedings of 24th AAAI International Conference on Artificial Intelligence, pp. 1399–1404 (2010)

24. Guo, J., Cheng, X., Xu, G., Shen, H.-W.: A structured approach to query recommendation with social annotation data. In: CKIM 2010: The Proceedings of 19th ACM International Conference on Information and Knowledge Management, pp. 619–628 (2010)
25. Song, Y., Zhou, D., He, L.: Post ranking query suggestion by diversifying search results. In: SIGIR 2011: The Proceedings of 34th International ACM SIGIR Conference on Research and Development in Information Retrieval, pp. 815–824 (2011)
26. Kunpeng, Z., Xiaolong, W., Yuanchao, L.: A new query expansion method based on query logs mining. Int. J. Asian Lang. Process. **19**(1), 1–12 (2009)
27. Craswell, N., Szummer, M.: Random walks on the click graph. In: SIGIR 2007: The Proceedings of 30th Annual International ACM SIGIR Conference on Research and Development in Information Retrieval, pp. 239–246 (2007)
28. Ma, H., King, I., Lyu, M.R.T.: Mining Web graphs for recommendations. IEEE Trans. Knowl. Data Eng. **24**(6), 1051–1064 (2012)
29. Hwang, H., Lauw, H.W., Getoor, L., Ntoulas, A.: Organizing user search histories. IEEE Trans. Knowl. Data Eng. **24**(5), 912–925 (2012)
30. Pass, G., Chowdhury, A., Torgenson, C.: A picture of search. In: The Proceedings of 1th International Conference on Scalable Information Systems, June 2006
31. Baeza-Yates, R., Tiberi, A.: Extracting semantic relations from query logs. In: KDD 2007: The Proceedings of 13th ACM SIGKDD International Conference on Knowledge Discovery and Data Mining, pp. 76–85 (2007)

Offline Writer Identification in Tamil Using Bagged Classification Trees

Sudarshan Babu[✉]

SVCE, Anna University, Chennai, Tamil Nadu, India
sudarshan.warft@gmail.com

Abstract. In this paper, we explore the effectiveness of bagged classification trees, in solving the writer identification problem in the Tamil
language. Unlike other languages, in Tamil the writer identification problem is mostly an unexplored problem. Novel feature extraction methods
tailored to better understand Tamil characters have been proposed. The
feature extraction methods used in this paper are chosen after analysing
the statistical spread of a feature across different handwriting classes.
We have also analysed how increasing the number of bagged classification trees would affect the classification accuracy. Our learning algorithm
is trained with hundred and forty four samples and is tested with twenty
different samples per handwriting style. In total the algorithm is trained
with ten different handwriting styles. Using the proposed features and
bagged classification trees, we achieve 76.4 % accuracy. The practicality
of the proposed method is also analysed using a few time consumption
measuring parameters.

Keywords: Image data analysis · Image understanding · Handwriting
recognition · Classification trees · Bagging

1 Introduction

Writer identification is an image understanding problem. Here the objective is
to identify the author of a given handwritten text. Here handwriting samples
have been scanned from paper and these scanned pictures of the paper are given
as input to the writer identification software. Due to the nature of the input it
is regarded as an offline method [25]. It is important not to confuse this with
OCR, where the objective is to recognize the text character.

References [1–3], along with other works have made huge improvements to the
state of the art in writer identification problems in English, Arabic and Chinese.
But very little work has been done towards solving the writer identification
problem in Tamil; the current work is one of the first attempts towards solving
the problem. Tamil, one of the oldest languages in the world, is widely spoken
in South East Asia. Owing to the widespread usage of Tamil language in the
region, there could be forensic and biometric applications that could require the
services of a writer identification software.

© Springer International Publishing Switzerland 2015
P. Perner (Ed.): MLDM 2015, LNAI 9166, pp. 342–354, 2015.
DOI: 10.1007/978-3-319-21024-7_23

In solving the writer identification problem, in languages like English, Chinese, Japanese, and Arabic, learning algorithms such as SVM, ANN, and HMM have been used quite popularly [3–5]. The results given by these algorithms compare favourably with the state of the art in their respective languages. Results provided by SVM, naive-bayes learner and sparse representation learner for the Tamil writer identification problem gave us relatively poorer results and moreover, previous works have not explored the effectiveness of ensembling their learners towards solving the writer identification problem. The improvement brought by the different techniques of ensembling was an added incentive in using bagged classification trees to solve this problem. Hence, in this work bagged classification trees were employed and the effect of varying the number of trees that were being used was studied.

The intricate structure of the script, makes describing the characters to the learning algorithm a challenging task. This calls for a feature set that is tailored for describing the Tamil characters. There are several feature extraction processes such as Zernike moments, wavelet transform, Fourier transforms that have been used in several image understanding problems [6–8]. These feature extraction methods have been used on other image recognition problems with a good measure of success, but performed poorly for this problem.

First experimentally, and then statistically, we found that feature extraction methods that describe the topology of the characters had better performances. Hence a few feature extraction methods such as circle features, corner features, black pixel coordinate product feature (BPCP feature) were proposed and their effectiveness was analysed using certain statistical parameters.

This paper is broken down into the following sections. Section 2, discusses the importance and widespread usage of Tamil as a language and describes why the writer identification problem is a hard problem and requires more attention. Then in Sect. 3, the contributions the paper has made is discussed. In Sect. 4, the lack of attention given to this problem and why the problem is a nascent problem is described. Moving on to Sect. 5, we describe the tailored features developed to solve the problem and in Sect. 6, the effectiveness of the features are analysed statistically. In Sect. 7, a brief background on constructing the classification tree and bagging is given. In Sect. 8, the algorithm proposed in the paper is given and the various sections of the algorithm are explained. Finally in Sect. 9, the paper projects and discusses the various performance attributes and this is followed by Sect. 10, which discusses the significance and possible extensions of the proposed work.

2 About the Tamil Language

Tamil is one of the oldest languages in the world and is widely spoken in southern India, Malaysia, Singapore, and Srilanka [9,10]. It is estimated that close to about 80 million people today use the language. The language's script consists of two hundred and forty seven letters. The large character set is one of the reasons why the problem is a challenging problem. Unlike English or other European languages, in Tamil there are instances where different characters have almost the same geometrical structure. To formulate an algorithm that will recognise these minute various across characters and perform writer classification is a hard problem to solve with high accuracy. The accuracy achieved in this paper could be a good start towards encouraging more works towards solving this problem.

3   Contributions by the Paper

This paper brings with it feature extraction methods tailored to solve the writer identification problem in Tamil. The effectiveness of these methods such as circle features, corner features, and black pixel coordinate product features (BPCP features) are analysed using a few statistical parameters and this is given in Sect. 6.

The paper also studies how increasing the number of bagged classification trees would affect the classification accuracy. The number of trees being used is an important parameter, as it helps users find a trade off between computational factors (time and memory) and accuracy.

4   Related Work

The Tamil writer identification problem is a nascent problem, and to the authors knowledge there is only one other paper that discusses the writer identification problem in Tamil. There are no standardised datasets for this problem. Hence comparing other works based on just performance would be unfair. Hence, we only discuss the methodology of the paper.

Reference [11] uses texture analysis to solve the problem. This work treats the entire text as one entity and performs classification. The solution proposed in the paper is text independent and is based on the features extracted from gray level co-occurrence matrix of the scanned image. There is no tailoring of feature extraction routines to done to extract more information form the characters.

5   Feature Extraction

The process of identifying and extracting a number of characteristics or properties that uniquely relate to an image is called feature extraction. These several characteristics obtained from a given image collectively form together what is called a feature vector. Framing the right feature vector is a challenging and a central problem in image understanding problems. For this specific problem, it was discovered that features that described the topology of the characters performed well. With this in mind the following features have been chosen to solve this problem.

5.1   Dimension Features

This feature captures the dimensions of the individual characters written by an author. Each handwriting expresses a character in certain dimensions, hence this feature was extracted.

5.2   Corner Features

Corners are formed at the intersection of different strokes that form the character. Since different writers write the same character differently, the points at which these corners are formed are different. Thus coordinates of the corners could be used in describing the character. Hence we sum up coordinates of the corners and use it as a feature. In this work the corners and subsequently their coordinates were identified using Harris corner detection algorithm [12].

5.3   Circle Features

The contours of the Tamil characters are characterized with circular or ellipsoidal boundaries that are present in different parts of the character. Since different writers write characters differently, the centers of these circles for the same character may be placed slightly apart for different writers. Thus coordinates of the circles could be used in describing the character. Hence we sum up coordinates of the circles and use it as a feature. In this work the circles and subsequently their centers were identified using Hough transform [13].

5.4   Black Pixel Coordinate Product Features(BPCP)

This feature is simply the products of the coordinates of the black pixels in the image. The black pixel coordinate is calculated for every black pixel in the image and then, these products are stacked one above the other to form the black pixel coordinate product feature. For ease of reference it is referred as BPCP in the remainder of the paper.

5.5   Line Features

The lines in an image are classified as horizontal, vertical, right diagonal, and left diagonal. The lengths of these lines in the image are measured. These lengths are used as features.

5.6   Zoning

It is the precursor step done before extracting zoned black pixel density features. This is a widely used image processing technique to split an image into smaller grids [14]. These grids contain portions of the image whose features are to be extracted. These grids are passed as inputs to the local features extractor programs. Since it is a common and well known procedure not much explanation is given about zoning process. These split zones are passed as parameters to the zoned black pixel density features algorithm.

5.7   Zoned Black Pixel Density Features

The black pixel to white pixel ratio is calculated for the split zones of an image. These ratios obtained from individual zones are stacked and are used as a ratio.

(a) Written by Author1.   (b) Written by Author2.

Fig. 1. This figure shows the Tamil character 'ka' written by two different authors.

Table 1. Table capturing various statistical differences brought about by the respective feature extraction methods.

Feature Extraction Method	Class	1st	median	mean	3rd
Circle Features	Author1	57.79	69.89	67.7 6	90.04
Circle Features	Author2	20.79	40.89	36.76	54.04
Corner Features	Author1	49	70	69.76	95
Corner Features	Author2	0	33	31	59
BPCP Features	Author1	1725	4030	4399	6570
BPCP Features	Author2	0	1040	1434	2646
Dimension Features	Author1	137	130.5	130.5	124
Dimension Features	Author2	90	87.5	87.5	85
Line Features	Author1	0.60	0.517	1.43	0.8
Line Features	Author2	0.40	0.53	0.96	0.8
Zoned Black Pixel Density Features	Author1	63.3	93.039	27.523	37.037
Zoned Black Pixel Density Features	Author2	73.1	41.669	38.3	58.7

6   Feature Analysis

In this section, we describe the effectiveness of a few feature extraction methods using statistical parameters like 1st quartile, median, mean, and 3rd quartile.

For this analysis, we ran the above mentioned feature extraction procedures on samples from two different authors only. Those two handwriting samples are given by Fig. 1a and b.

Table 1 shows how each feature extraction method generates a statistically different set of features for different handwritings, providing insights into the effectiveness of each feature extraction procedure. This effectiveness is shown by the various statistical parameters that have different values under their respective columns for author1 and author2. For example the mean for the circle features for author1 is 67.76 and the mean for the same feature for author2 is 36.8. This implies that the sum of coordinates of the center of the circles found in the character are scattered about 67.76 for author1 and 36.8 for author2. These inherent differences that are brought about by the proposed feature extraction procedures, make them very effective in solving this problem.

7   Classification Background

The identification problem is framed as a supervised learning problem. Here the learning algorithm is trained by providing it with features of labeled objects. From this training the learning algorithm is expected to determine the label of unknown objects. In the writer identification problem the object is a handwriting sample and the label is an author id. The objective of the method proposed in the paper is to determine the author id, given the handwriting samples of that unknown author.

7.1   Motivation to Use Bagged Classification Trees

From a purely applied perspective, works in other types of image understanding problems such as [15, 16] have shown us that ensembling different learning models have always given a improvement in the predication rates. With these works in mind we wanted to explore the effectiveness of ensembled learners in the Tamil writer identification problem. In this method, we used classification trees over other learning algorithms. This is because, though classification trees offer high variance, when a large number of such trees are combined by bagging, the variance provided by the entire learning model is lowered and hence makes it a strong learner. Reference [17, 18] provide theoretical background on the effectiveness of ensembling classification trees with bagging.

7.2   Classification Trees

Introduced by [19] classification trees have been one of the very successful learning algorithms. They are non parametric i.e., they do not expect the input to follow any structure [20]. They also produce white box solutions. So for the above reasons they have been popular over the years [20].

Growing a Classification Tree. The tree must be constructed such that variability within classes is minimised and the contrast across classes is maximised. Initially all images of all classes are assigned to the root node. Then a binary tree is grown from it by splitting the data into two children nodes based on certain conditions. This process of splitting is performed on the new children nodes till a terminating condition is achieved. A newly grown node may become a leaf when it contains images from only one class or images from one class dominate the population. The classification tree is constructed using the c4.5 algorithm. This is described in a more detailed fashion in [21].

Pruning. A central problem in growing trees is fixing the size of the tree. A large tree results in overfitting while a small tree fails to capture the difference across different classes. Pruning identifies portions of the tree that carry less information and removes those portions. This is done by cross validation. Pruning performs a trade-off between bias and variance. The process of pruning is explained more elaborately in [22].

Classification. The classification tree in its nodes has decision rules that are formed from the training examples. At first, the test image is checked with the decision rule in the root node. Based on this decision rule the test image then falls on to the left or the right child. This process of checking with the decision rule in each node, is repeated till the image reaches one of the leaves, at which point the tree assigns the test image a label based on the population in the respective leaf [19].

7.3 Bagging

Introduced by [17], bagging is one of the very popular ensembling techniques. N individual learners are trained with randomly selected portions of the training features called a bootstrap sample [18].

Now, based on its training, every individual learner predicts a label for the input test image. The label that has been predicted the most by the individual learners is chosen as the label of the test image.

8 Implementation of Writer Identification Software

This section dwells on the implementation aspects of the work proposed in the paper. The algorithm 1 enlists the different steps involved in building the Tamil writer identification software. At a higher level, the algorithm could be split into three sections, feature extraction, training, and testing.

8.1 Feature Extraction Code

The first for loop performs feature extraction where every training image is passed as an argument to the feature extraction procedure. The feature extraction procedure, extracts the different features described in Sect. 5 and the result is a vector. Hence there is a feature vector for every training image. These feature vectors are horizontally concatenated to form the feature matrix referred in the Algorithm 1 as features for all training images.

8.2 Training Code

The second phase is the training phase, where the ensembled classification tree is constructed. In this phase, the feature matrix formed in the previous step, along with the label vector, and the number of trees to be constructed variable are passed as input to the ensembled classification tree building program. The ensembled tree is then built by framing different decision rules. The model corresponding to the constructed ensembled setup is stored.

Data: 144 training images and 20 testing images per handwriting. In total 1440 training images and 200
  testing images
Result: Achieved a maximum accuracy of 76.4 %.
/* Feature Extraction   */
for *every training image* **do**
  $i^{th}\_Feature\_Vector = Feature\_Extraction\_Procedure(i^{th}\_Image)$;
  $Features\_For\_All\_Training\_Images =$
  $Horizontal\_Concatenate(Features\_For\_All\_Training\_Images, i^{th}\_Feature\_Vector)$;
end
/* Training Phase   */
$Train\_Label\_Vector = Generate\_Label\_Vector()$;
$Tree\_Model =$
$Classification\_Tree.Construct(Features\_For\_All\_Training\_Images, Train\_label\_vector,$
  $Num\_of\_Trees)$;
/* Testing Phase   */
for *every testing image* **do**
  $i^{th}\_Test\_Feature\_Vector = Feature\_Extraction\_Procedure(i^{th}\_Test\_Image)$;
  $i^{th}Test\_Image\_Predicted\_Class =$
  $Classification\_Tree.Predict(i^{th}\_Test\_Feature\_Vector, Treemodel)$;
  $Print(i^{th}Test\_Image\_Predicted\_Class)$
end

Algorithm 1. Offline writer identification in Tamil using bagged classification
trees.

Table 2. Table showing the performance analysis of the proposed work across different
handwritings.

Number of Classification Trees	100 Trees	200 Trees	300 Trees	400 Trees	500 Trees
Accuracy for Author1	75 %	75 %	80 %	85 %	85 %
Accuracy for Author2	60 %	65 %	70 %	70 %	70 %
Accuracy for Author3	65 %	70 %	70 %	70 %	75 %
Accuracy for Author4	70 %	65 %	75 %	75 %	75 %
Accuracy for Author5	70 %	75 %	75 %	70 %	75 %
Accuracy for Author6	45 %	50 %	60 %	55 %	65 %
Accuracy for Author7	75 %	80 %	75 %	85 %	80 %
Accuracy for Author8	70 %	80 %	75 %	75 %	80 %
Accuracy for Author9	80 %	80 %	80 %	80 %	85 %
Accuracy for Author10	70 %	80 %	70 %	75 %	75 %

8.3 Testing Code

In the third phase, we analyze the performance of the writer identification classification tree built in the second phase. Here the image whose class is to be determined is first passed as an argument to the feature extraction function. Then,
the feature vector corresponding to the unknown sample along with the stored
model of the ensembled classification tree setup is passed as input to the predict
function. The output of the predict function is the predicted class of the sample
determined by the proposed method. Then it is printed.

9 Results and Analysis

In this section, the experimentation methodology along with the environment
in which the experiments were conducted is described. Then the efficiency of

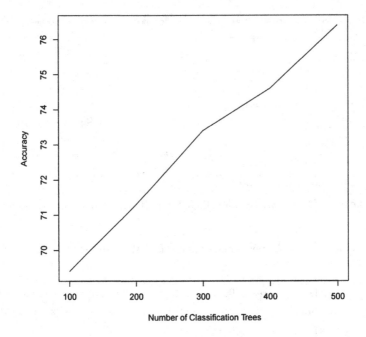

Fig. 2. Graph showing the effect of increasing the number of trees on accuracy.

the proposed method for each handwriting is analysed. This is then followed up with an analysis of how increasing the number of bagged classification trees would affect the classification accuracy. Then, parameters that capture the time consumption factors are analysed. Finally, issues and limitations of the proposed method are discussed.

9.1 Experimentation Methodology

The Tamil character database was obtained from [23]. From this database, we made a subset comprising of ten handwritings. We used 1440 images for training, with 144 images per class. For testing we used 200 images in total, with 20 images per class.

The project and all its associated experiments were executed on a three year old i-3 processor with 2.40 GHz clock. The entire project was written in MATLAB.

9.2 Analysing Performance per Handwriting Class

From the Table 2, we can observe that the algorithms accuracy rate falls within a certain numerical range for all handwritings with a few exceptions. The accuracy rates are mostly between 65 % and 85 % with a few aberrations. This shows the learning algorithm is robust to noise that different handwritings bring with them.

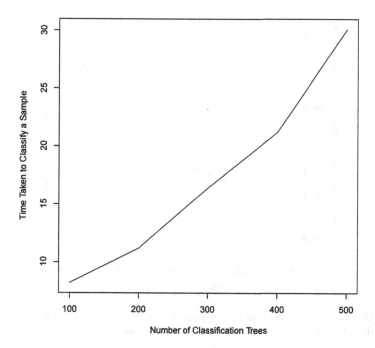

Fig. 3. Graph showing the increase in time taken to classify a sample with increase in number of trees

This also ascertains the effectiveness of the feature extraction methods, as they have identified characteristics unique to each handwriting, which is why almost equal efficiency across handwritings has been observed. It can also be seen that for a particular handwriting, with increase in number of trees the accuracy has increased gradually. Deviations in accuracy are seen only for author6. Generally all models have performed below their mean accuracy rate for author6.

9.3 Effect of Increase in Number of Classification Trees on Accuracy

We study the effect of increasing the number of classification trees on the accuracy of the proposed method. The graph in Fig. 2 shows the improvement in accuracy that is brought upon by increasing the number of trees. It is observed that there is a marginal improvement in accuracy with for every increase in 100 trees. We achieved our best result of 76.4 % by using five hundred classification trees.

9.4 Analysing Performance Parameters

In this section, the paper introduces two parameters into the discussion, time taken to classify a sample, and time taken to construct the entire ensembled

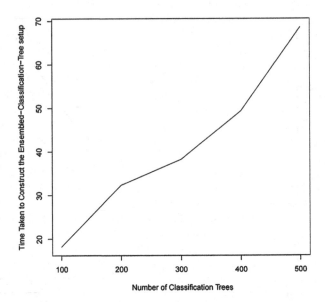

Fig. 4. Graph showing the increase in time taken to compute the entire ensembled classification-tree setup with in crease in number of trees.

classification tree setup. These two parameters give a sense of time consumed by the proposed method. Time taken to classify a sample is what that determines the run time performance of the proposed work. As, the ensembled-tree construction can be precomputed and the resultant model can be stored, it does not have any effect on the run time characteristics of the finished software.

As it can be seen in the graph given by Fig. 3, there is linear rise in time taken to classify a sample from one hundred trees to two hundred trees. Beyond this point, the rate of change of the graph increases further and remains constant till the four hundred trees mark. When five hundred trees are used, the rate of change of the graph becomes even more steeper. This increase rise in time taken, with increase in trees is because more number of nodes have to be traversed by the algorithm before finalising on the result. The time taken to classify a sample by the proposed method is in the order of seconds.

Time taken to construct the entire ensembled-classification-tree setup can consume more time, when compared to the time taken to classify a sample. As the entire ensembled-classification-tree can be precomputed it is acceptable to have a larger setup time. The graph given by Fig. 4, at first increases at a particular rate for hundred trees, and then for two hundred trees, the rate of change of the curve decreases sharply, before increasing sharply again for three hundred trees. This rise in time taken to construct the entire ensembled classification-tree setup is due to the increased number of trees to be constructed and then later ensembled. The time consumed for the construction of trees is in the order of minutes.

9.5   Issues and Limitations

The algorithm when given a handwriting sample from an user, that it has not been trained with, will miss-classify the sample as one of the already trained users, instead of classifying it as an unknown writer. This validation property is generally is not exhibited by classification trees and a few other machine learning algorithms. However, Lasso regression based learners and similar classifiers, inherently have this property in them [24]. This self-validation property is absent in the proposed work.

Though, increasing the number of trees to be ensembled has given us increased accuracy, we cannot expect to scale the algorithm in this manner, as increasing the number of trees places a substantial computational load and the software module becomes too slow to be put in use. This can be seen in the graphs given by Figs. 3 and 4. The amount of time taken to classify a image increases with increase in number of trees.

10   Conclusion

Again, we stress that the problem is a nascent problem and the proposed work is one of the first attempts to solve the problem.

This paper has come up with features customised to better describe Tamil characters to the classifier. To achieve more accuracy classifiers must also be customised to suit the problem. Developing classifiers to suit a specific problem is possible, but requires time and effort. The authors of the paper strongly believe this paper would direct a lot more research into solving this problem and this paper could act as a base line for further research.

Owing to the large population that uses Tamil language, solving the Tamil writer identification problem could open up several commercial avenues. Such a software could be put to use in banks for identification of a person, and in mobile phones and tablets to prevent unauthorised usage. It can be concluded that the proposed work has given a good perspective of the problem and the accuracy achieved in the paper is a good starting point for further research.

References

1. Bensefia, A., Paquet, T., Heutte, L.: A writer identification and verification system. Pattern Recogn. Lett. **26**(13), 2080–2092 (2005). Computer Vision ECCV 2010. Springer, Berlin Heidelberg, pp. 448–461 (2010)
2. Bulacu, M., Schomaker, L., Brink, A.: Text-independent writer identification and verification on offline arabic handwriting. In: Ninth International Conference on Document Analysis and Recognition, 2007, ICDAR 2007, vol. 2. IEEE (2007)
3. He, Z., You, X., Tang, Y.Y.: Writer identification of Chinese handwriting documents using hidden Markov tree model. Pattern Recogn. **41**(4), 1295–1307 (2008)
4. Justino, E.J.R., Bortolozzi, F., Sabourin, R.: A comparison of SVM and HMM classifiers in the off-line signature verification. Pattern recognition letters **26**(9), 1377–1385 (2005)

5. Quan, Z.-H., Liu, K.-H.: Online signature verification based on the hybrid HMM/ANN model. Int. J. Comput. Sci. Netw. Secur. **7**(3), 313–322 (2007)
6. Chong, C.-W., Raveendran, P., Mukundan, R.: Translation invariants of Zernike moments. Pattern Recogn. **36**(8), 1765–1773 (2003)
7. Bostanov, V.: BCI competition 2003-data sets Ib and IIb: feature extraction from event-related brain potentials with the continuous wavelet transform and the t-value scalogram. IEEE Trans. Biomed. Eng. **51**(6), 1057–1061 (2004)
8. Li, W., Zhang, D., Zhuoqun, X.: Palmprint identification by Fourier transform. Int. J. Pattern Recognit Artif Intell. **16**(04), 417–432 (2002)
9. Schiffman, H.F.: Linguistic culture and language policy. Psychology Press (1998)
10. Schiffman, H.F.: A reference grammar of spoken Tamil. Cambridge University Press, Cambridge (1999)
11. Jayanthi, S.K., Rajalakshmi, D.: Writer identification for offline Tamil handwriting based on gray-level co-occurrence matrices. In: Third International Conference onAdvanced Computing (ICoAC), 2011. IEEE (2011)
12. Harris, C., Stephens, M.: A combined corner and edge detector. Alvey vision conference, vol. 15 (1988)
13. Ballard, D.H.: Generalizing the Hough transform to detect arbitrary shapes. Pattern Recogn. **13**(2), 111–122 (1981)
14. Handbook of character recognition and document image analysis (1997)
15. Su, Y., Shan, S., Chen, X., Gao, W.: Hierarchical ensemble of global and local classifiers for face recognition. IEEE Trans. Image Process. **18**(8), 1885–1896 (2009)
16. Grabner, H., Bischof, H.: On-line boosting and vision. Computer Vision and Pattern Recognition, 2006 In: Conference on IEEE Computer Society, vol. 1. IEEE (2006)
17. Breiman, L.: Bagging predictors. Mach. Learn. **24**(2), 123–140 (1996)
18. Bryll, R., Gutierrez-Osuna, R., Quek, F.: Attribute bagging: improving accuracy of classifier ensembles by using random feature subsets. Pattern Recogn. **36**(6), 1291–1302 (2003)
19. Breiman, L., et al.: Classification and regression trees. CRC Press (1984)
20. http://www.clarklabs.org/applications/upload/Classification-Tree-Analysis-IDRI SI-Focus-Paper.pdf
21. Quinlan, J.R.: Bagging, boosting, and C4. 5. AAAI/IAAI, vol. 1 (1996)
22. Gelfand, S.B., Ravishankar, C.S., Delp, E.J.: An iterative growing and pruning algorithm for classification tree design. In: Conference Proceedings of IEEE International Conference on Systems, Man and Cybernetics, 1989. IEEE (1989)
23. HP-labs. Isolated Handwritten Tamil Character Dataset developed by HP india along with IISc (2006). http://lipitk.sourceforge.net/datasets/tamilchardata.htm (Accessed on 30 September 2010)
24. Wright, J., et al.: Robust face recognition via sparse representation. IEEE Trans. Pattern Anal. Mach. Intell. **31**(2), 210–227 (2009)
25. http://en.wikipedia.org/wiki/Handwriting_recognition

Applications of Data Mining

Data Analysis for Courses Registration

Nada Alzahrani[1]([✉]), Rasha Alsulim[1], Nourah Alaseem[1],
and Ghada Badr[1,2]

[1] King Saud University, Riyadh, Saudi Arabia
n.y.alzahrani@gmail.com, {r.m.z1433,nooone_2000}@hotmail.com
[2] IRI - The City of Scientific Research and Technological Applications,
Alex, Egypt
badrghada@hotmail.com

Abstract. Data mining is a knowledge discovery process to extract the interesting previously unknown, potentially useful and non-trivial patterns from large repositories of data. There is currently increasing interest in data mining in educational systems, making it into a growing new research community. This paper applies a frequent patterns extraction approach to analyzing the distribution of courses in universities, where there are core and elective courses. The system analyzes the data that is stored in the department's database. The objective is to consider if allocation of courses is appropriate when they are more likely to be taken in the same semester by most students. A workflow is proposed; where the data is assumed to be collected over many semesters for already graduated students. A case study is presented and results are summarized. The results show the importance of the proposed system to analyze the courses registration in a given department.

Keywords: Data mining · Courses registration · Education · Frequent patterns · Apriori · Frequent-pattern growth

1 Introduction

Data mining involves processes of extracting knowledge and discovering useful information from large repositories of data by combining artificial intelligence techniques and some statistical methods. Thus, useful knowledge can emerge about association, patterns, anomalies and changes in the data stored in a database, data warehouse, or any other repository [1,2]. Alternatively, another term from Data Mining is Knowledge Discovery from Data, or KDD, which is perfectly reflected in the detailed meaning of the data mining process. There is also another aspect which considers data mining as an essential step in knowledge discovery processes. But the term data mining is more popular in research and analysis reports and a shorter term than knowledge discovery from data [3].

In data mining, discovering the interesting correlations, frequent patterns, and relations between these hidden patterns help users to understand the data more than using simple query approaches. Therefore, the general data mining

© Springer International Publishing Switzerland 2015
P. Perner (Ed.): MLDM 2015, LNAI 9166, pp. 357–367, 2015.
DOI: 10.1007/978-3-319-21024-7_24

field has developed several techniques and functionalities to extract meaningful and interesting patterns such as association rules, classification, clustering, sequence analysis, prediction, and outlier analysis [1,4,5]. During recent decades, educational systems have been released from some of their traditional concepts; and innovations in technology have become useful in new areas of study. The emerging data mining techniques within educational systems has been an active field of research for conducting new approaches. Educational Data Mining can be defined as the process of using statistical, machine-learning, and data-mining algorithms to be applied to educational data. The main aim of educational data mining is to extract information from certain data and analyze the educational settings such as students and courses for better understanding their relationships [2,6]. Association rules - one of the most well studied data mining methods - and frequent patterns extraction have recently become very helpful data mining techniques applied in the education field.

In this paper we propose new data mining work flow to analyze the distribution of related and associated courses in a given department, where core courses and elective courses are provided. Moreover, we discuss if this allocation is reliable and in meaningful manner. To accomplish the goal of this study our system extracts frequent courses that are taken in the same semester by different students. We analyze these courses based on students choices through the last 10 years. After preprocessing the data, frequent patterns and association rules extraction techniques are applied using Apriori and Frequent-pattern growth (FP-Growth) algorithms. Association rules are helpful to find any relation or association between courses.

The rest of this paper is organized as follows: in Sect. 2, data mining techniques are discussed. In Sect. 3 data mining approaches that are used in the system are reviewed. Section 4 presents the course analysis work-flow. Our experiment and results are given in Sect. 5. Discussion is provided in Sect. 6. Finally, conclusions from our findings are presented in Sect. 7.

2   Data Mining Techniques

Among all data mining functions, association rule mining is considered one of the most important techniques. The aim of association rules is to discover relationships between attributes in a database [7]. Association rule mining is used in different areas such as market analysis and risk management, telecommunication networks, for instance, extracting interesting relations such as frequent patterns or associations [8].

Development of information technology in various fields has led to colossal data storage repositories in various formats such as those of records, documents, images, audio and video archiving, scientific data, and many new data formats in large repositories. The collected data of such types from different sources has required better methods to extract knowledge for better decision making. Knowledge discovery applies various methods and algorithms. Educational data mining aims to develop methods using data mining techniques for making discoveries within the unique kinds of data involved. Using the pattern discovered helps to better understand students and their activities as reflected in such data sets [9].

Extracting frequent patterns provides major assistance in constructing association rule mining. Mining frequent patterns from large scale databases has emerged as an important strategy in the data mining and knowledge discovery community. A number of algorithms have been proposed to discover frequent patterns. The Apriori Algorithm is the basic algorithm adopted in this field [8]. The FP-Growth Algorithm is considered an enhancement of the Apriori Algorithm.

The goal of frequent pattern analysis is to discover recurring associations in huge data repositories in an efficient way. Extracting frequent patterns is a fundamental step in association rule mining, classification, clustering, and other data mining tasks. Proposed by Agarwal [7], Frequent Pattern algorithms are divided into three types:

1. Candidate generation approach (e.g. Apriori algorithm).
2. Without candidate generation approach (e.g. FPgrowth algorithm).
3. Vertical layout approach (e.g. Eclat algorithm) [10].

3   Data Mining Approaches

3.1   Basic Algorithm (Apriori)

Apriori algorithm is the oldest and most used algorithm for extracting frequent patterns. It is used for association rule mining and was proposed by R.Agrawal and R.Srikant in 1994. It is an iterative algorithm used to produce frequent itemsets [7].

Apriori is considered an iterative approach level-wise, where k itemset is used to explore $(k+1)$ itemsets, under conditions where each k -itemset must be greater than or equal to the minimum support threshold to be frequent. In this algorithm there are two main iterative steps. The first step scans the data base to find 1-itemsets that frequently appear and contains only one item; then it generates a set of candidate itemsets. The second step ensures that each candidate set appears in the database and prunes all infrequent candidates [11]. This algorithm uses anti-monotonicity of itemsets, if an itemset is not frequent, any of its superset is never frequent.

Apriori algorithm is used for its simplicity and clarity but has some weakness. It will repeatedly generate many frequent items and scan the database to find candidate itemsets. In addition, the efficiency of the algorithm is reduced if memory is limited and there are large numbers of transactions [12]. Improvement of Apriori is presented in [12] by minimizing the time in scanning transactions for candidate itemsets. This improvement leads to reducing the number of scanned transactions. In comparison to the Apriori algorithm, the experimental results show that the enhanced method reduces the time by 67.38 %.

In [13] an enhanced Apriori algorithm based on a bottom up approach and a support matrix to identify frequent itemsets is presented. It uses minimum support with a functional model based on standard deviation. This enhancement also allows non-expert persons to work the data mining system.

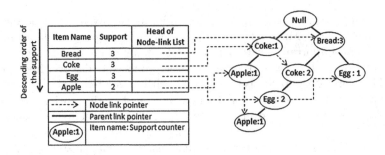

Fig. 1. An FP-tree and its header table [15].

3.2 Enhanced Algorithm (FP-Growth)

Frequent-pattern growth or simply FP-growth is one of the useful methods that discover itemsets without candidate generation. FP-growth was first proposed briefly by Han et al. (2000) [14]. It adopts a divide-and-conquer strategy [3]. This approach is entirely different from the generate-and-test approach of Apriori. The FP-growth algorithm consists of two main steps:

1. Construct a representation of the input transaction database as a tree (called an FP-tree).
2. Extract all frequent itemsets from the tree [15].

FP-tree abbreviates frequent-pattern tree. It consists of a root node labeled as null and child nodes that include an item name, support count, parent-link pointer and node-link pointer, as shown in Fig. 1 [15].

Today, current research concentrates on increasing performance of frequent patterns mining algorithms. Here, we show two enhancements of FP-tree. In [16] a more efficient algorithm than FP-tree is proposed and named the FP-split algorithm. FP-split is preferable compared to the FP-tree construction algorithm for three reasons. First, the proposed method scans the database only once where FP-tree scans twice. Second, filtering out and sorting the items descending depends on support as in FP-tree and each transaction record is no longer used in this method. Finally, header table and links will not be frequently examined, while a new node in the FP-split tree is added.

A new algorithm proposed in [17] is called Temporary Root growth and it is based on a Compressed FP-tree (TR-CFP). This algorithm uses a temporary root constructing concept during mining on a CFP tree instead of generating an FP-tree. Through its use, this approach can save a large amount of memory space and reduce the cost compared to FP-tree.

4   Proposed Course Analysis Work Flow

4.1   Acquiring Data

The data needed for this approach is all the information about graduated students in a university from a specific department. The more recent years included in the study, the more accurate the results to produce a comprehensive analysis. We recommend applying this methodology on at least ten years of students information. In addition, the number of semesters included in this methodology is important and at least 20 to 30 semesters are required for good analysis. The number of students in the data set affects the final analysis and, at least 100 students were included in each iteration of the study. The division of data into sets depends on the courses lists in the department.

4.2 Preprocessing

Pyle [18] suggested in 1999 that data preprocessing is one of the most important steps to the success of the project. Sixty percent of the total time required to complete a data mining project should be spent on data preprocessing as the preprocessing is a major element in data mining achieving a good result. It employs several steps including eliminating repeated information, outlier removal, detection of inconsistent information, data normalization and treatment of missing data.

Steps needed for this methodology is listed below:

- Save the Excel file as Comma Delimited (.csv) and then transform it to an Attribute Relation File Format (.arff) file which is better when extracting the knowledge in data mining systems.
- When the student registered in same course twice in two consecutive semesters. This state appears when the student drops the course in the first semester and takes it again in another semester. We consider the course is registered in the normal plan.
- Put aside any courses registered by only one student.
- If the department had an old list of courses and now a new one is considered; then reject any courses from the old list such that it is not available in the current list.
- Discard any courses registered after normal plan duration.
- Attribute selection is set to student id, course and semester.

4.3 Approach

The objective of this methodology is to analyze and discuss the distribution of courses in universities. To achieve this goal, frequent patterns extraction techniques should be used. The most commonly used algorithm in such a technique is Apriori. In addition, FP-Growth algorithm will provide good performance compared to basic Apriori. We suggest using the FP-Growth algorithm in the methodology due to its advantages over the Apriori algorithm as reviewed in the previous section. Finally, generating association rules will improve the analysis, so we recommend applying one of the association rules techniques.

5   Case Study

This case study focuses on female students studied in the regular master track who are graduated. Our data set is taken from the CS Department at King Saud University, Riyadh, Saudi Arabia. This data set contains 113 tuples of students information; each tuple includes Social Security Number, student ID, student name, gender, and a list of all courses taken by the student along with their grading. Our study was conducted based on data set analysis through ten years, which is approximately 20 semesters from 2003 to 2013.

In the CS department, the courses are divided into three lists, which are the core courses, list A courses, and list B courses are shown in Fig. 2. So, we divided the obtained data set into three separated sets that are compatible with the department lists.

(a)

Course Code	Course Title
CSC 512	Algorithm Analysis and Design
CSC 524	Computer Networks
CSC 541	Advanced Software Engineering
CSC 581	Advanced Data base Systems
CSC 595	Seminar and Decisions
CSC 597	Project 1
CSC 598	Project 2

(b)

Course Code	Course Title
CSC 519	Computer Security
CSC 543	Software Quality Management
CSC 551	Automata Calculability and Formal Languages
CSC 562	Artificial Intelligence
CSC 572	Advanced Computer Graphics
CSC 587	Web Databases and Information Retrieval
CSC 588	Data Warehouse and Mining Systems

(c)

Course Code	Course Title
CSC 520	Networking in the TSP/IP Environment
CSC 523	Distributed Systems
CSC 525	Distributed Real-Time Systems
CSC 526	Parallel Processing
CSC 527	Design and Implementation of Real –Time Systems
CSC 528	Interconnection Networks
CSC 529	Selected Topics in Computer Systems
CSC 530	High-Performance Computations
CSC 535	New Advances in Programming Languages
CSC 546	Designing Object-Oriented Software Systems
CSC 547	Software Measurements
CSC 548	Software projects Managements
CSC 549	Selected Topics in Software Engineering
CSC 558	Pattern recognition and Image Processing
CSC 561	Expert System and knowledge Engineering Applications
CSC 563	Neural Networks and Machine Learning Applications
CSC 566	Advanced Applications of pattern recognition and Machine Learning
CSC 567	E-Business and its Applications in large enterprises
CSC 569	Selected Topics in artificial intelligence
CSC 573	Numerical algorithms and their Applications in Computer science
CSC 574	Human-Machine communication and user-interface design
CSC 576	Graphics and multimedia Applications
CSC 578	Advances in multimedia Applications
CSC 579	Selected Topics in Computer Graphics
CSC 586	Hypermedia and Geographical Information Systems
CSC 589	Selected Topics in Database Systems
CSC 590	Selected Topics in Computer Applications
CEN 523	Fault-Tolerance Systems
CEN 545	Digital Image Processing

Fig. 2. Current courses lists in CS department. (a) Core courses list, (b) List A courses, (c) List B courses.

5.1   Acquiring Data

We acquired the data from the CS Department at King Saud University, in Excel sheet format. These data were about all graduated CS masters degree female students.

5.2   Preprocessing

The preprocessing step in this case study depends on the current list of courses in the CS department. Preprocessing steps were:

- The data sets saved in .arff file.
- Any courses registered twice by the same student, we had removed the first one and consider the second registration as normal plan. For example: when the

student registers the course in third semester then drop this semester and then he registers this course again in fourth semester, we consider the registration in fourth semester as the third semester.
- Any courses that had only one student registered were rejected.
- Any courses registered for by some student in the old CS department plan were rejected.
- Since the CS master project track is usually five semesters, any courses registered in the sixth semester were rejected.
- Student id, course name, and semester attributes.

5.3 Our Approach

Our new approach uses the FP-Growth data mining technique for extracting frequent patterns in the data sets. The original of the algorithm was obtained from Philippe Fournier-Verlag website [19]. We implemented it in NetBeans IDE 8.0.1 and integrated weka.jar file in the Java project. Then, this algorithm was applied on core courses, list A and list B files to find the courses most likely to be taken together in the same semester. Therefore, the performance of FP Growth algorithms is better for this application than Apriori based on the time. More discussion about the performance is provided in the discussion section.

5.4 Result Analysis

The data set is three files that our approach was applied on. Figure 3 illustrates the histogram of courses from list A. As shown, among all the seven courses that provided as list A there are three dominant courses; which are CSC 519, CSC 562 and CSC 572 taken by 107, 81, and 82 students respectively. We also notice that through ten years no student registered in CSC 543 or CSC 587 courses.

To analyze list B courses, Fig. 4 showed a histogram of this list. However, because of the large number of courses in list B (there were 29), there are no specific dominant courses. The most frequently taken courses involved 16. Students are more likely to register in these courses in fourth and fifth semesters. Figure 5(a) illustrates the output of our system using FP-Growth for three data sets. Figure 5(b) shows the histogram of second semester with its most frequent courses. These are two courses from list A which are CSC 562 and CSC 572 and one core course which is CSC 581. Moreover, Fig. 5(c) shows the histogram of third semester with its most frequent courses. These courses are CSC 519 from list A and CSC595 from core courses list.

6 Discussion

Comparison between Apriori and FP Growth algorithms can be discussed based on the memory usage and the time required for finding the result. We applied these algorithms on two different datasets of more than 350 records. One of the data sets is for list A courses and the other data set for list B. Table 1 below

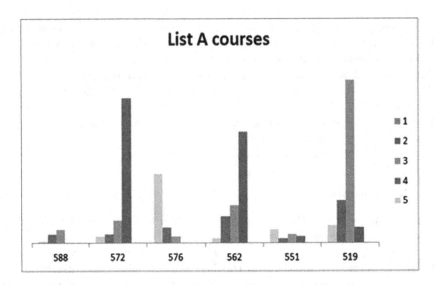

Fig. 3. Histogram of List A courses.

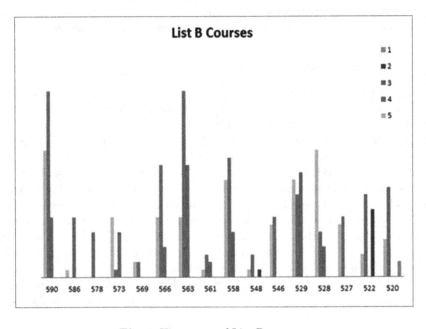

Fig. 4. Histogram of List B courses.

compares between Apriori and FP Growth algorithms on the time and memory usage.

As is shown, the Apriori algorithm required less memory to find the result. However, it takes a longer time than FP Growth algorithm for the same data

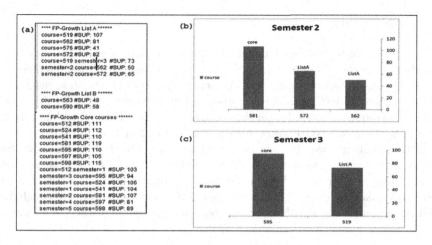

Fig. 5. Frequent courses in each semester. (a) Final output of the system, (b) Histogram of frequent courses in second semester, (c) Histogram of frequent courses in third semester.

Table 1. Comparison between Apriori and FP-Growth Algorithms

Data Set	Comparison Criteria	Apriori Algorithm	FP-Growth Algorithm
List A	Memory usage	5.13 MB	7.69 MB
	Time	70 ms	19 ms
List B	Memory usage	6.41 MB	9.61 ms
	Time	39 ms	17 ms

sets. For this application we find that FP-Growth is better. After applying our approach in this case study and analyzing the results, we have some recommendations for the CS department:

- Through ten years, there are two courses provided in list A for which no student registered in one of them. These were CSC 543: Software Quality Management and CSC 587: Web Databases and Information Retrieval.
- Among 113 students listed in the data sets there were 107 registered in CSC 519: Computer Security. Seventy-three of these students took CSC519 in the third semester with CSC595: Seminar and Discussion from the core courses list. Therefore, we recommend providing these two courses together.
- 50 students that registered in CSC 562: Artificial Intelligence in the second semester also chose to study CSC 572: Advanced Computer Graphics. We recommend to allow students to register in these two courses together in second or third semesters.

7 Conclusion

In this paper, we report on having conducted a new approach to analyze course distribution in the CS department. Certain related and associated courses are more likely to be taken together in the same semester. This system used data mining techniques to find frequent courses in the core courses list, list A courses and list B courses, provided from the department. The FP-Growth data mining algorithm was used. With respect to memory and run time we compare the performance of the two algorithms Apriori and FP-Growth. Results show that FP-Growth is much faster than Apriori but this is used the expense of more memory usage. After analyzing the data sets we found frequent courses that are frequently taken together by students during the same semester. We concluded with the recommendation that certain courses be available at the same time.

8 Future work

We will add more sophisticated association rules that find the relations between courses in list A and list B by joining these two data sets into one relation as a SQL database.

References

1. Delavari, N., Shirazi, M.R.A., Beikzadeh, M.R.: A new model for using data mining technology in higher educational systems. In: Information Technology Based Higher Education and Training. In: Proceedings of the Fifth International Conference on Information Technology Based Higher Education and Training pp. 319–324 (2004)
2. Abdullah, Z., Herawan, T., Ahmad, N., Deris, M.M.: Mining significant association rules from educational data using critical relative support approach. Procedia-Social Behav. Sci. **28**, 97–101 (2011)
3. Han, J., Kamber, M.: Data Mining. Concepts and Techniques. Morgan kaufmann, Southeast Asia Edition (2006)
4. Minaei-Bidgoli, B., Pang-Ning, T., Punch, W.F.: Mining interesting contrast rules for a web-based educational system. In: Proceedings of International Conference on Machine Learning and Applications, pp. 320–327 (2004)
5. Foster, I., Kesselman, C., Nick, J., Tuecke, S.: The Physiology of the Grid: an Open Grid Services Architecture for Distributed Systems Integration. Technical report, Global Grid Forum (2002)
6. Romero, C., Ventura, S.: Educational data mining: a review of the state of the art. IEEE Trans. Syst. Man Cybern. Part C Appl. Rev. **40**, 601–618 (2010)
7. Agrawal, R., Imielinsk, Swami, T.A.: Mining Association Rules between Sets of Items in Large Databases. In: Proceedings of the ACM SIGMOD International Conference on the Management of Data, pp. 207–216 (1993)
8. Kotsiantis, S., Kanellopoulos, D.: Association Rules Mining: A Recent Overview. GESTS Int. Trans. Comput. Sci. Eng. **32**, 71–82 (2006)
9. Shirwaikar, P.R., Rajadhyax, N.: Data Mining on Educational Domain (2012). arXiv preprint arXiv:1207.1535

10. Usha, D., Rameshkumar, D.: A complete survey on application of frequent pattern mining and association rule mining on crime pattern mining. Int. J. Adv. Comput. Sci. Technol. **3**, 264–275 (2014)

11. D.N, G., Anshu, C., Raghuvanshi, C.S.: An algorithm for frequent pattern mining based on Apriori. (IJCSE) Int. J. Comput. Sci. Eng. **2**, 942–947 (2010)

12. Al-Maolegi, M., Arkok, B.: An improved apriori algorith for association rules. Int. J. Nat. Lang. Comput. (IJNLC) **3**, 21–29 (2014)

13. Jha, J., Ragha, L.: Educational data mining using improved apriori algorithm. Int. J. Inf. Comput. Technol. **5**, 411–418 (2013)

14. Han, J., Pei, J., Yin, Y., Mao, R.: Mining frequent patterns without candidate generation: a frequent-pattern tree approach. Data Min. Knowl. Disc. **8**, 53–87 (2004)

15. Du, H.: Data Mining Techniques and Applications: An Introduction, Cengage Learning: Andover (2010)

16. Lee, C.-F., Shen, T.-H.: An FP-split method for fast association rules mining. In: 3rd International Conference on Information Technology: Research and Education. IEEE, pp.459–463 (2005)

17. Fei, C., Lin, S., Ming, L., Zhao-Qian, C., Shi-Fu, C.: Mining Frequent Patterns based on Compressed FP-tree without Conditional FP-tree Generation. In: IEEE International Conference on Granular Computing, pp. 478–481 (2006)

18. Pyle, D.: Data Preparation for Data Mining (1999). (The Morgan Kaufmann Series in Data Management Systems)

19. An Open-Source Data Mining Library. www.philippe-fournier-viger.com/spmf/

Learning the Relationship Between Corporate Governance and Company Performance Using Data Mining

Darie Moldovan[✉] and Simona Mutu

Faculty of Economics and Business Administration, Babeş-Bolyai University,
Theodor Mihali 58-60, 400591 Cluj-Napoca, Romania
{darie.moldovan,simona.mutu}@econ.ubbcluj.ro

Abstract. The objective of this paper is to identify the relationship between corporate governance variables and firm performance by employing data mining methods. We choose two dependent variables, Tobin's Q ratio and Altman Z-score, as measures for the companies' performances and apply machine learning techniques on the data collected from the components companies of three major stock indexes: S&P 500, STOXX Europe 600 and STOXX Eastern Europe 300. We use decision trees and logistic regressions as learning algorithms, and then we compare their performances. For the US components, we found a positive connection between the presence of women in the board and the company performance, while in Western Europe that it is better to employ a larger audit committee in order to lower the bankruptcy risk. An independent chairperson is a positive factor related to Altman Z-score, for the companies from Eastern Europe.

Keywords: Machine learning · Corporate governance · Firm performance · Classification · Logistic regression · Decision Trees

1 Introduction

The study of corporate governance has increased continuously in the last twenty years, benefiting from time to time boosts such as the large insolvencies of the early 2000s (Enron, WorldCom, Parmalat) or the effects of the recent financial crisis, which shown multiple weaknesses in the corporate governance systems of the companies.

The corporate governance rules are different from country to country, but major differences can be found between developed markets and emerging markets. The emerging markets countries implemented a series of reforms to their corporate governance systems in terms of legal measures, corporate governance codes or cross listings [13]. The integration of emerging markets within the global financial systems brought new investment opportunities, but also challenges in capital allocation. The financial crisis made portfolio managers to look more closely to the quality of corporate governance before taking their investment decisions.

© Springer International Publishing Switzerland 2015
P. Perner (Ed.): MLDM 2015, LNAI 9166, pp. 368–381, 2015.
DOI: 10.1007/978-3-319-21024-7_25

Our objective is to demonstrate the utility of data mining methods in identifying the relationships between corporate governance and performance. As a proxy for corporate performance we use two dependent variables: the company's performance evaluation (represented by the Tobin's Q ratio [45]) and the financial distress level (according to Altman Z-score [2]). Our testbed is represented by the data collected from three corporate governance systems: United States, Western Europe and Eastern Europe. We used the data from the companies composing the following three stock indexes: S&P500, STOXX Europe 600 and STOXX Eastern Europe 300. We selected 50 attributes, both financial indicators and corporate governance variables, in order to create a "competition" between them to explain the Tobin's Q ratio and Altman Z-score.

We used classification algorithms and regressions in order to demonstrate the utility of data mining techniques in creating portraits of different typologies for the companies analyzed. We employed the software Weka 3.7 to make our experiments, a powerful open source tool developed by the University of Waikato, New Zealand [25].

Our findings reveal that for each dataset the performances of algorithms are close to each other, with Adaboost M1 being the most consistent in terms of classification accuracy in most of the cases. In case of American companies the Tobin's Q ratio is positively influenced by the percentage women in the board, while an independent lead director and a financial leverage higher than 2.5 generates a higher risk of bankruptcy. For the Western European companies, the presence of an independent CEO and large percentage of women in the board could lead to weaker performances, being negatively correlated with Tobin's Q. As for the Eastern European companies, a smaller age range for the board members and a financial leverage less than 4 enhances companies' performance, while an independent chairperson or a woman as CEO reduces bankruptcy probability.

The obtained results can be useful either for investment decision or as a guide for board companies looking to improve their corporate governance policies, both in developed, but also in emerging markets.

The remaining of this paper is structured as follows. Section 2 is dedicated to describe the related work in this field. First we described the research approaches in using the Tobin's Q ratio and the Altman Z-score in relation with the corporate governance. Second, we review the literature concerning the application of data mining methods in the field of corporate governance, demonstrating the usefulness of the algorithms in solving research questions in this area. Section 3 describes the data used in this research. Section 4 describes in detail the experiments conducted. Section 5 presents the results. Finally, the conclusions of the paper are described in Sect. 6.

2   Related Work

Our paper connects several strands of literature. We relate to the studies that focus on the study of executives compensations influence on companies' performance and on the importance of the board members' independence [1,7,43]. The separation of the executive management from the stakeholders can lead sometimes to a tendency of the managers to accumulate wealth. The participation of the shareholders in the board can have a positive effect on the company's performance.

In other studies, Guest [24] found a strong negative impact of the board size on the profitability of the company and the Tobin's Q ratio; Li [34] determined that independent directors' presence is positively related with the performance of the company while Lückerath-Rovers [36] associated positively the presence of women on boards with the companies' performances.

To evaluate the firm performance, a widely accepted indicator as reference by many researchers in this field is the Tobin's Q. Several studies were made around the opportunity to use this ratio as a measure of intangible assets of the company [5,6,11,33], but also negative arguments were brought by Dybvig [17].

Regarding the nexus between governance and performance in emerging and developed markets, Gibson [23] suggests that corporate governance is ineffective in emerging markets. Claessens and Djankov [12] found that firms with concentrated and foreign ownership are more profitable, registering a higher labor productivity.

Further, our research is connected to the studies on companies' bankruptcy. The probability of default could be enhanced by the "dominant" leadership style [27], the proportion of outside directors [16], the average number of directorships in other firms held by outside directors [18]and by a small and younger board [39]. During crisis periods, [40] found that companies which change their CEO and with a low percentage of shares held by managers are more likely to fail than the others.

The literature proposes several methods to measure the companies' default like the Z-score of Altman [2], the O-score of Ohlson [38] and the structural model default probabilities of Merton [37], among others. However, a large number of studies advocate for the Altman Z-score [4,5,8]. Also, we link to the research on machine learning applications in this area. Related studies are represented by the works of Creamer [14,15] who uses boosting in order to examine the relationship between the Tobin's Q and variables representative for south American banks, but also for the S&P 500 component companies. Tsai [46] focuses on applying data mining in order to observe the intangible assets value on companies from South Korea. Lu [35] is using association rules to address the same problem.

Other authors are focusing on detecting credit risk, creating credit scoring systems with the help of data mining, and using the Altman Z-score, an indicator accepted on a large scale by researchers and professionals in the field [10,29,30].

We contribute to the existing literature in several ways. First, we analyze a unique sample of listed companies which are included in three major stock indexes: S&P 500, STOXX Europe 600 and STOXX Eastern Europe 300, representative for developed and emerging markets. Second, we explore a large and unique set of variables that characterize different corporate governance and accounting aspects. Third, we add to the literature on data mining and corporate performance, by applying machine learning techniques to investigate the impact of companies' governance and financial characteristics on their performance.

3 Data Preprocessing

We collected data for the companies composing three stock indexes: S&P500 (SPX), STOXX Europe 600 (SXXP) and STOXX Eastern Europe 300 (EEBP). The data source was Bloomberg, from where we extracted 52 variables for 1400 companies. The chosen dependent variables were Tobin's Q ratio, as a measure for the performance of the companies and Altman Z-score, as a measure of the financial distress.

Tobin's Q represents the ratio between the market value of a company and the replacement cost of the firm's assets and is computed using the following formula:

$$Tobin's\,Q = (Market\,Cap + Total\,Liabilities + Preferred\,Equity$$
$$+ Minority\,Interest)/Total\,Assets \qquad (1)$$

In order to find the relationship between these variables and the corporate governance, we chose 50 variables, both from the corporate governance and accounting area. The accounting indexes are well known for their predictive power, being used by researchers in detecting patterns inside the financial statements of the companies [2,30,42].

Among the corporate governance variables we can mention the board size, the number of independent directors, the presence of women on board, the percent of female executives, the average age of the board members or the percent of independent directors inside the board. Our interest is to find if there is a connection between these variables and the performance of the company, expressed by Tobin's Q ratio and, respectively the financial distress, expressed by the Altman Z-score.

The accounting variables chosen are among the most popular used in fundamental analysis of the companies. Indicators such as Operating Profit Margin, Return on Equity, Dividend Payout, Return on Capital or Price to Book ratio are all indicators frequently used to analyze the financial statements of a company. A list of some of the variables used in our study can be found in Appendix A.

During the preprocessing stage of the knowledge discovery process [20] we had to deal with the missing values inside the data set and removing outliers. We removed all the instances that had missing data on the attribute of the dependent variables or did not find enough information to calculate their values. We also removed the outliers for all the variables in order to eliminate any possible data errors.

The class variables Tobin's Q ratio and Altman Z-score were processed in order to obtain nominal variables. As suggested by Creamer [14], we discretized Tobin's Q ratio in order to obtain two classes, dividing each dataset according to its median value. In this way, a company that lies in the upper side of the median will be looked positively by the machine learning algorithms.

The Altman Z-score variable was divided into three classes, according to the definitions provided by Altman [2,3]. The indicator is computed considering the following formula:

$$Altman\,Z\,Score = 1.2 * (Working\,Capital/Tangible\,Assets)$$
$$+ 1.4 * (Retained\,Earnings/Tangible\,Assets)$$
$$+ 3.3 * (EBIT/Tangible\,Assets)$$
$$+ 0.6 * (Market\,Value\,of\,Equity/Total\,Liabilities)$$
$$+ (Sales/TangibleAssets) \qquad (2)$$

For the SPX and SXXP companies we used the classification values specific to the developed markets. In this case, a company that has a Z-score above 2.99 is considered to be in the "safe" zone, if the score is bellow 2.99 but above 1.81 it is considered to be in the "gray" zone and if it lies below 1.81 then it is in financial distress. The companies from the emerging markets are evaluated on a slightly different scale, where the "safe" zone begins at a score of 2.6, the "grey" zone is located between 1.1 and 2.6, while the "distress" zone is situated bellow the Z-score of 1.1.

After the preprocessing of the three datasets, the SPX dataset remained with 496 records, the SXXP dataset with 595 instances and the EEBP dataset with 297 instances.

4   Experiments

For each dataset, we performed an attribute evaluation in relation to the class, in order to eliminate all the attributes that do not offer an information gain with respect to the class. The attribute selection operation was performed only if the classification precision improved. If not, the dataset was maintained unaltered. Considering the benchmark realized by Hall [26] we used two alternative feature selection methods: CFS (Correlation based Feature Selection) and Information Gain Attribute Ranking. The first is a method that evaluates subsets of attributes rather than individual attributes. The algorithm takes into account the usefulness of the features to predict the class, considering the level of intercorrelation between them. The second method calculates the entropy before and after observing an attribute. The amount by which the entropy decreases represents the information gain provided by the attribute [41]. We used for our experiments four different algorithms: Alternating Decision Trees (ADTree), Adaboost M1 with ADTree, J48 and Simple Logistic. The Alternating Decision Tree [21,28] consists of decision nodes and prediction nodes; an instance will be classified by following all the paths for which the decision nodes are true and summing the values of the prediction nodes from the path.

The Boosting is a method introduced by Freund [22], that applies repeatedly an algorithm (named "weak learner") to different weightings of the same training set. It was first used for binary problems, but has new implementations that are capable to deal with multinomial classes. In our experiments we used the Adaboost M1 [22] implementation with ADTree as the weak learner algorithm.

Simple logistic is an algorithm designed by Sumner [44]and it works as a classifier for building linear logistic regression models. It uses LogitBoost [32] as base learners for fitting the logistic models.

Table 1. Tobin's Q as class – Algorithms performance – SPX dataset

Algorithm	Correctly classified instances	Coverage of cases (0.95 level)	Precision Class 0	Precision Class 1	ROC area
Adaboost M1	89.7177 %	91.3306 %	0.89	0.905	0.957
J48	85.2823 %	92.9435 %	0.841	0.866	0.854
Simple log	90.3226 %	98.5887 %	0.9	0.906	0.952
ADTree	88.1048 %	99.5968 %	0.865	0.898	0.941

Table 2. Tobin's Q as class – Algorithms performance – SXXP dataset

Algorithm	Correctly classified instances	Coverage of cases (0.95 level)	Precision Class 0	Precision Class 1	ROC area
Adaboost M1	88.2353 %	91.4286 %	0.891	0.874	0.946
J48	87.395 %	97.3109 %	0.871	0.877	0.874
Simple log	85.042 %	98.6555 %	0.845	0.856	0.927
ADTree	87.563 %	99.3277 %	0.881	0.87	0.948

Table 3. Tobin's Q as class – Algorithms performance – SXXP dataset

Algorithm	Correctly classified instances	Coverage of cases (0.95 level)	Precision Class 0	Precision Class 1	ROC area
Adaboost M1	81.8182 %	87.8788 %	0.823	0.813	0.889
J48	77.1044 %	100 %	0.783	0.76	0.836
Simple log	76.431 %	98.9899 %	0.78	0.75	0.861
ADTree	80.8081 %	99.6633 %	0.811	0.805	0.893

J48 is the implementation of Quinlan's C4.5 [41], and using the concept of information gain, or entropy splits the data into subsets. The greater the information gain brought by a certain attribute, the higher will be in the decision tree.

4.1 Tobin's Q Ratio as Class

The first round of experiments was conducted having the Tobin's Q ratio as class variable. Observing the robust results obtained by Creamer [15] in employing Adaboost implemented with Alternating Decision Trees, we chose the Weka algorithm Adaboost M1 with ADTree as a benchmark for our experiments related to Tobin's Q ratio. The other algorithms used were the J.48 decision tree with a 0.05 confidence factor, the Simple Logistic Regression and the ADTRee. The validation method chosen was 10-fold cross validation, as a useful way for validation when the dataset is not very large [31].

Table 4. Altman Z-score as class – Algorithms performance – SPX dataset

Algorithm	Correctly classified instances	Coverage of cases (0.95 level)	Precision Class 0	Precision Class 1	Precision Class 2	ROC area
Adaboost M1	82.9443 %	92.8187 %	0.843	0.725	0.865	0.942
J48	73.6086 %	91.2029 %	0.722	0.569	0.816	0.843
Simple log	79.1741 %	96.0503 %	0.793	0.655	0.843	0.907
LADTree	75.9425 %	94.3084 %	0.768	0.574	0.856	0.9

Table 5. Altman Z-score as class – Algorithms performance – SXXP dataset

Algorithm	Correctly classified instances	Coverage of cases (0.95 level)	Precision Class 0	Precision Class 1	Precision Class 2	ROC area
Adaboost M1	76.2681 %	93.1159 %	0.804	0.522	0.864	0.769
J48	74.2754 %	96.1957 %	0.788	0.531	0.803	0.856
Simple log	77.5362 %	98.3696 %	0.792	0.61	0.811	0.906
LADTree	73.7319 %	98.0072 %	0.793	0.53	0.807	0.893

Table 6. Altman Z-score as class – Algorithms performance – EEBP dataset

Algorithm	Correctly classified instances	Coverage of cases (0.95 level)	Precision Class 0	Precision Class 1	Precision Class 2	ROC area
Adaboost M1	63.7602 %	99.455 %	0.746	0.496	0.675	0.771
J48	65.6676 %	98.3651 %	0.79	0.615	0.598	0.8
Simple log	69.7548 %	99.1826 %	0.795	0.543	0.731	0.849
LADTree	64.5777 %	97.8202 %	0.796	0.485	0.683	0.823

To measure the performance of the algorithms we looked for the correctly classified instances and for the classification precision for every class. By observing the precision in every class (correctly classified instances/class) we monitor the consistency of the classification in detail, excluding the possibility that the global computed accuracy could hide the lack of performance in classifying a certain class. Supplementary, we considered the Coverage of cases (0.95 level) and the ROC area [19]. In Tables 1, 2 and 3 we present the compared performance of the algorithms after a tenfold cross-validation for each of the datasets.

4.2 Altman Z-Score as Class

The second part of our experiments was dedicated to the relation between the corporate governance variables and accounting indexes on one side, and the

Altman Z-score on the other side. The Altman Z-score being an indicator for bankruptcy, our objective was to find which of the variables could be considered red flags for distress, but also to identify the financial safety signals. The Altman Z-score variable had three classes: 0 for "distress", 1 for "gray", 2 for "safe". Because, generally, the number of instances in the "distress" and "gray" classes was much more smaller than those in the "safe" class, we used a re-sampling technique, named SMOTE, or Synthetic Minority Oversampling Technique [9]. This way, we re-balance the datasets by over-sampling the minority class, in order to achieve better classifier performance. This is achieved by creating synthetic minority class examples.

As the ADTree algorithm is not suitable for two class target variables, we used Adaboost M1 with LADTree [28], an adaptation of the original algorithm, in order to support multi-class. The other algorithms used are the logistic regression, implemented to support more than two values for the class variable, the J48 and the simple LADTree. In Tables 4, 5 and 6 we present the performances obtained after running the algorithms with the Altman Z-score variable as class.

5   Results Obtained

The results of the learning algorithms in terms of classification accuracy are different for each dataset used. While for the SPX and SXXP datasets the results do not differ too much, when the algorithms were applied to the EEBP set, the results were not so strong. The explanation for this behavior resides in the significant percent of missing data for the corporate governance variables collected from the Eastern Europe companies. The principles of corporate governance are not so well established for these companies and only a minority of them adopted the same good practices as the western ones.

In terms of algorithms comparison, we can note that for each dataset the performances are close to each other, with Adaboost M1 being the most consistent in terms of classification accuracy in most of the cases.

We can also note the Simple Logistic Regression algorithm that performed better than the others on the EEBP dataset with Altman Z-score as class variable (Table 6).

From a financial perspective we found interesting connections between the corporate governance variables and the two dependent variables, which was the main objective of our research.

For the American companies inside the S&P 500 index, we found a positive correlation between the percentage higher than 20 % of women in the board and the Tobin's Q ratio, but also the presence of an independent lead director in the company along with a financial leverage higher than 2.5 incur a higher risk of bankruptcy.

For the Western European companies, the presence of an independent lead director or a former CEO in the board could be a sign of weaker performances, being negatively correlated with Tobin's Q. A large percentage of women in the board could also affect negatively the performance. For the companies with large

```
J48 pruned tree
------------------

P/B <= 1.879789
|   Indep Lead Dir = Y
|   |   P/B <= 1.366616: 0 (62.37/4.78)
|   |   P/B > 1.366616
|   |   |   Fincl 1 <= 2.888611: 1 (15.34/4.0)
|   |   |   Fincl 1 > 2.888611: 0 (10.34/0.39)
|   Indep Lead Dir = N: 0 (161.23/8.09)
P/B > 1.879789
|   Asset <= 0.182254: 0 (23.79/2.21)
|   Asset > 0.182254
|   |   Fincl 1 <= 3.874863: 1 (254.57/18.0)
|   |   Fincl 1 > 3.874863
|   |   |   Oper ROE <= 29.051374
|   |   |   |   Board Size <= 13
|   |   |   |   |   Indep Directors <= 5: 0 (13.04/1.99)
|   |   |   |   |   Indep Directors > 5
|   |   |   |   |   |   Asset <= 0.98851: 0 (11.95/2.72)
|   |   |   |   |   |   Asset > 0.98851: 1 (8.5)
|   |   |   |   Board Size > 13: 0 (14.29)
|   |   |   Oper ROE > 29.051374: 1 (19.57/0.17)

Number of Leaves  :      11

Size of the tree :      21
```

Fig. 1. J48 Results for the SXXP dataset

financial leverage in order to be in the "safe" zone of the Altman Z-score it could be a good idea to adopt an Auditing Committee with more than four people.

When analyzing the Eastern European companies' data, we found that a smaller age range for the board members is positively related with the companies' performance and that a financial leverage less than 4 is needed in order to be on the upper side of the Tobin's Q ratio. To be on the "safe" zone of the Altman Z-score it is important to have an independent chairperson or even a woman as CEO.

We present in Fig. 1 an example of output from Weka, consisting in a decision tree for the SPX dataset with Tobin's Q dependent variable.

6   Conclusions

This paper develops a framework that applies four different machine learning algorithms (Adaboost, ADTree, J48 and Simple Logistic) to three data sets obtained from the component companies of the following stock market indexes: S&P500, STOXX Europe 600 and STOXX Eastern Europe 300. Our findings present several important features which have practical applicability both for the boards of the listed companies, as well as for investors. They can observe which corporate governance variables are closely connected to the performance indicator (Tobin's Q). The board members can learn what changes could be made in the corporate governance in order to align it with the best practices from the most successful companies. The investors can assess the changes in the corporate governance of a certain company and decide if it is heading the good direction. The relation between the corporate governance variables and the distress indicator Altman Z-score can serve to those interested in evaluating the bankruptcy risk of a company. While the indicators are computed using data from the past, a change in the corporate governance could be a flag in which direction the company is heading. By having the corporate governance variables categorized, one could take immediate action. In this respect, we obtained interesting relations between the corporate governance variables and the companies' performances, while observing different particularities of the three different world areas.

The performance of the algorithms is more solid for the S&P500 and STOXX Europe 600 indexes components because of the high availability of the data. For the Eastern Europe index the data related to corporate governance register missing values for many companies, mainly because they did not adopt all the good practices already established with the American and West European companies. The Adaboost M1 algorithm with ADTree and the Simple Logistic Regression obtained the best results, which is consistent with other research in this area [14].

The research could be improved by collecting more consistent data related to Eastern Europe companies, which could be achieved by questioning the targeted companies. Another direction for improvement is to collect historical data for the companies for several years in order to increase the datasets and to observe the variables evolution. In terms of learning algorithms, deep learning methods can be added to detect useful patterns.

Acknowledgments. This work was co financed from the European Social Fund through Sectoral Operational Program Human Resources Development 2007–2013, project number POSDRU/159/1.5/S/134197 and POSDRU/159/1.5/S/142115, Performance and excellence in doctoral and postdoctoral research in Romanian economics science domain" and from UEFISCDI under project JustASR - PN-II-PT-PCCA-2013-4-1644.

Appendix A

Variable	Description
Tax	Tax burden for the last 12 months
Interest	Interest burden for the last 12 months
Asset	Amount of sales or revenues generated per dollar of assets. The ratio is an indicator of the efficiency with which a company is deploying its assets
Fincl l	Financial leverage. Measures the average assets to average equity
Oper ROE	Normalized ROE. Returns on Common Equity based on net income excluding one-time charges
Dvd P/O	Dividend Payout Ratio. Fraction of net income a firm pays to its shareholders in dividends, in percentage
Board Size	Number of Directors on the company's board
% Non Exec Dir on Bd	Percentage of the board of directors that is comprised of non-executive directors
% Indep Directors	Independent directors as a percentage of total board membership
CEO Duality	Indicates whether the company's Chief Executive Officer is also Chairman of the Board, as reported by the company
Indep Chrprsn	Indicates whether the company chairperson was independent as of the fiscal year end
Indep Lead Dir	Indicates whether the company has an independent lead director within the board of directors
Frmr CEO or its Equiv on Bd	Indicates whether a former company chief executive officer (CEO) or person with equivalent role has been a director on the board
% Women on Board	Percentage of Women on the Board of Directors
Ind Dir Bd Mtg Att	Percentage of board meetings attended by independent directors
Unit or 2 Tier Bd Sys	Indicates whether the company's board has a Unitary (1) or Two Tier (2) system. Marked 2 when board system has separate boards for Supervisory/Commissioner board and Management board
Prsdg Dir	Indicates whether the company has a presiding director in its board of directors
% Feml Execs Feml CEO or Equiv	Number of female executives, as a percentage of total executives. Indicates whether the company Chief Executive Officer (CEO) or equivalent is female
Age Young Dir	Age of the youngest director on the company board in years
BOD Age Rng	Age range of the members of the company board in years, calculated by subtracting the age of the youngest director on the company board from the age of the oldest director on the company board
Age Old Dir	Age of the oldest director on the company board in years
Bd Avg Age	Average age of the members of the board
Board Duration	Length of a board member's term, in years
Board Mtgs #	Total number of corporate board meetings held in the past year
Exec Dir Bd Dur	Length of an executive director board member's term, in years

References

1. Acharya, V.V., John, K., Sundaram, R.K.: On the optimality of resetting executive stock options. J. Financ. Econ. **57**(1), 65–101 (2000)
2. Altman, E.I.: Financial ratios, discriminant analysis and the prediction of corporate bankruptcy. J. Financ. **23**(4), 589–609 (1968)
3. Altman, E.I., Saunders, A.: Credit risk measurement: developments over the last 20 years. J. Banking Financ. **21**(11), 1721–1742 (1997)
4. Becker, B., Strömberg, P.: Fiduciary duties and equity-debtholder conflicts. Rev. Financ. Stud. **25**(6), 1931–1969 (2012)
5. Bhagat, S., Bolton, B.: Corporate governance and firm performance. J. Corp. Financ. **14**(3), 257–273 (2008)
6. Bolton, P., Chen, H., Wang, N.: A unified theory of tobin's q, corporate investment, financing, and risk management. J. Financ. **66**(5), 1545–1578 (2011)
7. Brenner, M., Sundaram, R.K., Yermack, D.: Altering the terms of executive stock options. J. Financ. Econ. **57**(1), 103–128 (2000)
8. Bryan, D., Fernando, G.D., Tripathy, A.: Bankruptcy risk, productivity and firm strategy. Rev. Account. Financ. **12**(4), 309–326 (2013)
9. Chawla, N.V., Bowyer, K.W., Hall, L.O., Kegelmeyer, W.P.: Smote: synthetic minority over-sampling technique. J. Artif. Intell. Res. **16**(1), 321–357 (2002)
10. Chen, W., Xiang, G., Liu, Y., Wang, K.: Credit risk evaluation by hybrid data mining technique. Syst. Eng. Procedia **3**, 194–200 (2012)
11. Chung, K.H., Pruitt, S.W.: A simple approximation of Tobin's q. Financ. Manage. **23**, 70–74 (1994)
12. Claessens, S., Djankov, S.: Ownership concentration and corporate performance in the Czech Republic. J. Comp. Econ. **27**(3), 498–513 (1999)
13. Claessens, S., Yurtoglu, B.B.: Corporate governance in emerging markets: a survey. Emerg. Mark. Rev. **15**, 1–33 (2013)
14. Creamer, G., Freund, Y.: Learning a board balanced scorecard to improve corporate performance. Decis. Support Syst. **49**(4), 365–385 (2010)
15. Creamer, G.G., Freund, Y.: Predicting performance and quantifying corporate governance risk for Latin American adrs and banks. In: Financial Engineering and Applications. MIT, Cambridge (2004)
16. Daily, C.M., Dalton, D.R.: Bankruptcy and corporate governance: the impact of board composition and structure. Acad. Manage. J. **37**(6), 1603–1617 (1994)
17. Dybvig, P.H., Warachka, M.: Tobin's q does not measure performance: theory, empirics, and alternative measures. Unpublished Working paper, Washington University, Saint Louis, United Sates (2010)
18. Elloumi, F., Gueyie, J.P.: Financial distress and corporate governance: an empirical analysis. Corp. Governance Int. J. Bus. Soc. **1**(1), 15–23 (2001)
19. Fawcett, T.: An introduction to roc analysis. Pattern Recogn. Lett. **27**(8), 861–874 (2006)
20. Fayyad, U., Piatetsky-Shapiro, G., Smyth, P.: From data mining to knowledge discovery in databases. AL Mag. **17**(3), 37 (1996)
21. Freund, Y., Mason, L.: The alternating decision tree learning algorithm. In: ICML. vol. 99, pp. 124–133 (1999)
22. Freund, Y., Schapire, R.E., et al.: Experiments with a new boosting algorithm. In: ICML, vol. 96, pp. 148–156 (1996)
23. Gibson, M.S.: Is corporate governance ineffective in emerging markets? J. Financ. Quant. Anal. **38**(01), 231–250 (2003)

24. Guest, P.M.: The impact of board size on firm performance: evidence from the UK. Eur. J. Financ. **15**(4), 385–404 (2009)
25. Hall, M., Frank, E., Holmes, G., Pfahringer, B., Reutemann, P., Witten, I.H.: The weka data mining software: an update. ACM SIGKDD Explor. Newslett. **11**(1), 10–18 (2009)
26. Hall, M.A., Holmes, G.: Benchmarking attribute selection techniques for discrete class data mining. IEEE Trans. Knowl. Data Eng. **15**(6), 1437–1447 (2003)
27. Hambrick, D.C., D'Aveni, R.A.: Top team deterioration as part of the downward spiral of large corporate bankruptcies. Manage. Sci. **38**(10), 1445–1466 (1992)
28. Holmes, G., Pfahringer, B., Kirkby, R., Frank, E., Hall, M.: Multiclass alternating decision trees. In: Elomaa, T., Mannila, H., Toivonen, H. (eds.) ECML 2002. LNCS (LNAI), vol. 2430, pp. 161–172. Springer, Heidelberg (2002)
29. Kambal, E., Osman, I., Taha, M., Mohammed, N., Mohammed, S.: Credit scoring using data mining techniques with particular reference to sudanese banks. In: 2013 International Conference on Computing, Electrical and Electronics Engineering (ICCEEE), pp. 378–383. IEEE (2013)
30. Kirkos, E., Spathis, C., Manolopoulos, Y.: Data mining techniques for the detection of fraudulent financial statements. Expert Syst. Appl. **32**(4), 995–1003 (2007)
31. Kohavi, R., et al.: A study of cross-validation and bootstrap for accuracy estimation and model selection. In: IJCAI 2014, pp. 1137–1145 (1995)
32. Landwehr, N., Hall, M., Frank, E.: Logistic Model Trees. In: Lavrač, N., Gamberger, D., Todorovski, L., Blockeel, H. (eds.) ECML 2003. LNCS (LNAI), vol. 2837, pp. 241–252. Springer, Heidelberg (2003)
33. Lang, L.H., Stulz, R.M.: Tobin's q, corporate diversification and firm performance. Technical report, National Bureau of Economic Research (1993)
34. Li, H.X., Wang, Z.J., Deng, X.L.: Ownership, independent directors, agency costs and financial distress: evidence from chinese listed companies. Corp. Governance Int. J. Bus. Soc. **8**(5), 622–636 (2008)
35. Lu, Y.H., Tsai, C.F., Yen, D.C., et al.: Discovering important factors of intangible firm value by association rules. Int. J. Digital Account. Res. **10**, 55–85 (2010)
36. Lückerath-Rovers, M.: Women on boards and firm performance. J. Manage. Governance **17**(2), 491–509 (2013)
37. Merton, R.C.: On the pricing of corporate debt: the risk structure of interest rates*. J. Financ. **29**(2), 449–470 (1974)
38. Ohlson, J.A.: Financial ratios and the probabilistic prediction of bankruptcy. J. Account. Res. **18**(1), 109–131 (1980)
39. Parker, S., Peters, G.F., Turetsky, H.F.: Corporate governance and corporate failure: a survival analysis. Corp. Governance Int. J. Bus. Soc. **2**(2), 4–12 (2002)
40. Platt, H., Platt, M.: Corporate board attributes and bankruptcy. J. Bus. Res. **65**(8), 1139–1143 (2012)
41. Quinlan, J.R.: C4. 5: Programs for Machine Learning. Elsevier, Burlington (2014)
42. Ravisankar, P., Ravi, V., Rao, G.R., Bose, I.: Detection of financial statement fraud and feature selection using data mining techniques. Decis. Support Syst. **50**(2), 491–500 (2011)
43. Ryan, H.E., Wiggins, R.A.: Who is in whose pocket? Director compensation, board independence, and barriers to effective monitoring. J. Financ. Econ. **73**(3), 497–524 (2004)
44. Sumner, M., Frank, E., Hall, M.: Speeding up logistic model tree induction. In: Jorge, A.M., Torgo, L., Brazdil, P.B., Camacho, R., Gama, J. (eds.) PKDD 2005. LNCS (LNAI), vol. 3721, pp. 675–683. Springer, Heidelberg (2005)

45. Tobin, J.: A general equilibrium approach to monetary theory. J. Money Credit Bank. **1**(1), 15–29 (1969)
46. Tsai, C.F., Lu, Y.H., Yen, D.C.: Determinants of intangible assets value: the data mining approach. Knowl.-Based Syst. **31**, 67–77 (2012)

A Bayesian Approach to Sparse Learning-to-Rank for Search Engine Optimization

Olga Krasotkina[✉] and Vadim Mottl

Computing Center RAS Moscow, Moscow, Russia
{O.V.Krasotkina,vmott}@yandex.ru

Abstract. Search engine optimization (SEO) is the process of affecting the visibility of a web page in the engine's search results. SEO specialists must understand how search engines work and which features of the web-page affect its position in the search results. This paper employs machine learning ranking algorithms to constructing the rank model of a web-search engine. Ranking a set of retrieved documents according to their relevance to a given query has become a popular problem at the intersection of web search, machine learning and information retrieval. Feature selection in learning to rank has recently emerged as a crucial issue. Recent work on ranking, focused on a number of different paradigms, namely, point-wise, pair-wise, and list-wise approaches, for which several preprocessing feature section methods have been proposed. Unfortunately, only a few works have been focused on integrating the feature selection into the learning process and all of these embedded methods are based on l_1 regularization technique. Such type of regularization does not possess many properties, essential for SEO, such as unbiasedness, grouping effect and oracle property. In this paper we suggest a new Bayesian framework for feature selection in learning-to-rank problem. The proposed approach gives the strong probabilistic statement of shrinkage criterion for features selection. The proposed regularization is unbiased, has grouping and oracle properties, its maximal risk diverges to finite value. Experimental results show that the proposed framework is competitive on both artificial data and publicly available LETOR data sets.

Keywords: Search engine optimisation · Learning-to-rank · Feature selection · Rank regression · Ordinal regression · Support vector machines

1 Introduction

The web engines are the main instrument for information retrieval in the Internet. When a user enters a query into a search engine, the engine examines its index and provides a listing of best-matching web pages according to the relevance of documents to the entered query.

Users normally tend to visit websites that are at the top of this list as they perceive those to be more relevant to the query. If you have ever wondered why some of these websites rank better than the others then you must know that it is because of a powerful

P. Perner (Ed.): MLDM 2015, LNAI 9166, pp. 382–394, 2015.
DOI: 10.1007/978-3-319-21024-7_26

web marketing technique called search engine optimization (SEO). SEO is a technique which helps search engines find and rank your site higher than the millions of other sites in response to a particular search query.

In our works the SEO specialists face three kinds of problems. First, the modern ranking algorithms depend not only on the web pages index but on the query. There are many examples when the optimization technic for the same website worked with one query but work with another. Second problem is the crawling frequency of your website by search engine. The so-called deep crawl occurs roughly once a month. This extensive reconnaissance of the web content requires more than a week to complete and an undisclosed length of time after completion to build the results into the index. For this reason, it can take up to six weeks for a new page to appear in the search engine.

SEO algorithm is defined in terms of what should be done on the page, what should be done around the page, and what should be done to the page. So the third problem is that there is a very wide range of factors describing the page for each particular query and the SEO process is generally costly and can be time consuming. It would be better if SEO specialist has an instrument for advising the main factors affecting the rank of particular page for particular query.

But the search engine algorithms are highly secret, competitive things. No one knows exactly what each search engine weights and what importance it attaches to each factor in the ranking formula. So a new trend has recently arisen in search engine optimization, that is, to employ machine learning techniques to automatically reconstruct the ranking model. In web search engines, a large amount of search log data, such as click through data, is accumulated. This makes it possible to derive training data from search log data and automatically create the ranking model. Learning to rank refers to machine learning techniques for training the model in a ranking task. In fact, learning to rank has become one of the key technologies for modern web search.

Over the past decade, a large number of learning to rank algorithms has been proposed [1]. Based on how they treat sets of ratings, they can be effectively categorized into the following three groups: point-wise, pairwise, and list-wise approaches. For a given query and a set of retrieved documents, point-wise approaches try to directly estimate the relevance label for each query-document pair. It may seen that these approaches are consistent for a variety of performance measures, they ignore relative information within collections of documents. In other words, they ignore that a document with a mediocre rating may actually be desirable if all other documents carry even lower scores (i.e., the one- eyed may be the king among the blind). Pairwise approaches, as proposed by [2–4], take the relative nature of the scores into account by comparing pairs of documents. They ensure that we obtain the correct order of documents even in case when we may not be able to obtain a good estimate of the ratings directly. Finally, list-wise approaches, as proposed by [5–8], treat the ranking in total, and they exploit the fact that we may not even care about the entire ranking, and, furthermore, even among the retrieved subset of documents we may care considerably more about the topmost documents.

There has been significant discussion about the relative merit of these strategies, and the common wisdom is that the list-wise is better than the pair-wise, which outperforms the point-wise.

But during the past decade the number of features that can be used by the algorithms has increased, the issue of feature selection in learning to rank has emerged, for two reasons.

First, as more and more features are incorporated into algorithms, not only the models become more difficult to understand, but also, they potentially have to deal with more and more noisy and irrelevant features, and the generalization performance of learning-to-rank algorithms is essentially decreased. Second, the large number of features leads to the high computational complexity of algorithms. Reducing the number of features is a promising way to handle the problem of high computational cost.

Recent works have focused on the development of feature selection methods for learning-to-rank algorithms. All of them can be divided into three groups. First group, called "filters", can be some preprocessing step, second group, called "wrappers" [9–12], use a predictive model to score feature subsets. The third group of methods, called "embedded" methods [13–16], are incorporated into algorithm. The main advantages of the last group are the ability to use the properties of ranking algorithm in contrast to filters and the lower computational complexity then wrappers. These approaches introduce a sparse regularization term in the formulation of the optimization problem. Although sparse regularizations are widely used in classification to deal with feature selection, only a few attempts have been made to propose sparse- regularized learning to rank methods.

Sun et al. [14] implemented a sparse algorithm called RSRank to directly optimize the NDCG. A more recent work of Lai et al. [15] formulates the sparse learning to rank problem as a SVM problem and proposed an algorithm for learning to rank called FenchelRank. The authors formulate the sparse learning to rank problem in different ways but both with a l_1-regularization term. However, if the l_1 lasso penalty produces the good sparsity but, it has been shown to be, in certain cases, inconsistent for variable selection and biased [18]. Mothe et al. produced a set of the unbiased non-convex penalties based on non-convex function of l_1 penalty for the learning to rank problem. The proposed non-convex penalties provide good statistical properties such as unbiasedness and oracle inequalities [19]. But as it has been shown in [20] that any estimator satisfying a sparsity property has maximal risk that converges to the supremum of the loss function; in particular, the maximal risk diverges to infinity whenever the loss function is unbounded. Another disadvantage of the penalties based on the l_1 regularization is that such type of penalties has not grouping property that means impossibility to include into model correlated regressors. But then the main goal of SEO analysis is to understand the factors affecting the rank of particular page for particular query. It would be better that all correlated factors were included into model.

In this paper we propose a new Bayesian approach to variable selection in learning-to-rank problem. The proposed approach gives the strong probabilistic statement of shrinkage criterion for predictor selection. The proposed shrinkage criterion is unbiased, has grouping and oracle properties, its maximal risk diverges to finite value.

2   Hierarchical Bayesian Model for Sparse Learning to Rank

Let us start from the classical statement of dependency estimation problem. Let $\omega \in \Omega$ be a set of real-world objects naturally associated with a point in the linear feature space $\mathbf{x}(\omega) = (x_1(\omega), \ldots, x_n(\omega)) \in \mathbb{R}^n$ and a dependent rank variable $y \in \mathbb{Y}$, $\mathbb{Y} = \{y_1, \ldots, y_q\}$, $y_i \prec y_{i+1}$. The function $y(\omega): X(\Omega) \rightarrow \mathbb{Y}$ is known only within the bounds of a finite training set $\Omega^* \Rightarrow \{\mathbf{x}(\omega_j), y(\omega_j)\}_{j=1}^{N}$, where N is the number of observations. It is required to continue the function onto the entire set Ω, so that it would be possible to estimate the ranking for any subset of objects $\omega \in \Omega \backslash \Omega^*$.

We also assume a probability distribution in the set of observable feature values and hidden ranks $(x_1(\omega), \ldots, x_n(\omega), y(\omega)) \in (X_1 \times \ldots \times X_n \times Y)$ and that training set members are sampled independently. Let $S = \{(\mathbf{x}', \mathbf{x}'') : y' \prec y''\}$ be the set of object pairs from the training sample with proper ranks. Let the $\varphi(\mathbf{x}'\mathbf{x}''|\mathbf{w}, y', y'')$ for each pair from S be the parametric family of probability densities associated with ranking vector $\mathbf{w}$ expressing the assumption that the objects are projected on the direction $\mathbf{a}$ in the right order. We shall consider the improper densities for each pair of objects from S

$$\varphi(\mathbf{x}', \mathbf{x}''|\mathbf{w}, c) = \begin{cases} const, \mathbf{w}^T(\mathbf{x}' - \mathbf{x}'') < -1, \\ \exp(-c\mathbf{w}^T(\mathbf{x}' - \mathbf{x}')), \mathbf{w}^T(\mathbf{x}' - \mathbf{x}'') \geq -1, \end{cases} \tag{1}$$

where the $const = 1$ expresses the assumption that the random feature vectors of two objects are uniformly distributed if they are in right order with parameter c controlling their incorrect ranking. The conditional joint distribution of the observed training set is

$$\Phi(X|Y, \mathbf{w}, \mu_1, \ldots, \mu_m) = \prod_{(i,j) \in S} \varphi(\mathbf{x}_i, \mathbf{x}_j | y_i, y_j, \mathbf{w}).$$

Then, let the direction vector $\mathbf{w}$ be considered as a random vector distributed in accordance with a priori density $\Psi(\mathbf{w}|\mu_1, \ldots, \mu_m)$ parametrized by $(\mu_1, \ldots, \mu_m)$. Consequently, the a posteriori joint distribution density of the parameters of the ranking vector w.r.t. the training set is proportional to the product

$$P(\mathbf{w}|X, Y, \mu_1, \ldots, \mu_m) \propto \Psi(\mathbf{w}|\mu_1, \ldots, \mu_m) \ \Phi(X|Y, \mathbf{w}, \mu_1, \ldots, \mu_m)\sqrt{2} \tag{2}$$

It is natural to consider the maximum point of this a posteriori density as the object of training

$$(w_1, \ldots, w_n) \propto \arg\max [\ln \Psi(\mathbf{w}|\mu_1, \ldots, \mu_m) + \ln \Phi(X|Y, \mathbf{w}, \mu_1, \ldots, \mu_m)].$$

It is easy to show that, under these assumptions, we obtain the training criterion:

$$\begin{cases} -\ln \Psi(\mathbf{w}|\mu_1, \ldots, \mu_m) + c \sum_{(i,j) \in S} \delta_{i,j} \rightarrow \min(\mathbf{w}, \delta_{i,j}, (i,j) \in S), \\ \mathbf{w}^T(\mathbf{x}_j - \mathbf{x}_i) \leq 1 - \delta_{i,j}, \ \delta_{i,j} \geq 0, \ (i,j) \in S. \end{cases}$$

In regression and classification, a large panel of a priori densities $\Psi(\mathbf{w}|\mu_1,\ldots,\mu_m)$ has been developed to learn sparse models: Ridge, LASSO, Bridge, Elastic net, Adaptive LASSO, SCAD (Table 1).

A good a priori density should result in an estimator with six properties.

1. *Unbiasedness:* The resulting estimator is nearly unbiased when the true unknown parameter is large enough to avoid unnecessary modeling bias.
2. *Sparsity:* The resulting estimator can suppress the presence of the small estimated coefficients to reduce model complexity.
3. *Grouping effect:* Highly correlated variables will be in or out of the model together.
4. *Oracle properties:* The resulting estimator can correctly select the nonzero coefficients with probability converging to one, and that the estimators of the nonzero coefficients are asymptotically normal with the same means and covariance that they would have if the zero coefficients were known in advance.
5. *Finite maximum risk*: The resulting estimator has the finite risk in all points of parameter space and for all sample sizes.
6. *Continuity*: The resulting estimator is continuous in data to avoid instability in model prediction.

The overview of the above mentioned properties for all a priori densities is shown in Table 2.

As we can see from the literature there is no any a priory density possessing all the properties of a good estimator.

Table 1. A priory densities to learn sparse model

Penalty	$\psi(w_i|\mu_1,\ldots,\mu_m)$
Ridge	$\exp(-\mu w_i^2); \mu > 0$
LASSO	$\exp(-\mu|w_i|); \mu > 0$
Elastic net	$\exp(-\mu_1|w_i| + \mu_2 w_i^2); \mu_1, \mu_2 > 0$
Bridge	$\exp(-\mu|w_i|^p); \mu > 0, p < 1$
Adaptive LASSO	$\exp(-\mu_i w_i^2); \mu_i = 1/|w_i^{OLS}|^\gamma$
SCAD	$\exp(-p(|w_i|))$ $p'(|w_i|) = \mu_1\left\{1(|w_i| \leq \mu_1) + \frac{\mu_2\mu_1 - |w_i|}{\mu_2\mu_1 - \mu_1}1(|w_i| > \mu_1)\right\}$

Table 2. Properties of a priory densities to learn sparse model

Penalty	1	2	3	4	5	6
Ridge	−	−	+	−	+	+
LASSO	−	+	−	−	−	+
Elastic net	−	+	+	−	−	+
Bridge	+	+	−	−	−	−
Adaptive LASSO	+	+	−	+	−	+
SCAD	+	+	−	+	−	+

In this paper we proposed a general framework for sparse learning to rank possessing all the properties of a good estimator. We assume that the distribution density of the component w_i is the normal distribution with zero expectation and variance ρr_i

$$\psi(w_i|r_i) = (1/2\pi r_i)^{1/2} \exp\left[-w_i^2/(2r_i)\right].$$

In this case, the joint distribution of regression coefficients is

$$\Psi(w_1, \ldots, w_n|r_1, \ldots, r_n, \rho) \propto \left(\prod_{i=1}^n r_i\right)^{-1/2} \exp\left[-\sum_{i=1}^n w_i^2/(2r_i)\right].$$

Furthermore, we assume that inverse variances considered as independent priori gamma distributions $\gamma(1/r_i|\alpha, \beta) \propto (1/r_i)^{\alpha-1}\exp(-\beta/r_i)$. Joint priori distribution density of inverse variances $1/r_i$ is

$$G(1/r_1, \ldots, 1/r_n|\alpha, \beta) = \prod_{i=1}^n \left[(1/r)^{\alpha-1}\exp(-\beta/r_i)\right].$$

From the principle of maximum posteriori density by using Bayesian treatment

$$P(w_1, \ldots, w_n, r_1, \ldots, r_n \,|\, y_1, \ldots, y_N, \alpha, \beta, \rho) \propto \Phi(y_1, \ldots, y_N \,|\, w_1, \ldots, w_n, \rho) \times$$
$$\Psi(w_1, \ldots, w_n \,|\, r_1, \ldots, r_n, \rho) \times G(r_1, \ldots, r_n \,|\, \alpha, \beta)$$

so that results in the training criterion

$$J(w_1, \ldots, w_n, r_1, \ldots, r_n \,|\, \alpha, \beta) = c\sum_{(i,j)\in S} \delta_{i,j} + \frac{1}{2}\sum_{i=1}^n \frac{(w_i)^2}{r_i} +$$
$$+ \frac{1}{2}\sum_{i=1}^n \ln r_i + (\alpha-1)\sum_{i=1}^n \ln r_i + \beta\sum_{i=1}^n 1/r_i \rightarrow \min_{\substack{w_1,\ldots,w_n,\\ r_1,\ldots,r_n}}$$

$$(3)$$

under constraints

$$\mathbf{w}^T(\mathbf{x}_j - \mathbf{x}_i) \leq 1 - \delta_{i,j}, \quad \delta_{i,j} \geq 0, \quad (i,j)\in S.$$

We called the criterion of such a type as supervised selectivity ranking (Super-SelRank) criterion.

3 Group Coordinate Descent Procedure for Parameter Estimation

For finding the minimum point of the criterion, we apply the coordinate descent iteration to both groups of variables $w_i, i = 1, \ldots, n$ and $r_i, i = 1, \ldots, n$ starting with the initial values $\left(r_i^0 = 1, i = 1, \ldots, n\right)$.

Let $(\mathbf{w}^{(k)}, \mathbf{r}^{(k)})$ be the current approximation of the minimum point. The next values of regression coefficients and their variances $(\mathbf{w}^{(k+1)}, \mathbf{r}^{(k+1)})$ can be found as minimum points of partial derivatives

$$r_i^{(k+1)} = \arg\min_{r_i} \left[\sum_{i=1}^{n} \frac{(w_i^{(k)})^2}{2r_i} + \left(\alpha - \frac{1}{2}\right) \sum_{i=1}^{n} \ln r_i + \beta \sum_{i=1}^{n} 1/r_i \right], \qquad (4)$$

$$c^{(k+1)} = \underset{\substack{c, \\ w^T(x_j - x_i) \le 1 - \delta_{i,j}, \\ \delta_{i,j} \ge 0,\ (i,j) \in S}}{\arg\min} \left[c \sum_{(i,j) \in S} \delta_{i,j} + \frac{1}{2} \sum_{i=1}^{n} \frac{(w_i)^2}{r_i} \right]. \qquad (5)$$

So the variances $\mathbf{r}^{(k+1)}$ are defined by the following expression which is easy to prove

$$r_i^{(k+1)} = \frac{(w_i^k)^2 + 2\beta}{(2\alpha - 1)}, \qquad (6)$$

For optimization of the criterion (3) we use the SMO [21] method in the case of small training sample and algorithm PRSVM [22] for the large sample sizes.

We choose parameters of gamma-distributions by the following formula: $\alpha = 1 + 1/(2\mu)$ and $\beta = 1 + 1/(2\mu)$, so that $E(1/r_i) = (2\mu + 1)$ and $E(1/r_i^2) = (2\mu + 1)2\mu$. The parameter $0 < \mu < \infty$ plays the role of a parameter of supervised selectivity. If $\mu \to 0$ then $E(1/r_i) = 1$ and $E(1/r_i^2) = 0 \Rightarrow 1/r_i \cong \ldots \cong 1/r_n \cong 1$. If $\mu \to 0$ then $E(1/r_i) = \infty$ and $E(1/r_i^2) = \infty$ but $E(1/r_i^2)/E(1/r_i) = \mu \to \infty$. It means that if μ grows, the independent positive values may differ arbitrarily, because variances increase much faster than expectations. Convergence of the procedure occurs in 10–15 steps for typical problems. The appropriate value of the structural parameters μ, c is to be determined by cross-validation procedure.

4   Properties of Proposed Sparse Learning to Rank Approach

Substitution of the expression (4) into the training criterion (1) makes the expression for a priori density more convenient for the further analysis

$$\psi(w_i) \propto \exp(-p(w_i))\ p(w_i) = \left(\frac{\mu + 1}{2\mu}\right) \ln(\mu w_i^2 + 1) \qquad (7)$$

4.1 Unbiasedness

The resulting estimator is approximately unbiased if $\lim_{w_i \to \infty} p'(w_i) \to 0$. It is obvious that expression (5) satisfies this condition.

4.2 Selectivity

The criterion (1) penalizes the high values of coefficients variances and proposes the soft threshold for parameters controlled by selectivity parameter μ. In terms of the statistical learning theory decreasing values of μ determine a succession of "almost" nested classes of models. In other terms, different values of μ result in different "effective sizes" of the parameter space.

4.3 Grouping Effect

The resulting estimator possesses the grouping effect if highly correlated variables are included in model or are excluded out of the model together. As it can be proved proposed approach has grouping effect in the subspace $w_i \in [-1/\mu, 1/\mu]$, $i = 1, \ldots, n$.

4.4 Oracle Properties

Let $\mathbf{w}^0$ be true parameter vector consisted from nonzero part $\mathbf{w}^{10}$ and zero part $\mathbf{w}^{20}$. Let

$$a_N = \max\{p'_\mu(|w_{j0}|) : w_{j0} \neq 0\}, \quad b_N = \max\{p''_\lambda(|w_{j0}|) : w_{j0} \neq 0\}.$$

The resulting estimator possesses the oracle properties if it satisfies such conditions

$$b_N \to 0, \quad \sqrt{N}a_N \to 0, \quad N \to \infty$$

It is obvious that expression (5) satisfies these conditions.

4.5 Finite Maximum Risk

The resulting estimator has finite maximum risk if $\arg\min\{|w_i| + p'(w_i)\} = 0$ and $\min\{|w_i| + p'(w_i)\} = 0$. It is obvious that expression (5) satisfies these conditions.

4.6 Continuity

The resulting estimator is continuous in data if $\arg\min\{|w_i| + p'(w_i)\} = 0$. It is obvious that expression (5) satisfies this condition.

5   Experimental Study

We compared SuperSelRank with two other algorithms: the SVM based algorithm (RankSVM) [22] and the SVM based algorithm with l_1 regularization (FenchelRank) [15].

5.1   Groud-Truth Experiments

We tested the proposed approach, for obviousness sake, on a set of $\{(\mathbf{x}_j, y_j); j = 1, \ldots, N\}$ of $N = 300$ pairs consisting of randomly chosen feature vectors, $\mathbf{x}_j \in \mathbb{R}^n$, $n = 100$ and scalars obtained by the rule $y_j = a_1 x_{j,1} + a_2 x_{j,2} + \xi_j$ with $a_1 = a_2 = 1$ and ξ_j as normal white noise with zero mean value and some variance σ^2. So, only $n' = 2$ features of $n = 100$ were rational in the simulated data. In the experiment with learning to rank we took the set $\{\mathbf{x}_j, rank(y_j)\}_{j=1}^N$.

In the experiments, we randomly chose $N_{tr} = 20$ pairs for training. So, the size of the training set was ten times greater than the number of rational features, but five times less than the full dimensionality of the feature vector. The remaining $N_{test} = 280$ pairs we used as the test set.

We use discounted cumulative gain (NDCG) [23] and mean average precision (MAP) [23] as commonly adopted metrics that measure the quality of ranked list. As feature quality prediction metric we use the proportion

$$ratio = \min(|w_1|, |w_2|)/\max(w_i, i = 3, \ldots, n).$$

The comparative results of training are presented in Tables 3 and 4. As we can see, the proposed method of supervised selectivity ranking provides the better quality on artificial data in rank restoration and relevant feature restoration.

Real Data. We conduct our experiments on LETOR 3.0 collection [24]. These are benchmarks dedicated to learning to rank. LETOR 3.0 contains seven datasets: Ohsumed, TD2003, TD2004, HP2003, HP2004, NP2003 and NP2004. Their characteristics are summarized in Table 5. Each dataset is divided into five folds, in order to perform cross validation. For each fold, we dispose of train, test and validation set. The comparative results of training are presented in Table 6. SuperSelRank provides the higher value of MAP and NDCG for all datasets.

6   Conclusion

The main claim of this paper is a new variable selection technique for learning-to-rank that helps SEO specialists to reconstruct the ranking algorithms of the search engines and effectively to lift the web-pages in the search results. The focus is on the hierarchical Bayesian model with supervised selectivity which allows to obtain the asymptotic unbiased parameters estimates, possessing the grouping and oracle properties. We gave the strongly theoretical proofs of all properties for the proposed embedded feature

Table 3. Comparison of the three methods (RankSVM, FenchelRank and SuperSelRank) on rank restoration on test datasets

	NDCG @1	NDCG @3	NDCG @5	NDCG @7	NDCG @9	NDCG @10	MAP
$n = 100$							
RankSVM	0.967	0.898	0.883	0.882	0.862	0.856	0.786
Fenchel	0.981	0.981	0.981	0.982	0.981	0.982	0.941
SSelRank	**1**	**1**	**1**	**0.997**	**0.994**	**0.992**	**0.971**
$n = 200$							
RankSVM	0.433	0.610	0.627	0.627	0.648	0.658	0.758
Fenchel	0.883	0.937	0.944	0.948	0.940	0.944	0.919
SSelRank	**0.967**	**0.967**	**0.971**	**0.964**	**0.955**	**0.960**	**0.916**
$n = 500$							
RankSVM	0.483	0.558	0.553	0.541	0.548	0.550	0.722
Fenchel	0.648	0.677	0.694	0.712	0.719	0.800	0.617
SSelRank	**0.850**	**0.864**	**0.872**	**0.864**	**0.860**	**0.859**	**0.859**

Table 4. Comparison of the three methods (RankSVM, FenchelRank and SuperSelRank) on relevant features restoration

n	RankSVM	FenchelRank	Supervised selectivity
100	3.819	23.75	138.354
200	2.424	12.116	27.533
500	2.970	5.278	10.301

Table 5. Characteristics of LETOR 3.0 datasets

Table head	Number of		
	Features	Queries	Query-document pairs
TD_2003	64[a]	50	49058
NP_2003	64	150	148657
HP_2003	64	150	147606
TD_2004	64	75	74146
NP_2004	64	75	73834
HP_2004	64	75	74409
OHSUMED	45	106	16140

selection technic. The methodological power of the proposed supervised selectivity ranking algorithm has been demonstrated via carefully designed simulation studies and real data examples. Comparing with the version of newest ranking algorithms Rank-SVM and FenchelRank we used, it is far more specific in selecting important variables. As a result, it has much smaller absolute deviation in parameters estimation.

Table 6. Comparison of the three methods (RankSVM, FenchelRank and superselrank) on rank restoration on LETOR 3.0 datasets

	NDCG @1	NDCG @3	NDCG @5	NDCG @7	NDCG @10	MAP
TD_2003						
svm	0.74	0.791	0.808	0.816	0.812	0.765
fenchel	0.72	0.796	0.817	0.824	0.825	0.761
SSelRank	0.74	0.796	0.826	0.848	0.879	0.775
NP_2003						
svm	0.573	0.763	0.775	0.786	0.789	0.689
fenchel	0.553	0.716	0.752	0.762	0.768	0.668
SSelRank	0.587	0.761	0.770	0.778	0.785	0.693
HP_2003						
svm	0.32	0.355	0.366	0.358	0.357	0.265
fenchel	0.34	0.34	0.355	0.342	0.340	0.251
SSelRank	0.35	0.356	0.368	0.361	0.366	0.267
TD_2004						
svm	0.573	0.713	0.759	0.771	0.772	0.671
fenchel	0.667	0.796	0.818	0.823	0.827	0.745
SSelRank	0.673	0.828	0.870	0.888	0.889	0.881
NP_2004						
svm	0.56	0.724	0.772	0.786	0.795	0.675
fenchel	0.587	0.767	0.782	0.797	0.811	0.695
SSelRank	0.596	0.784	0.792	0.797	0.815	0.695
HP_2004						
svm	0.307	0.313	0.306	0.295	0.291	0.206
fenchel	0.36	0.353	0.338	0.317	0.311	0.228
SSelRank	0.393	0.362	0.345	0.395	0.395	0.244
OHSUMED						
svm	0.546	0.485	0.469	0.453	0.450	0.444
fenchel	0.581	0.501	0.479	0.469	0.458	0.451
SSelRank	0.588	0.507	0.499	0.477	0.468	0.466

Acknowledgment. The work is supported by grants of the Russian Foundation for Basic Research No. 11-07-00409 and No. 11-07-00634.

References

1. Liu, T.Y.: Learning to Rank for Information Retrieval. Now Publishers, Breda (2009)
2. Burges, C., Shaked, T., Renshaw, E., Lazier, A., Deeds, M., Hamilton, N., Hulldender, G.: Learning to rank using gradient descent. In: Proceedings of International Conference on Machine Learning (2005)
3. Joachims, T.: Optimizing search engines using clickthrough data. In: Proceedings of the ACM Conference on Knowledge Discovery and Data Mining (KDD). ACM (2002)
4. Zheng, Z., Zha, H., Chen, K., Sun, G.: A regression framework for learning ranking functions using relative relevance judgments. In: Proceedings of Annual International ACM SIGIR Conference on Research and Development in Information Retrieval (2007)
5. Cao, Z., Qin, T., Liu, T.-Y., Tsai, M.-F., Li, H.: Learning to rank: from pairwise approach to listwise approach. In: ICML 2007: Proceedings of the 24th International Conference on Machine Learning, pp. 129–136. ACM, New York (2007)
6. Weimer, M., Karatzoglou, A., Le, Q., Smola, A.: Cofi rank - maximum margin matrix factorization for collaborative ranking. In: Platt, J., Koller, D., Singer, Y., Roweis, S. (eds.) Advances in Neural Information Processing Systems 20. MIT Press, Cambridge (2008)
7. Taylor, M., Guiver, J., Robertson, S., Minka, T.: SoftRank: optimising non-smooth rank metrics. In: Proceedings of International ACM Conference on Web Search and Data Mining (2008)
8. Xia, F., Liu, T.Y., Wang, J., Zhang, W., Li, H.: Listwise approach to learning to rank - theory and algorithm. In: International Conference on Machine Learning (ICML) (2008)
9. Hua, G., Zhang, M., Liu, Y., Ma, S., Ru, L.: Hierarchical feature selection for ranking. In: Proceedings of 19th International Conference on World Wide Web, pp. 1113–1114 (2010)
10. Yu, H., Oh, J., Han, W.-S.: Efficient feature weighting methods for ranking. In: Proceedings of 18th ACM Conference on Information and Knowledge Management, pp. 1157–1166 (2009)
11. Pan, F., Converse, T., Ahn, D., Salvetti, F., Donato, G.: Feature selection for ranking using boosted trees. In: Proceedings of 18th ACM Conference on Information and Knowledge Management, pp. 2025–2028 (2009)
12. Dang, V., Croft, B.: Feature selection for document ranking using best first search and coordinate ascent. In: SIGIR Workshop on Feature Generation and Selection for Information Retrieval (2010)
13. Pahikkala, T., Airola, A., Naula, P., Salakoski, T.: Greedy RankRLS: a linear time algorithm for learning sparse ranking models. In: SIGIR 2010 Workshop on Feature Generation and Selection for Information Retrieval, pp. 11–18 (2010)
14. Sun, Z., Qin, T., Tao, Q., Wang, J.: Robust sparse rank learning for non-smooth ranking measures. In: Proceedings of 32nd International ACM SIGIR Conference on Research and Development in Information Retrieval, pp. 259–266 (2009)
15. Lai, H., Pan, Y., Liu, C., Lin, L., Wu, J.: Sparse learning-to-rank via an efficient primal-dual algorithm. IEEE Trans. Comput. **99**(PrePrints), 1221–1233 (2012)
16. Lai, H.-J., Pan, Y., Tang, Y., Yu, R.: Fsmrank: feature selection algorithm for learning to rank. IEEE Trans. Neural Netw. Learn. Syst. **24**(6), 940–952 (2013)
17. Zou, H.: The adaptive lasso and its oracle properties. J. Amer. Stat. Assoc. **101**(476), 1418–1429 (2006)

18. Mothe, J.: Non-convex regularizations for feature selection in ranking with sparse SVM. **X** (X): 1

19. Zhang, C.-H.: Nearly unbiaised variable selection under minimax con-cave penalty. Ann. Stat. **38**(2), 894–942 (2010)

20. Leeb, H., Pötscher, B.M.: Sparse estimators and the oracle property, or the return of Hodges' estimator. J. Econometrics **142**(1), 201–211 (2008)

21. Platt, J.: Fast training of support vector machines using sequential minimal optimization. In: Scholkopf, B., Burges, C., Smola, A. (eds.) Advances in Kernel Methods - Support Vector Learning. MIT Press, Cambridge (1998)

22. Chapelle, O., Keerthi, S.S.: Efficient algorithms for ranking with SVMs. Inf. Retrieval J. **13** (3), 201–215 (2010)

23. Moon, T., Smola, A.J., Chang, Y., Zheng, Z.: IntervalRank: isotonic regression with listwise and pairwise constraints. In: WSDM 2010, pp. 151–160

24. http://research.microsoft.com/en-us/um/beijing/projects/letor/

Data Driven Geometry for Learning

Elizabeth P. Chou[(⊠)]

Department of Statistics, National Chengchi University, Taipei, Taiwan
eptchou@nccu.edu.tw

Abstract. High dimensional covariate information provides a detailed description of any individuals involved in a machine learning and classification problem. The inter-dependence patterns among these covariate vectors may be unknown to researchers. This fact is not well recognized in classic and modern machine learning literature; most model-based popular algorithms are implemented using some version of the dimension-reduction approach or even impose a built-in complexity penalty. This is a defensive attitude toward the high dimensionality. In contrast, an accommodating attitude can exploit such potential inter-dependence patterns embedded within the high dimensionality. In this research, we implement this latter attitude throughout by first computing the similarity between data nodes and then discovering pattern information in the form of Ultrametric tree geometry among almost all the covariate dimensions involved. We illustrate with real Microarray datasets, where we demonstrate that such dual-relationships are indeed class specific, each precisely representing the discovery of a biomarker. The whole collection of computed biomarkers constitutes a global feature-matrix, which is then shown to give rise to a very effective learning algorithm.

Keywords: Microarray · Semi-supervised learning · Data cloud geometry · biDCG

1 Introduction

Under the high dimensionality, it becomes unrealistic to build learning algorithms based on required smoothness of manifolds or distributions to typical real world datasets. After recognizing the fact of that, it is clearly essential to extract authentic data structure in a data-driven fashion. Ideally if such computed structures can be coherently embedded into a visible geometry, then the developments of learning algorithm would be realistic and right to the point of solving the real issues in hand.

Microarrays are examples of the high dimensional datasets. Microarrays provide a means of measuring thousands of gene expression levels simultaneously. Clustering genes with similar expression patterns into a group can help biologists obtain more information about gene functioning [5,10]. In addition, clustering subjects into groups by their gene expression patterns can help medical professionals determine people's clinical diagnosis status [3,4,14]. Machine learning

© Springer International Publishing Switzerland 2015
P. Perner (Ed.): MLDM 2015, LNAI 9166, pp. 395–402, 2015.
DOI: 10.1007/978-3-319-21024-7_27

has been discussed extensively in this setting because it can help researchers investigate medical data in a more efficient way. Therefore, many methods for classifying microarray data have been developed and reviewed by researchers [17,20,22,24].

Many studies have shown that the logistic regression approach is a fast and standardizable method for data classification [9,25]. Regardless of its extensive use, it might not be appropriate for dealing with gene expression data [19,23,26]. Since most of the microarray data are in a large p small n setting, a subset of the genes is selected through some methods and the regression prediction is performed with these genes. However, it is difficult to determine the size of the gene subset that will be chosen. If too few genes are included, the prediction error may be large. If too many genes are used, the model may be overestimated and either fail to converge or yield an unstable result. It is difficult to find a reliable method for both selecting the genes and performing logistic regression. Although logistic regression can be extended to a multi-class classification problem, a suitable method for multi-class classification with gene expression is needed [2,6,8,21].

Multicollinearity may be another problem in regression analysis on gene expression data. Since gene expression is highly correlated to the expression of other genes, the classification line that we obtain to separate the data may be unstable. Another problem may be sparseness. The regression model may not reach convergence under these conditions. When the sample size is too small, logistic regression may not provide enough power for performing the prediction.

Cross-validation is a measure for checking the performance of a predicted model. However, in such high dimensional microarray data, it may not be efficient and may yield a range of predicted results.

Two-way clustering was introduced to microarray clustering decades ago. Researchers tried to narrow down the numbers of genes and of subjects and found features for a small subset of genes and a small subset of subjects [1,13]. The two-way method overcomes the problems identified above and also decreases the noise from irrelevant data. Feature selections can improve the quality of the classification and clustering techniques in machine learning. Chen et al. [7] developed an innovative iterative re-clustering procedure biDCG through a DCG clustering method [12] to construct a global feature matrix of dual relationships between multiple gene-subgroups and cancer subtypes.

In this research, we attempt to take the accommodating attitude toward the high covariate dimensionality, and to make use computational approaches to uncover the hidden inter-dependence patterns embedded within the collection of covariate dimensions. The essential component is to include all covariate information when constructing the DCG tree geometry. It is important because the geometry pertaining to a subset of covariate data might be significantly different from the geometry pertaining to the whole. The DCG tree is better to be based all involving covariate information of labeled and unlabeled subjects. The computed pattern information would be used as the foundation for constructing learning algorithm. So that the theme of "machine learning" here is a data-driven discovery in the computational and experimental enterprize in contrasting to heavily handed statistical modeling endeavors. This data-driven discovery theme is detailed as follows.

Consider n subjects indexed by $i = 1, .., n$, and each subject is encoded with class-category number and is equipped with p-dimensional covariate information. Let a $n \times p$ matrix collectively record all covariate information available. Here we assume that an empirical distance among the n row vectors, and another empirical distance for the p column vectors are available. By using either one of empirical distances, we calculate a symmetric distance matrix. Then we apply the Data Cloud Geometry (DCG) computational algorithm, developed by Fushing and McAssey [11,12], to build an Ultrametric tree geometry T_S on the subject space of n p-dimensional row vectors, and another Ultrametric tree geometry T_C on the covarite space of p n-dimensional column vectors.

In our learning approach, we try to make simultaneous use of computed pattern information in the Ultrametric tree geometries T_S and T_C. The key idea was motivated by the interesting block patterns seen by coupling these two DCG tree geometries on the $n \times p$ covariate matrix. The coupling is meant to permute the rows and columns according to the two rooted trees in such a fashion that subject-nodes and covariate-nodes sharing the core clusters are placed next to each other, while nodes belonging to different and farther apart branches are placed farther apart. This is the explicit reason why a geometry is needed in both subject and covariate spaces. Such a block pattern indicates that each cluster of subjects has a tight and close interacting relationship with a corresponding cluster of covariate dimensions. This block-based interacting relationship has been discovered and explicitly computed in [7], and termed a "dual relationship" between a target subject cluster and a target covariate cluster. Functionally speaking, this dual relationship describes the following fact: By restricting focus to a target subject cluster, the target covariate cluster can be exclusively brought out on the DCG tree as a standing branch. That is, this target covariate cluster is an entity distinct from the rest of the covariate dimensions with respect to the target subject cluster. Vice versa, by focusing only on the target covariate cluster, the target subject cluster can be brought out in the corresponding DCG tree.

Several real cancer-gene examples are analyzed here. Each cancer type turns out to be one target subject cluster. And interestingly, a cancer type has somehow formed more than one dual relationship with distinct target covariate (gene) clusters. If an identified dual relationship constitutes the discovery of a biomarker, then multiple dual relationships mean multiple biomarkers for the one cancer type. Further, the collection of dual relationships would constitute a global-feature matrix of biomarkers. A biomarker for a cancer type not only has the capability to identify such a cancer type, but at the same time it provides negative information to other cancer types that have no dual relationships with the biomarker. Therefore, a collection of dual-relation-based blocks discovered on the covariate matrix would form a global feature identification for all involved cancer types. An effective learning algorithm is constructed in this paper.

2   Method

2.1   Semi-supervised Learning

Step 1. Choosing a particular cancer type (which includes target labeled subjects and all unlabeled subjects) to cluster genes into groups.

Step 2. Classifying whole labeled and unlabeled subjects by each gene-subgroup. Finding a particular gene-subgroup that can classify the target cancer type. Repeating the procedures to whole the cancer types. These procedures yield the first dual relationship between the gene-subgroups and cancer subtypes. The cancer subtypes here may contain some unlabeled subjects within the cluster.

Step 3. Classifying genes again by a particular cancer subtype and the unknown ones that are in the same cluster as in step 2 yields the second gene-subgroups. Then, with these new gene-subgroups, classifying all subjects will yield the second dual-relationship.

Step 4. The calculation of

$$\frac{V_i V_{i'}}{||V_i|| ||V_{i'}||} = cos\,\theta_{ii'}, \ i, i' = 1, .., n$$

is performed using the 2nd dual relationship to calculate. Here V_i is a vector for the unlabeled subject's data and $V_{i'}$ is a vector for the other target labeled subject's data..

Step 5. Plotting the density function of $cos\,\theta_{ii'}$ for each cancer subtype determines the classification with the function having the largest density mode.

By the method above, we can obtain clusters of the unlabeled data and labeled data. We will not lose any information from the unlabeled data. By repeating the re-clustering procedure, we can confirm that the unlabeled subjects have been correctly classified.

2.2   Datasets

We applied our learning algorithm to several datasets. The first dataset is the one from [7]. The dataset contains 20 pulmonary carcinoids (COID), 17 normal lung (NL), and 21 squamous cell lung carcinomas (SQ) cases. The second dataset was obtained from [18], containing 83 subjects with 2308 genes with 4 different cancer types: 29 cases of Ewing sarcoma (EWS), 11 cases of Burkitt lymphoma (BL), 18 cases of neuroblastoma (NB), and 25 cases of rhabdomyosarcoma (RMS). The third gene expression dataset comes from the breast cancer microarray study by [16]. The data includes information about breast cancer mutation in the BRCA1 and the BRCA2 genes. Here, we have 22 patients, 7 with BRCA1 mutations, 8 with BRCA2 mutations, and 7 with other types. The fourth gene expression dataset comes from [15]. The data contains a total of 31 malignant pleural mesothelioma (MPM) samples and 150 adenocarcinoma (ADCA) samples, with the expression of the 1626 genes for each sample. A summary of the datasets can be found in Table 1.

Table 1. Data description

Data	Number of labels	Number of subjects in each label	Dimensions
Chen	3	20 COID, 17 NL, 21 SQ	58×1543
Khan	4	29 EWS, 11 BL, 18 NB, 25 RMS	83×2308
Hedenfalk	3	7 BRCA1, 8 BRCA2, 7 others	22×3226
Gordon	2	31 MPM, 150 ADCA	181×1626

Table 2. Data description in semi-supervised setting

Data	Number of unlabeled subjects	Number of subjects in each label
Chen	15	15 COID, 12 NL, 16 SQ
Khan	20	23 EWS, 8 BL, 12 NB, 20 RMS
Hedenfalk	6	5 BRCA1, 6 BRCA2, 5 others
Gordon	20	21 MPM, 140 ADCA

Table 3. Accuracy rates for different examples - semi-supervised learning

Data set	Accuracy
Chen	15/15
Khan	1/20
Hedenfalk	4/4
Gordon	20/20

3 Results

We made some of the subjects unlabeled to perform semi-supervised learning. For the Chen dataset, we took the last 5 subjects in each group as unlabeled. For the Khan dataset, unlabeled data are the same as those mentioned in [18]. Since the sample size for Hedenfalk dataset is not large, we unlabeled only the last 2 subjects in BRCA1 and the last 2 subjects in BRCA2. We unlabeled 10 subjects for each group for the Gordon dataset. The number of labeled subjects and unlabeled subjects can be found in Table 2. The predicted results can be found in Table 3. However, we could not find the distinct dual-relationship for the second dataset.

4 Discussion

In the present study, we have proposed a semi-supervised data-driven learning rule based on the biDCG algorithm. Through our learning rule, we have efficiently classified most of the datasets with their dual relationships. In addition, we incorporated unlabeled data into the learning rule to prevent misclassification and the loss of some important information.

A large collection of covariate dimensions must have many hidden patterns embedded in it to be discovered. The model-based learning algorithm might capture the aspects allowed by the assumed models. We made use computational approaches to uncover the hidden inter-dependence patterns embedded within the collection of covariate dimensions. However, we could not find the dual relationships for one dataset, as demonstrated in the previous sections. For that dataset, we could not predict precisely. The reason is that the distance function used was not appropriate for a description of the geometry of this particular dataset. We believe that the measuring of similarity or distance for two data nodes plays an important role in capturing the data geometry. However, choosing a correct distance measure is difficult. With high dimensionality, it is impossible to make assumptions about data distributions or to get *a priori* knowledge of the data. Therefore, it is even more difficult to measure the similarity between the data. Different datasets may require different methods for measuring similarity between the nodes. A suitable selection of measuring similarity will improve the results of clustering algorithms

Another limitation is that we have to decide the smoothing bandwidth for the kernel density curves. A different smoothing bandwidth or kernel may lead to different results. Therefore, we can not make exact decisions. Besides, when the size of gene is very large, a great deal of computing time may be required.

By using the inner product as our decision rule, we know that, when two subjects are similar, the angle between the two vectors will be close to 0 and cos_θ will be close to 1. The use of cos_θ makes our decision rule easy and intuitive. The performance of the proposed method is excellent. In addition, it can solve the classification problem when we have outliers in the dual relationship.

The contributions of our studies are that the learning rules can specify gene-drug interactions or gene-disease relations in bioinformatics and can identify the clinical status of patients, leading them to early treatment. The application of this rule is not limited to microarray data. We can apply our rule of learning processes to any large dataset and find the dual-relationship to shrink the dataset's size. For example, the learning rules can also be applied to human behavior research focusing on understanding people's opinions and their interactions.

Traditional clustering methods assume that the data are independently and identically distributed. This assumption is unrealistic in real data, especially in high dimensional data. With high dimensionality, it is impossible to make assumptions about data distributions and difficult to measure the similarity between the data. We believe that measuring the similarity between the data nodes is an important way of exploring the data geometry in clustering. Also, clustering is a way to improve dimensionality reduction, and similarity research is a pre-requisite for non-linear dimensionality reduction. The relationships among clustering, similarity and dimensionality reduction should be considered in future research.

References

1. Alon, U., Barkai, N., Notterman, D.A., Gish, K., Ybarra, S., Mack, D., Levine, A.J.: Broad patterns of gene expression revealed by clustering analysis of tumor and normal colon tissues probed by oligonucleotide arrays. Proc. Nat. Acad. Sci. **96**(12), 6745–6750 (1999)
2. Bagirov, A.M., Ferguson, B., Ivkovic, S., Saunders, G., Yearwood, J.: New algorithms for multi-class cancer diagnosis using tumor gene expression signatures. Bioinformatics **19**(14), 1800–1807 (2003)
3. Basford, K.E., McLachlan, G.J., Rathnayake, S.I.: On the classification of microarray gene-expression data. Briefings Bioinf. **14**(4), 402–410 (2013)
4. Ben-Dor, A., Bruhn, L., Laboratories, A., Friedman, N., Schummer, M., Nachman, I., Washington, U., Washington, U., Yakhini, Z.: Tissue classification with gene expression profiles. J. Comput. Biol. **7**, 559–584 (2000)
5. Ben-Dor, A., Shamir, R., Yakhini, Z.: Clustering gene expression patterns. J. Comput. Biol. **6**(3–4), 281–297 (1999)
6. Bicciato, S., Luchini, A., Di Bello, C.: PCA disjoint models for multiclass cancer analysis using gene expression data. Bioinf. **19**(5), 571–578 (2003)
7. Chen, C.P., Fushing, H., Atwill, R., Koehl, P.: biDCG: a new method for discovering global features of dna microarray data via an iterative re-clustering procedure. PloS One **9**(7), 102445 (2014)
8. Chen, L., Yang, J., Li, J., Wang, X.: Multinomial regression with elastic net penalty and its grouping effect in gene selection. Abstr. Appl. Anal. **2014**, 1–7 (2014)
9. Dreiseitl, S., Ohno-Machado, L.: Logistic regression and artificial neural network classification models: a methodology review. J. Biomed. Inf. **35**(5–6), 352–359 (2002)
10. Eisen, M.B., Spellman, P.T., Brown, P.O., Botstein, D.: Cluster analysis and display of genome-wide expression patterns. PNAS **95**(25), 14863–14868 (1998)
11. Fushing, H., McAssey, M.P.: Time, temperature, and data cloud geometry. Phys. Rev. E **82**(6), 061110 (2010)
12. Fushing, H., Wang, H., Vanderwaal, K., McCowan, B., Koehl, P.: Multi-scale clustering by building a robust and self correcting ultrametric topology on data points. PLoS ONE **8**(2), e56259 (2013)
13. Getz, G., Levine, E., Domany, E.: Coupled two-way clustering analysis of gene microarray data. Proc. Natl. Acad. Sci. USA **97**(22), 12079–12084 (2000)
14. Golub, T.R., Slonim, D.K., Tamayo, P., Huard, C., Gaasenbeek, M., Mesirov, J.P., Coller, H., Loh, M.L., Downing, J.R., Caligiuri, M.A., Bloomfield, C.D., Lander, E.S.: Molecular classification of cancer: Class discovery and class prediction by gene expression monitoring. Science **286**(5439), 531–537 (1999)
15. Gordon, G.J., Jensen, R.V., Hsiao, L.L., Gullans, S.R., Blumenstock, J.E., Ramaswamy, S., Richards, W.G., Sugarbaker, D.J., Bueno, R.: Translation of microarray data into clinically relevant cancer diagnostic tests using gene expression ratios in lung cancer and mesothelioma. Cancer Res. **62**(17), 4963–4967 (2002)
16. Hedenfalk, I.A., Ringnér, M., Trent, J.M., Borg, A.: Gene expression in inherited breast cancer. Adv. Cancer Res. **84**, 1–34 (2002)
17. Huynh-Thu, V.A., Saeys, Y., Wehenkel, L., Geurts, P.: Statistical interpretation of machine learning-based feature importance scores for biomarker discovery. Bioinformatics **28**(13), 1766–1774 (2012)

18. Khan, J., Wei, J.S., Ringnér, M., Saal, L.H., Ladanyi, M., Westermann, F., Berthold, F., Schwab, M., Antonescu, C.R., Peterson, C., Meltzer, P.S.: Classification and diagnostic prediction of cancers using gene expression profiling and artificial neural networks. Nat. Med. **7**(6), 673–679 (2001)

19. Liao, J., Chin, K.V.: Logistic regression for disease classification using microarray data: model selection in a large p and small n case. Bioinformatics **23**(15), 1945–1951 (2007)

20. Mahmoud, A.M., Maher, B.A., El-Horbaty, E.S.M., Salem, A.B.M.: Analysis of machine learning techniques for gene selection and classification of microarray data. In: The 6th International Conference on Information Technology (2013)

21. Nguyen, D.V., Rocke, D.M.: Multi-class cancer classification via partial least squares with gene expression profiles. Bioinformatics **18**(9), 1216–1226 (2002)

22. Saber, H.B., Elloumi, M., Nadif, M.: Clustering Algorithms of Microarray Data. In: Biological Knowledge Discovery Handbook: Preprocessing, Mining, and Postprocessing of Biological Data, pp. 557–568 (2013)

23. Shevade, S.K., Keerthi, S.S.: A simple and efficient algorithm for gene selection using sparse logistic regression. Bioinformatics **19**(17), 2246–2253 (2003)

24. Thalamuthu, A., Mukhopadhyay, I., Zheng, X., Tseng, G.C.: Evaluation and comparison of gene clustering methods in microarray analysis. Bioinformatics **22**(19), 2405–2412 (2006)

25. Wasson, J.H., Sox, H.C., Neff, R.K., Goldman, L.: Clinical prediction rules. Applications and methodological standards. New Engl. J. Med. **313**(13), 793–799 (1985). PMID: 3897864

26. Zhou, X., Liu, K.Y., Wong, S.T.: Cancer classification and prediction using logistic regression with bayesian gene selection. J. Biomed. Inform. **37**(4), 249–259 (2004)

Mining Educational Data to Predict Students' Academic Performance

Mona Al-Saleem[1,3]([✉]), Norah Al-Kathiry[1], Sara Al-Osimi[1],
and Ghada Badr[1,2]

[1] King Saud University, Riyadh, Saudi Arabia
{Zamil.mona,norahabdulaziz.cs}@gmail.com,
{srmo2008,badrghada}@hotmail.com
[2] IRI-the City of Scientific Research and Technological Applications,
Alexandria, Egypt
badrghada@hotmail.com
[3] Qassim University, Qassim, Buraydah, Saudi Arabia

Abstract. Data mining is the process of extracting useful information from a huge amount of data. One of the most common applications of data mining is the use of different algorithms and tools to estimate future events based on previous experiences. In this context, many researchers have been using data mining techniques to support and solve challenges in higher education. There are many challenges facing this level of education, one of which is helping students to choose the right course to improve their success rate. An early prediction of students' grades may help to solve this problem and improve students' performance, selection of courses, success rate and retention. In this paper we use different classification techniques in order to build a performance prediction model, which is based on previous students' academic records. The model can be easily integrated into a recommender system that can help students in their course selection, based on their and other graduated students' grades. Our model uses two of the most recognised decision tree classification algorithms: ID3 and J48. The advantages of such a system have been presented along with a comparison in performance between the two algorithms.

1 Introduction

According to Wook (2009): [1] "data mining is the process of automatically extracting useful information and relationships from immense quantities of data". In this sense data mining is not about acquiring specific information; rather, it is about answering a question or approving theory. Another definition states: "Data mining algorithms can predict future trends and behaviours and enable businesses to make proactive, knowledge-driven decisions" [2]. These definitions argue that data mining could be very beneficial in every aspect of our lives.

One of the most famous applications of data mining is its use to support and improve higher education. In this context, data mining techniques can have a direct impact on the workforce being provided to industry and, as a result, can directly affect the economy. However, academic failure among university students is one of the

© Springer International Publishing Switzerland 2015
P. Perner (Ed.): MLDM 2015, LNAI 9166, pp. 403–414, 2015.
DOI: 10.1007/978-3-319-21024-7_28

biggest problems affecting higher education. One of the solutions involves predicting a student's performance in order to help the student to choose the right course and increase their performance. Using data mining we can study and analyse information from previous students (which is the main subject of this paper). The various data mining techniques, which include classification, clustering and relationship mining, can be applied in educational settings to predict the performance of a student.

In this paper we present a data mining technique that analyses previous students' data in order to predict future students' performance on specific courses. To achieve this and support the development of higher education in our country, we use ID3 and J48 decision tree classification algorithms. The data set used in this paper is student data from the Computer Science Department, Computer and Information Sciences College, King Saud University of Saudi Arabia (KSU).

The rest of this paper is organised as follows. In Sect. 2, we present a brief background about the paper. Research related to our objectives is reviewed in Sect. 3. In Sect. 4, the methodology and experimental results are discussed. Conclusions are drawn in Sect. 5.

2   Decision Tree Classification Algorithms

Classification generally refers to the mapping of data items into predefined groups and classes [4]. It is also referred to as supervised learning and involves learning and classification. In the learning phase, training data are analysed by classification algorithm; during the classification phase, test data are used to estimate the accuracy of the classification rules [5]. Data mining involves various classification techniques such as decision tree algorithms, Bayesian Classification, and classification by Back Propagation, Support Vector Machine and K Nearest Neighbour.

Decision tree algorithms are used for gaining information for the purpose of decision-making and are one of the most widely used algorithms in data mining. The decision tree starts with a root node from which users take actions. From this first node, users split further nodes recursively according to the decision tree learning algorithm [6]. The final result is a decision tree in which each branch represents a possible scenario of decision and its outcome. The two widely used decision tree learning algorithms discussed in this paper are: ID3 and J48 (which is an implementation of C4.5).

2.1   ID3 (Iterative Dichotomiser 3) [6]

Quinlan Ross introduced ID3 in 1986. It is based on Hunt's algorithm. The tree is constructed in two phases: tree building and pruning. ID3 uses information gain measure to choose the splitting attribute. It only accepts categorical attributes in building a tree model.

To build the decision tree, information gain is calculated for each and every attribute, with the attribute containing the highest information gain designated as a root node. The attribute is labelled as a root node and the possible values of the attribute are represented as arcs. Then all possible outcome instances are tested to check whether they fall under the same class or not. If all the instances fall under the same class, the node is represented with a single class name; if not, the splitting attribute is chosen to classify the instances.

2.2 J48 (Implementation of C4.5)

This algorithm is a successor to ID3 developed by Quinlan Ross. Unlike ID3 it handles both categorical and continuous attributes. It splits the attribute values into two partitions based on the selected threshold so that all the values above the threshold sit as one child and the remaining sit as another child. It uses Gain Ratio as an attribute selection measure to build a decision tree.

3 Data Mining in Higher Education

Universities process a vast amount of data each year, including information on the enrolment of students and academic performance. Therefore, these institutions require a suitable data mining tool to process past data and come up with solutions to resolve current situations. When data mining was introduced, the application of data mining techniques boosted many industries such as business, telecommunications, banking and education. In the education sector, data mining was defined as "the process of converting raw data from educational systems to useful information that can be used to inform design decisions and answer research questions" [3]. The prediction and prevention of academic failure among students has long been debated within each higher education institution. It has become one of the most important reasons for using data mining in higher education.

Performance prediction is a prominent field of research in educational data mining. There are many researchers who have addressed this problem, drawing similar conclusions to ours but using different techniques and tools. Vialardi et al. [7] developed a recommender system to support new students to choose better academic itineraries based on the acquisition of knowledge from past students' academic performance. Their system was based on using the C4.5 algorithms to provide the rules used in the recommender system to decide if the enrolment of the student on a specific course has a good probability of success or not. Experiments on the performance of the system give a success rate for 80 % of cases in terms of predicting the results for each student.

Another piece of research [8] developed a model to predict students' academic performance by taking into account socio-demographic and academic variables. The model, which is based on using the Naïve Bayes classifier and the Rapid miner software, reached accuracy of 60 % correct classification. Bharadwaj and Pal [9] conducted a study on student performance by selecting 300 students from five different colleges. By means of Bayesian classification method, which examines 17 attributes, it was found that factors including students' living location and medium of teaching were highly interlinked with student academic performance.

From the range of data mining techniques that exist, classification, and in particular decision tree algorithms, are the most effective modelling functions since they can be used to find the relationship between a specific variable, target variable and other variables. Al-Radaideh et al. [10] proposed a prediction approach that used data mining classification techniques to evaluate data that may affect student performance. They used three different classification methods: ID3, C4.5 and the Naïve Bayes. The results indicated that the decision tree model had better prediction accuracy than the other two models.

Another research study used decision tree algorithms to predict the results of students in their final semester based on the marks they obtained in previous semesters. They used two algorithms, J48 (implementation of C4.5) and Random Tree, to predict the result of the fifth semester based on the marks obtained by the students in the previous four semesters. They applied the two algorithms to the students' records and found that Random Tree is more accurate in predicting performance than the J48 algorithm.

4   Proposed Workflow

Data mining educational information is a vibrant and important field of research. Generally, the majority of research for the prediction of student academic performance shares the same methodology: data collection and preprocessing followed by the implementation of the classification algorithm and the analysis of results.

4.1   Data Collection and Preprocessing

Data preparation is the most complex step in the data mining process as it involves the collection of data and preprocessing. The collection of data includes the identification of data resources and the gathering of relevant data from these resources. Following this initial step, the collected data should be preprocessed to make it suitable for data mining algorithm implementation, which will also provide a greater understanding of the data collected. Preprocessing includes four important stages: data cleaning, which includes studying the data to remove noise (which are mostly the errors and biased values that can affect the accuracy of results) and dealing with unknown or missing data values; removal of duplicate values; smoothing noisy data and resolving inconsistencies.

Preprocessing also includes the integration of data from multiple resources, data values transformation from one domain to another and normalisation. Another preprocessing task is data reduction, which might be in the form of a feature selection to reduce the data dimensionality (which eliminates features that are not much use in model construction and produce better performance by reducing the number of used objects). Preprocessing also involves data discretisation, which reduces the number of values for a given continuous attribute by dividing its range into intervals. This helps to replace the numerous values of a continuous attribute by a small number of interval labels and supports concept hierarchy generation, which is used for discretisation.

Finally, the resulted data should be transferred into a suitable file format for the desired application, for example the WEKA file format (.arff).

4.2 Algorithm Implementation

After preparing the data, the dataset is divided into training and test datasets. The training dataset is used to construct the classification model and the test dataset is used either to test the performance of the constructed classification model or to compare predictions to the known target values [12]. Different algorithms are used for the classification and prediction of student performance; however, the decision tree classification is the most frequently used algorithm for this type of application.

4.3 Analysis of Results

After applying the algorithm, the resultant data patterns should undergo further processing. The post-processing stage includes pattern evaluation to measure different parameters such as accuracy. Post-processing also includes pattern selection because not all of the resultant patterns are of interest so a further selection may be beneficial. Finally, it is important that the result is interpreted correctly so that it is effectively viewed and understood.

5 Case Study

Our proposed system is modelled on predicting students' grades using former students' grades as a basis. Our methodology consists of four steps:

a. Collecting and preparing educational data.
b. Building the models of ID3 and J48 classification.
c. Evaluating the models using one of the evaluation methods.
d. The best model will be chosen as the model for online prediction.

We used NetBeans IDE as the programming environment and JAVA programming language in implementation. WEKA classes were integrated as a jar library.

5.1 Data Preparation

The dataset used in this work was obtained from King Saud University of Saudi Arabia, Computer Science Department, Computer and Information Sciences College. It was a sample of CS Master student records from session 2003 to 2011. Microsoft Excel was used to save the collected data for preprocessing; it was then stored as an .arff file for data mining tasks.

Preprocessing the data was a time consuming process; the original data were stored in an un-normalised database table with a row each for student ID, term, course and grade for each course. As part of the preprocessing stage we:

1. Filtered all the unneeded columns (like registration term and registration place), as those have no influence on the required results.
2. Removed the old courses that are no longer available in the department course plan.
3. Filled the empty cells where values were missing with the average of the grades taken in the same course by other students.
4. Transformed data to a transactional format where each row was indexed by student ID and each column indexed by course. At the intersection of each row and column, a value of the grade taken for each course is recorded.
5. Finally, our data form included 112 records and 37 attributes (one student ID, four core courses and 32 elective courses). We split our data into 82 records for training data and 30 records for test data. These training data are used to construct the model while the test data is used in evaluating. Table 1 shows an example of the normalised data.

Table 1. Example of the used dataset

	CS 541	CS 524	CS 512	CS 581	CS 595	CS 597
424221334	B+	B	A+	B+	B+	B+
426221091	A	B	A+	B	B+	A+
426221091	B	A	C	C+	A	A+
426221091	B+	A	B	C+	B+	A+
426221091	A	A	A+	B+	A	A+
426221091	A	A	C+	C+	A+	A
426221091	A+	A+	A	A	B+	A+

5.2 Offline Classification

We selected two classification models: ID3 and J48. The performance of the two models was compared to decide on the best approach for online prediction. Each model was built for each elective course as a class attribute based on the grades in the core courses. The average accuracy in prediction grades for all elective courses was computed to decide the best model.

Table 2 shows the analysis of the classification algorithms and the accuracy percentages for each elective course.

Table 2 shows that the performance of both algorithms is satisfactory; however, higher overall accuracy (83.75 %) was attained through J48 implementation compared to ID3, which had 69.27 % accuracy.

Table 2. Comparison of classification algorithms

Elective Courses	Classification Algorithms	
	J48	ID3
CS 595	50 %	50 %
CS 597	76.66 %	63 %
CS 598	96.66 %	73.33 %
CS 519	60 %	40 %
CS 551	70 %	70 %
CS 562	63.33 %	13.33 %
CS 572	36.66 %	26.66 %
CS 588	80 %	56.66 %
CS 520	90 %	73.33 %
CS 522	100 %	60 %
CS 523	70 %	46.66 %
CS 526	96.66 %	90 %
CS 527	86.66 %	63.33 %
CS 528	76.66 %	73.33 %
CS 529	100 %	100 %
CS 530	100 %	100 %
CS 546	73.33 %	60 %
CS 547	100 %	100 %
CS 548	100 %	100 %
CS 549	100 %	100 %
CS 558	80 %	53.33 %
CS 561	100 %	86.66 %
CS 563	56.66 %	6.66 %
CS 566	96.66 %	83.33 %
CS 579	100 %	100 %
CS 573	100 %	76.66 %
CS 574	96.66 %	96.66 %
CS 576	66.66 %	46.66 %
CS 578	100 %	80 %
CS 586	83.33 %	83.33 %
CS 589	96.66 %	83.33 %
CS 590	76.66 %	60 %
Accuracy Average	83.75 %	69.27 %

5.3   Online Classification

Based on the results shown in Table 2, the J48 algorithm was used for online prediction of students' grades. Our small system allows students to ask for predicted grades for one or more courses. The classification models are built using training data and specific attributes, using students' records to get a prediction. Attributes used to construct the model include the elective courses passed by a student together with core courses.

The system also recommends the best courses for students based on two criteria: maximum accuracy value and maximum grade value for all predicted courses. The goal is to help the students to choose a suitable course that enhances their academic attainment level.

The remainder of this section presents some examples of how the system interacts with students, how the results are displayed and an explanation of how the system works.

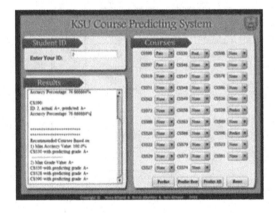

Fig. 1. Example 1

In Fig. 1, the student selected CS595 and CS597 as pass courses and CS530, CS528 and CS590 as predicted courses. Here, three classification models were built by using the training student's dataset after he or she pressed the predict button.

After building the models, the system predicted the student's grade in CS530, CS528 and CS590. In addition, it gave an accuracy percentage for this prediction by comparing predicted grades with actual grades:

CS530:
Predicted Grade: A+
Accuracy Percentage: 100.0%

CS528:
Predicted Grade: A+
Accuracy Percentage: 76.666664%

CS590:
Predicted Grade: A+
Accuracy Percentage: 76.666664%

The system also recommended the best courses based on maximum accuracy value and maximum grade value:

Recommended Courses Based on:
1) Max Accuracy Value: 100.0%
CS530 with predicting grade: A+

2) Max Grade Value: A+
CS530 with predicting grade: A+
CS528 with predicting grade: A+
CS590 with predicting grade: A+

Here, the system has 100 % accuracy in predicting course CS530, so it is recommended for the student. Moreover, all three courses gave an A+ prediction, which means that all three courses are recommended.

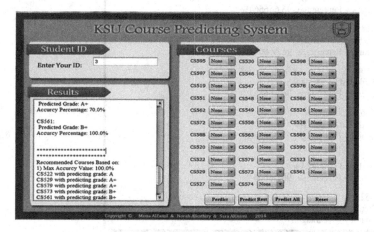

Fig. 2. Example 2

In Fig. 2, a separate student selected some pass courses and obtained a prediction for the rest of the courses. The attributes used to construct these models were the pass courses together with four core courses and each of the rest of the courses, which were used as class attributes for each model. The system gave a predicted grade for each course along with an accuracy percentage:

CS588:
  Predicted Grade: A
Accuracy Percentage: 80.0%

CS522:
  Predicted Grade: A
Accuracy Percentage: 100.0%

CS520:
  Predicted Grade: A
Accuracy Percentage: 90.0%

CS529:
  Predicted Grade: A+
Accuracy Percentage: 100.0%

CS527:
  Predicted Grade: B+
Accuracy Percentage: 86.666664%

CS579:
  Predicted Grade: A+
Accuracy Percentage: 100.0%

CS573:
  Predicted Grade: B+
Accuracy Percentage: 100.0%

CS574:
  Predicted Grade: B
Accuracy Percentage: 96.666664%

CS589:
  Predicted Grade: B+
Accuracy Percentage: 96.666664%

CS590:
  Predicted Grade: A+
Accuracy Percentage: 76.666664%

CS563:
  Predicted Grade: B+
Accuracy Percentage: 56.666668%

CS566:
  Predicted Grade: A+
Accuracy Percentage: 96.666664%

CS523:
  Predicted Grade: A+
Accuracy Percentage: 70.0%

The system also recommended the best courses for students based on maximum accuracy value and maximum grade value:

```
Recommended Courses Based on:
1) Max Accuracy Value: 100.0%
      CS522 with predicting grade: A
      CS529 with predicting grade: A+
      CS579 with predicting grade: A+
      CS573 with predicting grade: B+
      CS561 with predicting grade: B+
      ---------------------
2) Max Grade Value: A+
      CS529 with predicting grade: A+
      CS579 with predicting grade: A+
      CS590 with predicting grade: A+
      CS566 with predicting grade: A+
      CS523 with predicting grade: A+
```

6 Conclusion

To support the aims of higher education institutions, we proposed a model that would predict student performance in future courses based on the grade of previous students in core and elective courses. The workflow for our system was presented. A case study was also presented, using student records obtained from the department of Computer Science in KSU, where programmes consist of core and elective courses. The prediction system is based on the decision tree classification methods, specifically ID3 and J48. First, a preprocessing step was applied to the data and then the model was built using training data for each elective course as a class attribute based on the grades in the core courses. By calculating the average accuracy for each algorithm we found that J48 achieved a better accuracy performance of 83.75 % when compared to an accuracy of 69.27 % for the ID3 algorithm. On that basis we used the J48 algorithm for online prediction to enable the students to get a prediction for one or more courses.

References

1. Wook, M., Yahaya, Y.H., Wahab, N., Isa, M.R.M., Awang, N.F., Hoo Yann, S.: Predicting NDUM student's academic performance using data mining techniques. In: Second International Conference on Computer and Electrical Engineering, ICCEE, pp. 357–361 (2009)
2. Hoe, A.K., Ahmad, M.S., Tan Chin, H., Shanmugam, M., Gunasekaran, S.S., Cob, Z.C., Ramasamy, A.: Analyzing students records to identify patterns of students' performance. In International Conference on Research and Innovation in Information Systems (ICRIIS), pp. 544–547 (2013)
3. Heiner, C., Baker, R., Yacef, K.: Preface. In: Workshop on Educational Data Mining at the 8th International Conference on Intelligent Tutoring Systems (ITS), Jhongli, Taiwan (2006)

4. Dunham, M.H.: Data Mining: Introductory and Advanced Topics. Pearson Education India, Delhi (2006)
5. Taruna, S., Pandey, M.: An empirical analysis of classification techniques for predicting academic performance. In: IEEE International Advance Computing Conference (IACC), pp. 523–528 (2014)
6. Bunkar, K., Singh, U.K., Pandya, B., Bunkar, R.: Data mining: prediction for performance improvement of graduate students using classification. In: Ninth International Conference on Wireless and Optical Communications Networks (WOCN), pp. 1–5 (2012)
7. Vialardi, C., Bravo, J., Shafti, L., Ortigosa, A.: Recommendation in higher education using data mining techniques. In: International Working Group on Educational Data Mining (2009)
8. Garcia, E.P.I., Mora, P.M.: Model prediction of academic performance for first year students. In: 10th Mexican International Conference on Artificial Intelligence (MICAI), pp. 169–174 (2011)
9. Bhardwaj, B.K., Pal, S.: Data mining: a prediction for performance improvement using classification. Int. J. Comput. Sci. Inf. Secur. (IJCSIS) 9, 136–140 (2012)
10. Al-Radaideh, Q.A., Al-Shawakfa, E.M., Al-Najjar, M.I.: Mining student data using decision trees. In: International Arab Conference on Information Technology (ACIT), Yarmouk University, Jordan (2006)
11. Anupama Kumar, S., Vijayalakshmi, M.N.: Mining of student academic evaluation records in higher education. In: International Conference on Recent Advances in Computing and Software Systems (RACSS), pp. 67–70 (2012)
12. Krishna Kishore, K.V., Venkatramaphanikumar, S., Alekhya, S.: Prediction of student academic progression: a case study on Vignan University. In: International Conference on Computer Communication and Informatics (ICCCI), pp. 1–6 (2014)

Patient-Specific Modeling of Medical Data

Guilherme Alberto Sousa Ribeiro[1]([✉]), Alexandre Cesar Muniz de Oliveira[1],
Antonio Luiz S. Ferreira[1], Shyam Visweswaran[2], and Gregory F. Cooper[2]

[1] Departament of Informatics, Federal University of Maranhao,
Av. of Portugueses, São Luís, MA 1966, Brazil
{guilherme.ufma,aluizsf22}@gmail.com, acmo@deinf.ufma.br
[2] Department of Biomedical Informatics, 5607 BAUM, University of Pittsburgh,
Boulevard BAUM 423, Pittsburgh, PA 15206-3701, USA
{shv3,V.gfc}@pitt.edu

Abstract. Patient-specific models are instance-based learn algorithms
that take advantage of the particular features of the patient case at hand
to predict an outcome. We introduce two patient-specific algorithms
based on decision tree paradigm that use AUC as a metric to select
an attribute. We apply the patient specific algorithms to predict outcomes in several datasets, including medical datasets. Compared to the
standard entropy-based method, the AUC-based patient-specific decision path models performed equivalently on area under the ROC curve
(AUC). Our results provide support for patient-specific methods being
a promising approach for making clinical predictions.

Keywords: Classification problems · Approach instance-based · Area
under the roc curve

1 Introduction

Clinical decision-making may be improved by using predictive models [1]. Predicting patient outcomes under uncertainty constitute an important healthcare
problem. Enhanced decision models can lead to better patient outcomes as well
as efficient use of healthcare resources.

The typical paradigm in predictive modeling is to learn a single model from
a database of patient cases, which is then used to predict outcomes for future
patient cases [2]. This approach is known as *population-wide model* because it
is intended to be applied to an entire population of future cases. Examples of
popular population-wide methods are decision trees, logistic regression, neural
networks and Bayesian networks.

In contrast to that general approach, a *patient-specific model* consists of learning models that are tailored to the particular features of a given patient. Thus,
a patient-specific model is specialized to the patient case at hand, and it is optimized to predict especially well for that case. Moreover, patient-specific models can also be seen as examples of instance-based learning schemes, of which
k-nearest neighbor, local regression and lazy decision trees are examples.

© Springer International Publishing Switzerland 2015
P. Perner (Ed.): MLDM 2015, LNAI 9166, pp. 415–424, 2015.
DOI: 10.1007/978-3-319-21024-7_29

Instance-based algorithms learn a distinct model for a test instance and take advantage of the features in the test instance to learn the model [3]. Typically, the instance-based algorithms are lazy, since no model is constructed a priori before a test instance becomes available, and a model is learned only when a prediction is needed for a new instance [4]. In contrast, algorithms that learn population-wide models are eager since such explicitly build a model form the training data before a test instance becomes available.

There are several advantages of patient-specific models over population-wide methods. For instance, patient-specific models may have better predictive performance for taking advantage of any particular characteristic of the case at hand, whereas population-wide methods converge to an average method, derived for an entire population. Second, a patient-specific model structure is usually simpler than that of its population-wide counterpart. Thus, a patient-specific model can provide a more succinct explanation of its decision. Third, the construction patient-specific models may be computationally faster, though this advantage is offset by the observation that a patient-specific method has to construct a distinct model for each patient case of interest while a population-wide method has to construct just a single model. Finally, the task of handling of missing features is simplified on patient-specific approach.

In this paper, we investigate the performance of two patient-specific methods, based on the lazy decision tree approach. We compare the performance of the AUC-based patient-specific methods with the entropy-based model. We focus on the discriminative performance of the three methods and evaluate them using the area under the ROC curve (AUC) [5].

The remainder of this paper is organized as follows. Section 2 presents background and related work on instance-based methods. Section 3 provides details of the patient-specific decision path algorithms that we have developed. Section 4 describes the datasets, experimental methods and presents and discusses the results of the patient-specific decision path algorithm on several datasets. Section 5 presents our conclusions.

2   Background

The canonical instance-based method is a kind of nearest-neighbor technique, that is, the most similar training instance to a given the test instance is located its target value is returned as the prediction [6].For a test instance, the k-Nearest Neighbor (KNN) method, for example, selects the k most similar training instances and either averages or takes a majority vote of their target values. Modified version of kNN have been applied successfully to medical databases for diagnosis and knowledge extraction [7]. Other instance-based methods are not as reliant on a similarity measure as is the case for the nearest-neighbor methods, taking advantage of the values of the predictors in the test instance to learn a model.

Friedman et al. have described the LazyDT method [4] that searches for the best CART-like decision tree for a test instance. When compared to standard population-wide methods, for inducing decision trees, as ID3 and C4.5, LazyDT does not perform pruning, handles only discrete variables, and has higher accuracies on average.

Zheng et al. have developed an instance-based method called the Lazy Bayesian Rules (LBR) learner that builds a rule tailored to the values of the predictors of the test instance, used to classify it [8]. A LBR rule consists of: (1) a conjunction of the features (predictor-value pairs) present in the test instance as the antecedent, and (2) a consequent nave Bayes classifier that consists of the target variable as the parent of all other predictors that do not appear in the antecedent. The classifier parameters are estimated from those training instances that satisfy the antecedent. A greedy step-forward search selects the optimal LBR rule for a test instance to be classified. LBR uses values of predictors in the test instance to drive the search for a suitable model in the model space and, when compared to a variety of population-wide methods, LBR has reached higher accuracies on average.

Visweswaran et al. have developed and applied an instance-based algorithm that performs Bayesian model averaging over LBR models [2,9], using the features of the test case to drive the search. The prediction for the test case target variable is obtained by combining the predictions of the selected models weighted by their posterior probabilities. This method has obtained higher accuracies than LBR on a range of non-medical datasets and also performed better than several population-wide methods on a pneumonia dataset, when evaluated within a clinically relevant range of the ROC curve. Furthermore, instance-based algorithms that use the test instance to drive the search over a space of Bayesian network models have been developed and applied to patient-specific modeling with good results [10,11].

Ferreira et al. developed patient-specific decision path (PSDP) algorithms that can build a decision path to predict patient outcome [12]. Given a patient for whom the values of the features are known, these algorithms construct a decision path using a subset of those features. Two selection criteria were investigated for selecting features: balanced accuracy and information gain. Results obtained with those algorithms using clinical datasets were compared with CART using the AUC metric.

3   Patient-Specific Decision Path Algorithms Based on AUC

The proposed patient-specific decision path algorithm uses AUC as a metric to select patient's features that will compose the path [13]. The Area under the ROC Curve (AUC) is a widely used measured of performance of supervised classification rules. It has the attractive property of circumvent the need to specify misclassification costs.

The use of AUC as a metric to select attributes in a decision tree was introduced by Ferri and colleagues [14,15]. Based on the optimal choice of possible labellings of the tree, the AUC-split criterion leads to good AUC values, without

```
PSDP-AUC-Split (labels, dataset, testset)
    INPUT: labels contain a set of possible labels to predict,
              dataset is database,
              testset is a test case to predict
    OUTPUT: an array with the label, estimates probability of the label and the path.

    LOOP: until dataset to be empty, all cases in dataset have the same target or the set of atributes to be empty

        FOR each testeset(i)
            subset = getSubSet(dataset, testset(i))
            diff_subset = dataset - subset

            partition(1) = counting of positive and negative values of subset
            partition(2) = counting of positive and negative values of diff_subset

            coordinates = sort(partition, 'descend')

            fpr = false positive rate based on negative coordinates matrix
            tpr = true positive rate based on positive coordinates matrix

            AUC = trapz(fpr, tpr) %this function calculate the area of trapezoid formed by coordinates

            v_AUC(i) = AUC

            IF AUC < BestAUC THEN
                  BestAUC = AUC
            END IF
        END FOR

        path = attribute selected according BestAUC
        dataset = dataset according attribute selected
        dataset = remove atribute selected
        eProb_path = calculate probability of predicted label to belong to positive or negative class
        predicted_label = max(eProb_path)
    END LOOP
    RETURN eProb_path, predicted_label and path
```

Fig. 1. Pseudocode for PSDP-AUC-Split algorithm.

compromising the accuracy if a single labelling is chosen. Thus, for a two class classification problem and a tree with n leaves, there are 2^n possible labellings, of which $n + 1$ are optimal. That optimal labelling gives the convex hull for the considered leaves. Figure 1 shows the pseudocode of the patient-specific decision path that uses the AUC-split proposed by Ferri.

This algorithm receives as parameters the dataset, the label of each dataset instance and the test instance in which you want to classify. As shown in Fig. 1, the first step of the algorithm is to select an attribute of the test instance by times, calculate the partitions according to the value of this attribute in this instance using as reference the training dataset in order setting the partitions of each one of subsets (number of positive and negative cases). The matrix *coordinates* receives the matrix *partition* sorted in decreasing so that the AUC value obtained is the maximum. The calculation of the AUC is done by the function *trapz* [16], written in Matlab [17], using as parameters the false positive rate and true positive rate which are calculated based on *coordinates* matrix. After this calculation, the value is stored in *v_AUC* vector. All these steps will be repeated for the other attributes. At the end of the calculation for all, the attribute with best AUC value will be selected and so that one of the stopping conditions is true, the path will be complete together the predicted label.

In contrast, another alternative way to use the AUC metric to select patient features requires a prediction for class probability or some other score as proposed by Fawcett. In this case, a leave-one-out cross-validation scheme was employed in order to generate that probability estimate and further calculating a Mann-Whitney-Wilcoxon test statistic, which direct relates to the AUC. To avoid overfitting, Laplace smoothing [18] was employed when estimating class probabilities. The patient-specific decision path that uses this standard approach just described is shown in the pseudo-code of Fig. 2.

The main difference between PSDP-STD-AUC and PSDP-AUC-Split algorithms are the steps prior to the calculation of the AUC. An operation probability estimates is performed for all cases of dataset. However, these estimates are calculated according to the subset of training instances that have the same value of the test instance and the subset of training instances that have different values of the test instance. After the calculations of these estimates according to Laplace smoothing, the AUC is calculated by the function *colAUC* [19] written in Matlab [17], using as parameters all probability the true class of the training set, accompanied by the labels of each instance. The following steps of this algorithm are identical to the steps described in the previous paragraph for the PSDP-AUC-Split algorithm.

4   Experimental Results

In this section we presented the datasets on which the algorithms were evaluated, the preprocessing of the datasets, the performance measures used in the evaluation, and the experimental settings used for the algorithms. Amongst those datasets, there are clinical datasets including two on heart disease, two on diabetes and one cancer patients. Brief descriptions of the datasets are given in Table 1.

4.1   Datasets

All datasets used to test the algorithms were taken from the *UCI Repository* and are used on classification tasks. The continuous variables were discretized using the entropy-based method developed by Fayyad and Irani [20]. Although big part of data presented in the datasets are numeric, they are not used to calculate the area under the curve. The AUC is calculated receiving as parameters the positive probabilities estimated for each instance of the dataset, with the respective labels of each of these instances. The discretization process is conducted in order to facilitate the handle of data by the algorithm, creating ranges of values and using

```
PSDP-STD-AUC (labels, dataset, testset)
    INPUT: labels contain a set of possible labels to predict,
                dataset is database,
                testset is a test case to predict
    OUTPUT: an array with the label, estimates probability of the label and the path.

    LOOP: until dataset to be empty, all cases in dataset have the same target or the set of atributes to be empty

        eProb = calculate the probability of happen two class for each instance of dataset

        FOR each testeset(i)
            subset = getSubSet(dataset, testset(i))
            diff_subset = dataset - subset

            eProb(subset) = calculate probability according number of testset
            eProb(diff_subset) = calculate probability different of testset

            AUC = colAUC(true_probability_of_dataset, labels)

            v_AUC(i) = AUC

            IF AUC < BestAUC THEN
                    BestAUC = AUC
            END IF
        END FOR

        path = attribute selected according BestAUC
        dataset = dataset according attribute selected
        dataset = remove atribute selected
        eProb_path = calculate probability of predicted label to belong to positive or negative class
        predicted_label = max(eProb_path)
    END LOOP
    RETURN eProb_path, predicted_label and path
```

Fig. 2. Pseudocode for PSDP-STD-AUC algorithm.

a single number or character that represents each interval. Missing values were imputed using an iterative non-parametric imputation algorithm described by Caruana [21] which has previously been applied to fill in missing predictor values for medical datasets with good results.

4.2 Test Settings

The proposed patient-specific decision path algorithms were implemented in MATLAB (R2013b version). We evaluated the algorithms using 20-fold cross-validation. This method randomly divided each dataset into 20 approximately equal sets such that each set had a similar proportion of individuals who developed the positive outcome. For each algorithm, we combined 19 sets and evaluated it on the remaining test set, and we repeated this process once for each possible test set. We thus obtained a prediction for the outcome variable for every instance in a dataset. The final result of the algorithms will be presented

Table 1. Brief description of used UCI datasets

Dataset	Instances	Attributes	Positive cases (%)
Australian	690	14	44%
Breast	699	9	34%
Cleveland	296	13	45%
Corral	128	6	44%
Crx	653	15	45%
Diabetes	768	8	35%
Flare	1066	10	17%
Glass 2	163	9	47%
Heart	270	13	44%
Pima	768	8	35%
Sonar	208	60	53%
Tic-tac-toe	958	9	65%
Vote	435	16	61%

in terms of AUC and processing time. The algorithms that used the AUC measures to select an attribute were compared with the entropy-based algorithm proposed in [12].

All experiments were performed on a PC with a processor Intel Core i5 two cores with frequency of 2.5GHz, 8GB of RAM and running the operating system Mac OS X 64-bit Yosemite.

4.3   Results

Table 2 shows the AUCs obtained by the three algorithms. For each dataset, we present the mean AUC based on the 20-fold cross-validation and the respective confidence intervals at the 0.05 level. Overall, the two AUC based split algorithms perform comparably to the entropy method. Except for the Crx and Tic-tac-toe datasets, there is no statistically significant difference between the three methods.

In the Crx dataset, the entropy based model was statistically significant better than the other two methods with an mean AUC of 0.92. As for the Tic-tac-toe dataset, the PSDP-STD-AUC performed better, with a mean AUC of 0.98. ANOVA analysis [22] was performed and we verify that there is not statistical significant different between the three methods ($p >> 0.05$).

Table 3 shows the execution time of the proposed algorithms (means pm standard deviation). Each dataset was run 30 times for each model. Because the entropy based algorithm requires less operations, it presented better run time methods. The PSDP-STD-AUC requires an estimation of the class probabilities, which demands several computational operations. Even though the PSDP-AUC-Split does not require class probability estimation. Sorting of the leaves is

Table 2. AUCs for the datasets in Table 1. For each algorithm the table gives the mean AUC obtained from 20-fold cross-validation along with 95 % confidence intervals. Statistically significant mean AUC are in bold.

Datasets	Entropy	PSDP-STD-AUC	PSDP-AUC-Split
Australian	0.919 [0.910,0.928]	0.896 [0.877,0.916]	0.889 [0.869,0.909]
Breast	0.984 [0.980,0.988]	0.985 [0.978,0.992]	0.983 [0.975,0.991]
Cleveland	0.862 [0.837,0.888]	0.845 [0.779,0.911]	0.834 [0.773,0.895]
Corral	1.000 [1.000,1.000]	1.000 [1.000,1.000]	1.000 [1.000,1.000]
Crx	**0.920** [0.911,0.929]	0.879 [0.855,0.903]	0.885 [0.861,0.909]
Diabetes	0.827 [0.812,0.842]	0.815 [0.781,0.850]	0.820 [0.785,0.855]
Flare	0.730 [0.717,0.745]	0.718 [0.680,0.757]	0.715 [0.681,0.749]
Glass 2	0.831 [0.795,0.867]	0.870 [0.794,0.946]	0.865 [0.785,0.945]
Heart	0.879 [0.862,0.898]	0.877 [0.832,0.921]	0.877 [0.831,0.923]
Pima	0.825 [0.810,0.840]	0.813 [0.769,0.858]	0.811 [0.781,0.841]
Sonar	0.889 [0.867,0.911]	0.862 [0.818,0.907]	0.887 [0.835,0.939]
Tic-tac-toe	0.960 [0.952,0.969]	**0.989** [0.978,1.000]	0.977 [0.968,0.986]
Vote	0.986 [0.982,0.990]	0.985 [0.970,1.000]	0.985 [0.975,0.995]

Table 3. Results of execution time for proposed algorithms

Datasets	Entropy	PSDP-STD-AUC	PSDP-AUC-Split
Australian	3.346 ±0.046	35.821 ±1.464	7.388 ±0.138
Breast	1.140 ±0.010	10.102 ±0.048	2.115 ±0.025
Cleveland	1.091 ±0.011	10.827 ±0.052	2.466 ±0.026
Corral	0.170 ±0.005	1.651 ±0.019	0.341 ±0.009
Crx	3.342 ±0.032	34.111 ±0.094	7.620 ±0.229
Diabetes	1.974 ±0.013	17.305 ±0.094	4.002 ±0.040
Flare	4.348 ±0.021	36.206 ±1.179	8.019 ±0.322
Glass 2	0.427 ±0.007	3.298 ±0.038	0.920 ±0.014
Heart	1.042 ±0.011	10.301 ±0.057	2.220 ±0.019
Pima	1.973 ±0.012	19.175 ±0.093	3.781 ±0.137
Sonar	3.730 ±0.029	34.608 ±0.134	8.040 ±0.038
Tic-tac-toe	2.527 ±0.013	26.731 ±0.124	4.647 ±0.029
Vote	1.366 ±0.017	14.300 ±0.089	2.686 ±0.018

necessary to obtain the convex hull, for each split, also demanding extra CPU time. As per an ANOVA analysis, we verified that there is significant difference on execution time, since the entropy-based model runs faster ($p << 0$).

5   Conclusions

We have introduced PSDP, a new patient-specific approach for predicting outcomes, based on the AUC metric to select patient attributes. We evaluated this method on several datasets, including medical data. The results show that the PSDP-AUC based methods performs equivalently to the standard information based method. These results are encouraging and provide support for further investigation into the PSDP methods and more extensive evaluation on a wide range of datasets.

In future work, we plan to examine the complexity of the models generated by the PSDP methods and also explore other criteria for selecting predictors. Moreover, the presentation of patient-specific decision paths as IF-THEN rules to a domain expert may provide insight into patient populations.

The current PSDP methods have several limitations. One limitation is that they handle only discrete variables and continuous variables have to be discretized. A second limitation is the execution time on large data samples.

References

1. Abu-Hanna, A., Lucas, P.J.: Prognostic models in medicine. AI and statistical approaches. Methods Inf. Med. **40**(1), 1–5 (2001)
2. Visweswaran, S., Cooper, G.F.: Patient-specific models for predicting the outcomes of patients with community acquired pneumonia. In: AMIA Annu Symposium Proceedings (2005)
3. Mitchell, T.M.: Machine Learning, 1st edn. McGraw-Hill Inc, New York (1997)
4. Friedman, J.H.: Lazy decision trees. In: Proceedings of the Thirteenth National Conference on Artificial Intelligence, (AAAI 1996), v.1, pp. 717–724. AAAI (1996)
5. Foster J.P., Fawcett, T., Kohavi, R.: The case against accuracy estimation for comparing induction algorithms. In: Proceedings of the Fifteenth International Conference on Machine Learning (ICML 1998), pp. 445–453, San Francisco, CA, USA (1998)
6. Cover, T., Hart, P.: Nearest neighbor pattern classification. IEEE Trans. Inf. Theory **13**(1), 21–27 (1967)
7. Gagliardi, F.: Instance-based classifiers applied to medical databases: diagnosis and knowledge extraction. Artif. Intell. Med. **52**(3), 123–139 (2011)
8. Zheng, Z.J., Webb, G.I.: Lazy learning of Bayesian rules. Mach. Learn. **41**(1), 53–84 (2000)
9. Visweswaran, S., Cooper,: G.F.: Instance-specific Bayesian model averaging for classification. In: Proceedings of the Eighteenth Annual Conference on Neural Information Processing Systems. Vancouver, Canada (2004)
10. Visweswaran, S., et al.: Learning patient-specific predictive models from clinical data. J. Biomed. Inform. **43**(5), 669–685 (2010)
11. Visweswaran, S., Cooper, G.F.: Learning instance-specific predictive models. J. Mach. Learn. Res. **11**, 3333–3369 (2010)
12. Ferreira, A., Cooper, G.F., Visweswaran, S.: Decision path models for patient-specific modeling of patient outcomes. In: Proceedings of the Annual Symposium of the American Medical Informatics Association, pp. 413–21 (2013). PMID: 24551347, PMCID: PMC3900188

13. Hand, D.J., Till, R.J.: A simple generalisation of the area under the roc curve for multiple class classification problems. Mach. Learn. **45**(2), 171–186 (2001)
14. Ferri, C., Flach, P., Hernndez-Orallo, J.: Learning decision trees using the area under the ROC curve. In: Proceedings of the 19th International Conference on Machine Learning, pp. 139–146, Sydney, NSW, Australia, 8–12 July 2002
15. Ferri, C., Flach, P., Hernndez-Orallo, J.: Rocking the ROC Analysis within Decision Trees. Technical report, 20 December 2001
16. MATLAB: Trapz: Trapezoidal numerical integration. The Mathworks Inc. (2006). Accessed 17 February 2015
17. MATLAB: Version 8.4.0 (R2014b) The Mathworks Inc. (2014). Accessed 17 February 2015
18. Zadrozn, B., Elkan, C.: Obtaining calibrated probability estimates from decision trees and naive Bayesian classifiers. In: Proceedings of the Eighteenth International Conference on Machine Learning (ICML 2001), pp. 609–616, San Francisco, CA, USA, (2001)
19. colAUC: Calculates Area under ROC curve (AUC). The Mathworks Inc. (2011). Accessed 17 February 2015
20. Fayyad, U.M., Irani, K.B.: Multi-interval discretization of continuous-valued attributes for classification learning. In: IJCAI, pp. 1022–1029 (1993)
21. Caruana, R.: A non-parametric EM-style algorithm for imputing missing values. In: Proceedings of Artificial Intelligence and Statistics (2001)
22. Janez, Demar: Statistical comparisons of classifiers over multiple data sets. J. Mach. Learn. Res. **7**, 1–30 (2006)

A Bayesian Approach to Sparse Cox Regression in High-Dimentional Survival Analysis

Olga Krasotkina and Vadim Mottl$^{(\boxtimes)}$

Computing Center RAS, Moscow, Russia
{O.V.Krasotkina,vmottl}@yandex.ru

Abstract. Survival prediction and prognostic factor identification play an important role in machine learning research. This paper employs the machine learning regression algorithms for constructing survival model. The paper suggests a new Bayesian framework for feature selection in high-dimensional Cox regression problems. The proposed approach gives a strong probabilistic statement of the shrinkage criterion for feature selection. The proposed regularization gives the estimates that are unbiased, possesses grouping and oracle properties, their maximal risk diverges to a finite value. Experimental results show that the proposed framework is competitive on both simulated data and publicly available real data sets.

Keywords: Survival analysis · Cox regression · Proportional hazard model · Regularization · Feature selection · Oracle properties · Bayesian approach

1 Introduction

Survival analysis is a commonly adopted method for the analysis of failure time, for instance, biological death, mechanical failure, or credit default [1,2]. The objects in this area are patients, technical devices, clients, etc. Within this context, death or failure is also referred to as an event. The Machine Learning theory considers the survival analysis as modeling the relationship between survival time and one or several predictors, usually termed covariates in the survival-analysis literature. In supervised learning, estimating an unknown dependency between the covariates and the survival time is based on a limited number of observations called the training set. The primary objective is to predict the survival time for new objects represented the sets of their features (covariates). But another and more interesting aspect is assessing the importance of certain prognostic factors, such as age, gender, or race, in predicting the survival outcome.

Survival data have two aspects that are difficult to handle by the traditional statistical and machine learning methods: time as the goal variable and the presence of censored observations, namely, observations, for which the failure

The work is supported by grants of the Russian Foundation for Basic Research No. 11-07-00409 and No. 11-07-00634.

time is not registered exactly. The censored data arise when an object's life length is known to occur only within a certain period of time. In other words, we have to take into account also the objects for which the event of interest has not occurred during the observation period but might have occurred after it. Possible censoring schemes are said to be right censoring, if all what is known is that the respective individual is still alive at a given time, left censoring when all what is known is that the individual has experienced the event of interest prior to the start of the study, or interval censoring if the only information is that the event has occurred within the observation interval. In this paper, we consider only the right censoring scheme.

Many models of survival prediction have been proposed in the statistical literature. The most popular one is the accelerate failure time (AFT) model [8,9]. AFT is a linear regression model in which the response variable is the logarithm or a known monotone transformation of the failure (death) time. Even though the semi-parametric estimation of an AFT model with an unspecific error distribution has been studied extensively in the literature, the model has not been widely used in practice, mainly due to the difficulties in computing model parameters [10]. Another one is Cox proportional hazards model [3–5], in which the model parameters are estimated by way of partial log likelihood maximization. However, it is much more difficult to apply these methods to survival data of very high dimension.

Recent technological advances have made it possible to collect a huge amount of covariate information while observing survival information via bioimaging technology, such as microarray, proteomic and SNP data. However, it is quite likely that not all available covariates are associated with the survival time. In fact, typically a small fraction of covariates is associated with the goal variable. This is the phenomenon of sparsity that calls for the identification of important risk factors along with the evaluation of their risk contributions, when analyzing time-to-event data with many predictors.

Recent publications have focused on the development of feature selection methods for survival analysis problems. All of them can be divided into three groups. The methods of first group, called filters, are meant to be applied at the preprocessing step, the second group methods [11,12], called wrappers, use a predictive model to score the feature subsets. The third group methods, called embedded methods [13–18], are incorporated into the respective algorithm. The main advantages of the last group are the ability to use the properties of the respective learning algorithm, in contrast to filters, and the lower computational complexity than that of wrappers. These approaches are based on the presence of a sparse regularization term in the formulation of the optimization problem. Although several kinds of sparse regularization are widely used in classification to deal with feature selection, only a few attempts have been made to propose sparse-regularized survival analysis methods.

Tibshirani [13] applied the LASSO penalty to survival analysis. Fan and Li [5] considered survival analysis with the SCAD and other folded concave penalties. Zou, and Hastie [15] used the elastic net penalty for Cox regression. Later Zou [16] proposed the adaptive LASSO penalty while studying time-to-event data. The available theory and empirical results show that these penalization approaches work well only with a moderate number of covariates. The proposed non-convex penalties provide good statistical properties such as unbiasedness and oracle inequalities [16]. But, as it is shown in [19], any estimator that satisfies the sparsity property pays for this by the maximal risk that converges to the supremum of the loss function. In particular, the maximal risk tends to infinity whenever the loss function is unbounded. Another disadvantage of the penalties based on the l_1 regularization is that this type of penalties has no grouping property, i.e., displays its inability to include into the model correlated regressors. However, the main goal of survival analysis is to understand the factors affecting the survival time. It would be better that all correlated factors were included into the model.

In this paper, we propose a new Bayesian approach to variable selection in the Cox regression problem. The proposed approach gives a strong probabilistic formulation of the shrinkage criterion for predictor selection. The proposed shrinkage criterion is unbiased, has grouping and oracle properties, its maximal risk diverges to a finite value.

This rest of the paper is organized as follows. Section 2 specifies the general Bayesian formulation of Cox regression for survival analysis, gives an overview of variable selection methods via penalization of the training criterion, and presents the mathematical framework of the proportional hazard model with supervised selectivity as a new more effective hierarchical Bayesian Cox regression model. The respective group coordinate descent algorithm for feature-selective parameter estimation is outlined in Sect. 3. Section 4 investigates in full details the main properties of our Cox regression technique with supervised variable selectivity. Experimental results on simulated and real-world data in Sect. 5 demonstrate its effectiveness in both the variable selection and survival time prediction. Finally, the conclusions are formulated in Sect. 6.

2 The Hierarchical Bayesian Model with Supervised Selectivity for Sparse Cox Regression

Let us start withm the classical statement of the dependency estimation problem. Let $\omega \in \Omega$ be a set of real-world objects each of which is naturally associated with a point in the linear feature space $\mathbf{x}(\omega) = (x_1(\omega), ..., x_p(\omega)) \in R^p$ and a value of the time variable $t(\omega) \in T = [0, \infty)$. The function

$$\big(\mathbf{x}(\omega), t(\omega)\big) : \Omega \to R^p \times T \tag{1}$$

is known only within the bounds of a finite training set $\big\{\big(\mathbf{x}(\omega_i), t(\omega_i)\big) = (\mathbf{x}_i, t_i, \delta_i), \, i = 1, ..., N)\big\}$, where N is the number of observations and $\delta_i \in \{0, 1\}$ is the censoring indicator. If $\delta_i = 1$ (no censoring) t_i is survival time $t(\omega_i) = t_i$, if $\delta_i = 0$ then t_i is censoring time $t(\omega_i) \geq t_i$, i.e., it is only known that object ω_i was safe, at least, up to t_i. It is required to continue the function (1) onto the entire set Ω, so that it would be possible to estimate the survival time for new objects $\omega \in \Omega$ not represented in the training set.

We also assume a probability distribution in the joint range of observable features and hidden survival time $(\mathbf{x}(\omega), t(\omega)) \in R^p \times T$, and consider the training-set members as independently sampled. The conditional distribution function $F(t|\mathbf{x}, \boldsymbol{\beta}) = P\big[t(\omega) < t|\mathbf{x}(\omega) = \mathbf{x}, \boldsymbol{\beta}\big]$ describes the probability that the survival time of an object $\mathbf{x}(\omega) = \mathbf{x}$ is less then the fixed value t, where the regression parameter vector $\boldsymbol{\beta}$ specifies a fixed distribution function within the parametric family.

If the survival time is a continuous random variable, the density function is defined $p(t|\mathbf{x}, \boldsymbol{\beta}) = dF(tx, \boldsymbol{\beta})/dt$. Let $S(t|\mathbf{x}, \boldsymbol{\beta}) = 1 - F(t|\mathbf{x}, \boldsymbol{\beta}) = P\big[t(\omega) \geqslant t|\mathbf{x}(\omega) = \mathbf{x}, \boldsymbol{\beta}\big])$ be the survival function, namely, the conditional probability for a random object to survive the time point t. Let also $h(t|\mathbf{x}, \boldsymbol{\beta}) = p(t|\mathbf{x}, \boldsymbol{\beta})/S(t|\mathbf{x}, \boldsymbol{\beta})$ be the hazard function defining the probability density of a random object to die at the time moment t if it has survived up to this moment. Under these assumptions, the complete likelihood of the observed data set is given by the criterion

$$\rho(T|X, \boldsymbol{\beta}) = \prod_{i:\delta_i=0} S(t_i|\mathbf{x}_i, \boldsymbol{\beta}) \prod_{i:\delta_i=1} h(t_i|\mathbf{x}_i, \boldsymbol{\beta}) =$$

$$\prod_{i:\delta_i=0} \exp\left[-\int_0^{t_i} h(t_i|\mathbf{x}_i, \boldsymbol{\beta})\right] \prod_{i:\delta_i=1} h(t_i|\mathbf{x}_i, \boldsymbol{\beta}). \quad (2)$$

This criterion uses the survival function as the likelihood for the censored objects and the hazard function as the likelihood for the non-censored ones. The presence of the integrals in the likelihood function makes the parameters estimation process very complicate. To overcome this obstacle David Cox [3] suggested to replace the general criterion (2) by the partial likelihood criterion. If there are no any simultaneous events in the training data the partial likelihood of the observed data set can be written as

$$L(T|\boldsymbol{X}, \boldsymbol{\beta}) = \prod_{i=1}^{N}\left[h(t_i|\mathbf{x}_i, \boldsymbol{\beta})/\sum_{j \in R_i} h(t_j|\mathbf{x}_j, \boldsymbol{\beta})\right]^{\delta_i}, \quad (3)$$

where the risk set $R_i = \{j: t_j \geq t_i\}$ includes subjects for whom the event has not yet occurred or who have yet to be censored. The second key assumption in the Cox model is that of proportional hazards. In this model, the hazard function for individual i is written as $h(t|\mathbf{x}, \boldsymbol{\beta}) = q(t)\exp\big(\boldsymbol{x}^T\boldsymbol{\beta}\big)$, where $q(t)$ is a baseline hazard function for the population Ω and $\exp\big(\boldsymbol{x}^T\boldsymbol{\beta}\big)$ is the relative risk of an object. Under the proportional risk assumption the partial likelihood is a product over the uncensored failure times written as

$$L(T|\boldsymbol{X}, \boldsymbol{\beta}) = \prod_{i=1}^{N}\left[\frac{\exp\big(\boldsymbol{x}_i^T\boldsymbol{\beta}\big)}{\sum_{j \in R_i} \exp\big(\boldsymbol{x}_j^T\boldsymbol{\beta}\big)}\right]^{\delta_i} \quad (4)$$

Furthermore, let the parameter vector $\boldsymbol{\beta}$ be considered as random vector distributed in accordance with a priori density $\Psi(\boldsymbol{\beta}|\mu_1, ..., \mu_m)$ parametrized by $(\mu_1, ..., \mu_m)$. Consequently, the a posteriori joint distribution density of the parameters of the parameter vector w.r.t. the training set is proportional to the product $P(\boldsymbol{\beta}|\boldsymbol{X}, T, \mu_1, ..., \mu_m) \propto \Psi(\boldsymbol{\beta}|\mu_1, ..., \mu_m) L(T|\boldsymbol{X}, \boldsymbol{\beta}, \mu_1, ..., \mu_m)$. It is natural to consider the maximum point of this a posteriori density as the object of training

$$(\beta_1, ..., \beta_p) \propto \arg\max[\ln \Psi(\boldsymbol{\beta}|\mu_1, ..., \mu_m) + \ln L(T|\boldsymbol{X}, \boldsymbol{\beta}, \mu_1, ..., \mu_m)]. \quad (5)$$

It is easy to show that, under these assumptions, we obtain the training criterion:

$$J(\boldsymbol{\beta}, |(T, X), \boldsymbol{\mu}) = -\ln \Psi(\boldsymbol{\beta}|\mu_1, ..., \mu_m) -$$

$$\sum_{i=1}^{N} \delta_i \left(\boldsymbol{x}_i^T \boldsymbol{\beta} - \ln \sum_{j \in R_i} \exp\left(\boldsymbol{x}_j^T \boldsymbol{\beta}\right) \right) \to \min(\boldsymbol{\beta}). \quad (6)$$

In regression and classification, a large panel of a priori densities $\Psi(\boldsymbol{\beta}|\mu_1, ..., \mu_m)$ has been developed to learn sparse models: Ridge, LASSO, Bridge, Elastic Net, Adaptive LASSO, SCAD (Table 1).

Table 1. A priory densities to learn sparse model

Penalty	$\psi(\beta_i|\mu_1, ..., \mu_m)$
Ridge	$\exp(-\mu\beta_i^2); \mu > 0$
LASSO	$\exp(-\mu|\beta_i|); \mu > 0$
Elastic Net	$\exp(-\mu_1|\beta_i| + \mu_2\beta_i^2); \mu_1, \mu_2 > 0$
Bridge	$\exp(-\mu|\beta_i|^p); \mu > 0, p < 1$
Adaptive LASSO	$exp(-\mu_1|\beta_i| + \mu_2\beta_i^2); \mu_1, \mu_2 > 0$
SCAD	$\exp(-p(|\beta_i|)), p'(|\beta_i|) = \mu_1\left\{1(|\beta_i| \leq \mu_1) + \frac{\mu_2\mu_1 - |\beta_i|}{\mu_2\mu_1 - \mu_1}1(|\beta_i| > \mu_1)\right\}$

A good a priori density should result in an estimator with six properties [6, 19]:

1. *Unbiasedness*: The resulting estimator is nearly unbiased when the true unknown parameter is large to avoid unnecessary modeling bias.
2. *Sparsity*: The resulting estimator can suppress the presence of the small estimated coefficients to reduce model complexity
3. *Grouping effect*: Highly correlated variables will be in or out of the model together.
4. *Oracle properties*: The resulting estimator can correctly select the nonzero coefficients with probability converging to one, and that the estimators of the nonzero coefficients are asymptotically normal with the same means and covariance that they would have if the zero coefficients were known in advance.
5. *Finite maximum risk*: The resulting estimator has the finite risk in all point of parameter space and for all sample sizes.
6. *Continuity*: The resulting estimator is continuous in data to avoid instability in model prediction.

Table 2. Properties of a priory densities to learn sparse model

Penalty	1	2	3	4	5	6
Ridge	No	No	Yes	No	Yes	Yes
LASSO	No	Yes	No	No	No	Yes
Elastic Net	No	Yes	Yes	No	No	Yes
Bridge	Yes	Yes	No	No	No	No
Adaptive LASSO	Yes	Yes	No	Yes	No	Yes
SCAD	Yes	Yes	No	Yes	No	Yes

The overview of above mentioned properties for all a priori densities is shown in Table 2. As we can see from the literature [6,19] there is no any a priory density possessing by all properties of a good estimator. In the next section we proposed a general framework for sparse Cox regression possessing all properties of a good estimator.

We assume that the distribution density of the component β_i is the normal distribution with zero expectation and variance r_i

$$\phi(\beta_i|r_i) \propto (r_i)^{-1/2} \exp\left(-\beta_i^2/(2r_i)\right).$$

In this case, the joint distribution of regression coefficients is

$$\Phi(\beta_1, ..., \beta_p|r_1, ..., r_p,) \propto \prod_{i=1}^{p} (r_i)^{-1/2} \exp\left[-\sum_{i=1}^{p} w_i^2/(2r_i)\right].$$

Furthermore, we assume that inverse variances considered as independent priori gamma distributions

$$\gamma\left((1/r_i)|\alpha, \theta\right) \propto (1/r_i^{\alpha-1}) \exp\left(-\theta(1/r_i)\right).$$

Joint a priori distribution density of inverse variances $1/r_i$ is

$$G(1/r_1, \ldots, 1/r_p|\alpha, \theta) \propto \left(\prod_{i=1}^{p}(1/r_i^{\alpha-1}) \exp\left(-\theta(1/r_i)\right).\right.$$

Joint a priori distribution density of regression coefficients $1/r_i$ and their variances is

$$\Psi(\beta, \mathbf{r}|\alpha, \theta) = \Phi(\beta|\mathbf{r})G(\mathbf{r}|\alpha, \theta) \propto$$
$$\propto \prod_{i=1}^{p} (r_i)^{(1/2-\alpha)} \exp\left[-\sum_{i=1}^{p} \left(\beta_i^2/(2r_i) + \theta/r_i\right)\right]. \quad (7)$$

From the principle of maximum posteriori density by using Bayesian treatment

$$P(\beta, \mathbf{r}|(T, X), \alpha, \theta) \propto L(T|X, \beta)\Phi(\beta|\mathbf{r})G(\mathbf{r}|\alpha, \theta), \quad (8)$$

so that results in the training criterion

$$J(\boldsymbol{\beta}, \mathbf{r}|(T, X), \alpha, \theta) = \underbrace{(1/2) \sum_{i=1}^{p} (\beta_i^2)/(r_i) + (\alpha - 1/2) \sum_{i=1}^{p} \ln r_i + \theta \sum_{i=1}^{p} 1/r_i}_{p(\beta, \mathbf{r}|\alpha, \theta) = \ln \Psi(\beta, \mathbf{r}|\alpha, \theta)}$$

$$\underbrace{- \sum_{i=1}^{n} \delta_i \left(\boldsymbol{x}_i^T \boldsymbol{\beta} - \ln \sum_{j \in R_i} \exp \left(\boldsymbol{x}_j^T \boldsymbol{\beta} \right) \right)}_{l(\boldsymbol{\beta})}. \quad (9)$$

We called the criterion of such type as Supervised Selectivity Cox (SuperSelCox) criterion.

3 Group Coordinate Descent Procedure for Parameter Estimation

We will optimize the criterion (9) by the group coordinate descent algorithm. For finding the minimum point of the criterion, we apply the coordinate descent iteration to both groups of variables $(\boldsymbol{\beta}, \mathbf{r})$. We start from $(r_i^{(0)} = 1)_{i=1}^{n}$. Let $(\mathbf{r}^{(k)})$ be the current approximation of the variances. The next values of regression coefficients $(\boldsymbol{\beta}^{(k+1)})$ can be found as

$$\boldsymbol{\beta}^{(k+1)} = \arg \min_{\boldsymbol{\beta}} \left(-l(\boldsymbol{\beta}) + (1/2) \sum_{i=1}^{p} \beta_i^2/(r_i^{(k)}) \right), \quad (10)$$

where

$$l(\boldsymbol{\beta}) = \sum_{i=1}^{n} \delta_i \left(\boldsymbol{x}_i^T \boldsymbol{\beta} - \ln \sum_{j \in R_i} \exp \left(\boldsymbol{x}_j^T \boldsymbol{\beta} \right) \right). \quad (11)$$

For optimization the criterion (10) we use the Newton-Raphson procedure. Let $[\tilde{\beta}_i = \beta_i/\sqrt{r_i^{(k)}}, \tilde{x}_{i,j} = x_{i,j}\sqrt{r_i^{(k)}}]_{i=1}^{p}$, $\tilde{X} = \{\tilde{x}_{i,j}\}, \tilde{\eta} = \tilde{X}\tilde{\beta}$. We use the two term Taylor series expansion of the log-partial likelihood centered at point $\tilde{\beta} = \beta^k$. Lets denote g_i and $h_{i,i}$ the elements of gradient and Hessian for the log-partial likelihood with respect to η_i

$$\begin{cases} g_i = \frac{\partial l}{\partial \tilde{\eta}_i} = \delta_i - \exp(\tilde{\eta}_i) \sum_{k \in C_i} \left(1/\sum_{j \in R_k} \exp(\eta_i) \right) \\ h_{i,i} = \frac{\partial^2 l}{\partial \tilde{\eta}_i^2} = -\exp(\tilde{\eta}_i) \sum_{k \in C_i} \left(1/\sum_{j \in R_k} \exp(\eta_i) \right) + \\ \qquad + \exp(2\tilde{\eta}_i) \sum_{k \in C_i} \left(1/\sum_{j \in R_k} \exp(\eta_i) \right)^2 \\ C_i = \{k : i \in R_k\}, i = 1, ..., n. \end{cases}$$

Under these notations the Newton-Raphson procedure is

1. Initialize $\tilde{\beta}$ and $\tilde{\eta}$.
2. Compute $w(\tilde{\eta})_i = h_{i,i}, z(\tilde{\eta})_i = \eta_i - g_i/h_{i,i}$.

3. Find $\hat{\beta} = \arg\min_{\tilde{\beta}} M(\tilde{\beta})$, where $M(\tilde{\beta}) = \sum_{i=1}^{n} w(\tilde{\eta})_i (z(\tilde{\eta})_i - \tilde{x}_i^T \tilde{\beta})^2 +$ $(1/2)\sum_{i=1}^{p}(\tilde{\beta}_i)^2/n$. It can be easily shown that

$$\hat{\beta} = \frac{\sum_{i=1}^{n} w(\tilde{\eta}_i) z(\tilde{\eta}_i)\tilde{x}_{ik} - \sum_{i=1}^{n} w(\tilde{\eta}_i) z(\tilde{\eta}_i)\tilde{x}_{ik} \sum_{j\neq k} \tilde{x}_{ij}\tilde{\beta}_k}{1/n + \sum_{i=1}^{n} w(\tilde{\eta}_i)\tilde{x}_{ik}^2}.$$

4. Set $\tilde{\beta} = \hat{\beta}$.
5. Repeat steps 2-4 until the convergence of $\hat{\beta}$.

The next values of the variances $r^{(k+1)}$ can be found as minimum point

$$r_i^{(k+1)} = \arg\min_{r_i} \left[\sum_{i=1}^{p} \frac{(\beta_i^{(k+1)})^2}{2r_i} + \left(\alpha - \frac{1}{2}\right)\sum_{i=1}^{p}\ln r_i + \theta\sum_{i=1}^{p} 1/r_i \right]. \qquad (12)$$

So the variances $\mathbf{r}^{(k+1)}$ are defined by the following expression which is easy to prove

$$r_i^{(k+1)} = ((\beta_i^{(k+1)})^2 + 2\theta)/(2\alpha - 1). \qquad (13)$$

We choose parameters of gamma-distributions α, θ by the following formula: $\alpha = 1 + 1/(2\mu), \theta = 1/(2\mu)$, so that $E(1/r_i) = 2\mu + 1$ and $E(1/r_i^2) = (2\mu+1)2\mu$. The parameter $0 < \mu < \infty$ plays the role of a parameter of supervised selectivity. If $\mu \to 0$ then $E(1/r_i) = 1$ and $E(1/r_i^2) = 0 \Rightarrow 1/r_1 \cong ...1/r_p \cong 1$. If $\mu \to \infty$ then $E(1/r_i) = \infty$ and $E(1/r_i^2) = \infty$ but $E(1/r_i^2)/E(1/r_i) = \mu \to \infty$. It means that if μ grows, the independent positive values $1/r_i$ may differ arbitrarily, because variances increase much faster than expectations.

The appropriate value of the structural parameter μ is determined by cross-validation procedure. We use a technique proposed in van Houwelingen et al. [7]. We split our data into k parts. Our goodness of fit estimate for a given part i and μ is

$$CV_i(\mu) = \sum_{k=1}^{K} \left(l(\hat{\beta}_{(-k)}(\mu)) - l_{(-k)}(\hat{\beta}_{(-k)}(\mu)) \right)$$

where $l_{(-k)}(\hat{\beta}_{(-k)}(\mu))$ is the log-partial likelihood excluding part i of the data and $\hat{\beta}_{(-k)}$ is the optimal β for the non-left out data, $l(\hat{\beta}_{(-k)}(\mu))$ is the log-partial likelihood full sample. Our total criterion $CV(\mu)$ is the sum of $CV_i(\mu)$. We choose the μ value which maximize the $CV(\mu)$.

4   Properties of Proposed Sparse Cox Regression with Supervised Selectivity

For investigating the properties of proposed penalty we exclude the variances from criterion (9). The optimal values of the variances $\mathbf{r}^{(k)}$ can be found from expression $\partial J(\beta, \mathbf{r}|(T, X), \alpha, \theta)/\partial r_i = 0$ as

$$r_i = ((\beta_i)^2 + 2\theta)/(2\alpha - 1). \qquad (14)$$

Substitution of the expression (14) into (7) makes the expression for a priori density more convenient for the further analysis

$$\psi(\beta_i) \propto \exp(-p(\beta_i)), p(\beta_i) = \left(\frac{\mu + 1}{2\mu}\right) \ln\left(\mu\beta_i^2 + 1\right) \tag{15}$$

1. *Unbiasedness.* The resulting estimator is approximate unbiased [6] if

$$\lim_{\beta_i \to \infty} p'(\beta_i) \to 0.$$

It is obvious that expression (15) satisfies this condition.
2. *Selectivity.* The criterion (15) penalized the high values of coefficients variances and propose soft threshold for parameters controlled by selectivity parameter μ. In terms of the Statistical Learning Theory decreasing values of μ determine a succession of almos nested classes of models. In other terms, different values of μ result in different effective sizes of the parameter space.
3. *Grouping effect.* The resulting estimator possesses the grouping effect if highly correlated variables are in or out of the model together.
   Theorem 1. Let $x^{(i)} = x^{(j)}$ for some features $i, j \in 1, ..., p$. Let $\hat{\beta}$ is the minimum point of the criterion (9) and $\beta_l \in [-1/\mu, 1/\mu], l = 1, ..., p.$. Then $\beta_i = \beta_j$.
   Theorem 2. Let $\hat{\beta}$ is the minimum point of the criterion (9), $\beta_l \in [-1/\mu, 1/\mu], l = 1, ..., p.$ and $sign(\beta_i) = sign(\beta_j)$. Let transformed response vector $z(\tilde{\eta})$ and covariate matrix $\tilde{X}$ be mean-centered and standardized. Without loss generality, assume $\rho = cor(\tilde{x}^{(i)}, \tilde{x}^{(j)}) > 0$. Then for fixed μ

$$\frac{\hat{\beta}_i}{\mu\hat{\beta}_i^2 + 1} - \frac{\hat{\beta}_j}{\mu\hat{\beta}_j^2 + 1} \le \frac{\sqrt{2(1 - \rho)}}{\mu + 1} \tag{16}$$

If $\rho \to 1$, then $\left(\frac{\hat{\beta}_i}{\mu\hat{\beta}_i^2+1} - \frac{\hat{\beta}_j}{\mu\hat{\beta}_j^2+1}\right) \to 0$ and $\hat{\beta}_i - \hat{\beta}_j \to 0$. Derivation of Theorems 1 and 2 is provided in the Appendix.
4. *Oracle properties.* Let β^0 be true parameter vector. Let $a_N = \max\{p'_\mu(|\beta_{j0}|) : \beta_{j0} \ne 0\}$, $b_N = \max\{p''_\lambda(|\beta_{j0}|) : \beta_{j0} \ne 0\}$. As it was shown in [6] the resulting estimator possesses the oracle properties if it satisfies such conditions

$$b_N \to 0, \sqrt{N}a_N \to 0, N \to \infty. \tag{17}$$

t is obvious that expression (15) satisfies these conditions.
6. *Finite maximum risk.* The resulting estimator has finite maximum risk if

$$\arg\min\{|\beta_i| + p'(\beta_i)\} = 0$$

and

$$\min\{|\beta_i| + p'(\beta_i)\} = 0.$$

It is obvious that expression (15) satisfies these conditions.
6. *Continuity.* The resulting estimator is continuous in parameter space if

$$\arg\min\{|\beta_i| + p'(\beta_i)\} = 0.$$

It is obvious that expression (15) satisfies this condition.

Table 3. Comparison of the four methods (Ridge Regression, Lasso, Elastic Net, Supervised Selectivity) on relevant features restoration

p	Lasso	Ridge	ElasticNet	SuperSelCox
100	0.020	0.023	0.0163	**0.015**
500	0.031	0.055	0.023	**0.019**
1000	0.085	0.107	0.070	**0.055**

5   Experimental Study

We compared SuperSelCox with three other algorithms: the Cox Regression with Ridge, Lasso and Elastic Net penalties [21].

5.1   Ground-Throuth Experiments

We considered the simulation model by Simon et al. [21]. We generated standard Gaussian predictor data $X(N, p)$: $N(0, 1)$. with $n = 25$ observations and number of features p changed from 100 to 1000, only 5 coefficients are non-zero and generated as $\beta_i = (-1)^i \exp\left((1 - i)/10\right)$, $i = 1, ..., 5$, $\beta_i = 0, i > 5$ The true survival time Y is obtained from exponential distribution with parameters $X\beta$. Censoring times C are generated according to uniform distribution in interval $[0, \xi]$, where ξ - censoring parameter. Censoring indicator is $\delta = 1$, if $C \geq Y$ and $\delta = 0$, if $C < Y$ The recorded survival time is $t = \min\{C, Y\}$. Table 3 shows mean square error in estimating true regression coefficients for all three methods. We repeated the process 100 times and reported the average precision. As we can see the proposed Supervised Selectivity Cox regression algorithm provides the better quality on artificial data in relevant feature restoration.

5.2   Real Data

The DLBCL dataset of Rosenwald et al. [22] was employed as a main illustration of application of our proposed dimension reduction method. These data consist of measurements of 7399 genes from 240 patients obtained from customized cDNA microarrays (lymphochip). Among those 240 patients, 222 patients had the IPI recorded, and they were stratified to three risk groups indicated as low, intermediate and high. A survival time was recorded for each patient, ranging between 0 and 21.8 years. Among them, 127 were deceased (uncensored) and 95 were alive (censored) at the end of the study. A more detailed description of the data can be found in [22] .

We divided the patients into a training group of 160 samples and a testing group of 80 samples. Additionally, a nearest neighbor technique was applied to fill in the missing values for the gene expression data. The comparative results of training are presented in Table 4. SuperSelRank provides the higher value of likelihood for this dataset and gene subset from 30 gene is more suitable for prognosis according to the literature.

Table 4. Comparison of the four methods (Ridge Regression, Lasso, Elastic Net, Supervised Selectivity) on partial likelihood restoration

Lasso	ElasticNet	SS
4.38	4.41	4.52

6 Conclusion

The main claim of this paper is a new variable selection technique for survival analysis that helps to identify the covariates that affect the survival time. The focus is on the hierarchical Bayesian model with supervised selectivity which allows to obtain the asymptotic unbiased parameters estimates, possessing the grouping and oracle properties. We gave the strongly theoretical proofs of all properties for the proposed embedded feature selection technic. The methodological power of the proposed Supervised Selectivity Cox Regression Algorithm has been demonstrated via carefully designed simulation studies and real data examples. Comparing with the version of newest Cox regression algorithms we used, it is far more specific in selecting important variables. As a result, it has much smaller absolute deviation in parameters estimation.

Appendix: Proofs

Proof of Theorem 1. The function $l(\boldsymbol{\beta})$ (11) is strictly convex [5]. The penalty term (15) is strictly convex in subspace $\beta_l \in [-1/\mu, 1/\mu], l = 1, ..., p$. In fact, the second derivatives of $p(\boldsymbol{\beta})$ are

$$\frac{\partial^2 p(\boldsymbol{\beta})}{\partial \beta_i \partial \beta_j} = \begin{cases} \frac{1-\mu\beta_i^2}{(\mu\beta_i+1)^2}, i = j \\ 0, i \neq j. \end{cases} \tag{18}$$

So the Hessian of $p(\boldsymbol{\beta})$ non-negative defined in subspace $\beta_l \in [-1/\mu, 1/\mu], l = 1, ..., p$. Define estimator $\hat{\boldsymbol{\beta}}^*$: let $\hat{\beta}_k^* = \hat{\beta}_k$ for all $i \neq j$, otherwise let $\hat{\beta}_k^* = a\hat{\beta}_i + (1-a)\hat{\beta}_j$ for $a = 1/2$. Since $\boldsymbol{x}^{(i)} = \boldsymbol{x}^{(j)}$, $\tilde{\boldsymbol{X}}\hat{\boldsymbol{\beta}}^* = \tilde{\boldsymbol{X}}\hat{\boldsymbol{\beta}}$ and $|\tilde{\boldsymbol{z}} - \tilde{\boldsymbol{X}}\hat{\boldsymbol{\beta}}^*| = |\tilde{\boldsymbol{z}} - \tilde{\boldsymbol{X}}\hat{\boldsymbol{\beta}}|$. However, the penalization function is convex in $\beta_l \in [-1/\mu, 1/\mu], l = 1, ..., p$., that

$$p(\hat{\boldsymbol{\beta}}^*) = p(a\hat{\beta}_i + (1-a)\hat{\beta}_j) < ap(\hat{\beta}_i) + (1-a)p(\hat{\beta}_j) < p(\hat{\boldsymbol{\beta}}).$$

Because $p(\hat{\boldsymbol{\beta}}^*) = p(\hat{\boldsymbol{\beta}})$ and because $p(.)$ is additive, $p(\hat{\boldsymbol{\beta}}^*) < p(\hat{\boldsymbol{\beta}})$ and therefore cannot be the case that $\hat{\boldsymbol{\beta}}$ is a minimizer. Hence $\hat{\beta}_i = \hat{\beta}_j$.

Proof of Theorem 2. By definition,

$$\frac{\partial J(\boldsymbol{\beta}|\mu)}{\partial \beta_k} \Big|_{\beta=\hat{\beta}} = 0. \tag{19}$$

By (19) (for non-zero $\hat{\beta}_i$ and $\hat{\beta}_j$),

$$- 2\tilde{\boldsymbol{x}}_i^T (\tilde{\boldsymbol{z}} - \tilde{\boldsymbol{X}}\hat{\boldsymbol{\beta}}) + (1 + 1/\mu) \frac{2\mu\hat{\beta}_i}{\mu\hat{\beta}_i^2 + 1} = 0 \qquad (20)$$

and

$$- 2\tilde{\boldsymbol{x}}_j^T (\tilde{\boldsymbol{z}} - \tilde{\boldsymbol{X}}\hat{\boldsymbol{\beta}}) + (1 + 1/\mu) \frac{2\mu\hat{\beta}_j}{\mu\hat{\beta}_j^2 + 1} = 0 \qquad (21)$$

Hence

$$\frac{\hat{\beta}_i}{\mu\hat{\beta}_i^2 + 1} - \frac{\hat{\beta}_j}{\mu\hat{\beta}_j^2 + 1} = \frac{1}{1 + \mu}(\tilde{\boldsymbol{x}}_i - \tilde{\boldsymbol{x}}_j)^T (\tilde{\boldsymbol{z}} - \tilde{\boldsymbol{X}}\hat{\boldsymbol{\beta}}) \leq \frac{1}{1 + \mu}|\tilde{\boldsymbol{x}}_i - \tilde{\boldsymbol{x}}_j||\tilde{\boldsymbol{z}} - \tilde{\boldsymbol{X}}\hat{\boldsymbol{\beta}}|. \quad (22)$$

Also, note that $J(\hat{\boldsymbol{\beta}}|\mu) \leq J(\hat{\boldsymbol{\beta}} = \boldsymbol{0}|\mu)$, so $|\tilde{\boldsymbol{z}} - \tilde{\boldsymbol{X}}\hat{\boldsymbol{\beta}}| \leq |\tilde{\boldsymbol{z}}| = 1$, since $\tilde{\boldsymbol{z}}$ is centered and standardize . Hence,

$$\frac{\hat{\beta}_i}{\mu\hat{\beta}_i^2 + 1} - \frac{\hat{\beta}_j}{\mu\hat{\beta}_j^2 + 1} \leq \frac{1}{1 + \mu}|\tilde{\boldsymbol{x}}_i - \tilde{\boldsymbol{x}}_j| = \frac{\sqrt{2(1 - \rho)}}{\mu + 1}. \qquad (23)$$

References

1. Aalen, O., Borgan, O., Gjessing, H., Gjessing, S.: Survival and Event History Analysis: A Process Point of View, ser. Statistics for Biology and Health. Springer-Verlag, New York (2008)
2. Klein, J.P., Moeschberger, M.L.: Survival Analysis, 2nd edn. Springer, New York (2005)
3. Cox, D.R.: Regression models and life-tables (with discussion). J. Roy. Stat. Soc. B **34**, 187–220 (1972)
4. Gui, J., Li, H.: Penalized Cox regression analysis in the high-dimensional and low-sample size settings, with applications to microarray gene expression data. Bioinformatics **21**, 3001–3008 (2005)
5. Fan, J., Li, R.: Variable selection for Coxs proportional hazards model and frailty model. Ann. Stat. **30**, 74–99 (2002)
6. Fan, J., Li, R.: Variable selection via nonconcave penalized likelihood and its oracle properties. J. Am. Stat. Assoc. **96**, 1348–1360 (2001)
7. Van Houwelingen, H.C., et al.: Cross-validated Cox regression on microarray gene expression data. Stat. Med. **25**, 3201–3216 (2006)
8. Lin, D.W., Porter, M., Montgomery, B.: Treatment and survival outcomes in young men diagnosed with prostate cancer: a Population-based Cohort Study. Cancer **115**(13), 2863–2871 (2009)
9. Ying, Z.L.: A large sample study of rank estimation for censored regression data. Ann. Stat. **21**, 76–99 (1993)
10. Jin, Z., Lin, D.Y., Wei, L.J., Ying, Z.L.: Rank-based inference for the accelerated failure time model. Biometrika **90**, 341–353 (2003)
11. Sauerbrei, W.: The use of resampling methods to simplify regression models in medical statistics. Apply Stat. **48**, 313–339 (1999)

12. Sauerbrei, W., Schumacher, M.: A bootstrap resampling procedure for model building: application to the cox regression model. Stat. Med. **11**, 2093–2109 (1992)
13. Tibshirani, R.: The lasso method for variable selection in the Cox model. Stat. Med. **16**, 385–395 (1997)
14. Zhang, H.H., Lu, W.: Adaptive lasso for Coxs proportional hazards model. Biometrika **94**, 691–703 (2007)
15. Zou, H., Hastie, T.: Regularization and variable selection via the elastic net. J. Roy. Stat. Soc. B **67**, 301–320 (2005)
16. Zou, H.: The adaptive lasso and its oracle properties. J. Am. Stat. Assoc. **101**, 1418–1429 (2006)
17. Zou, H., Li, R.: One-step sparse estimates in nonconcave penalized likelihood models (with discussion). Ann. Stat. **36**(4), 1509–1566 (2008)
18. Fan, J., Samworth, R., Wu, Y.: Ultrahigh dimensional feature selection: beyond the linear model. J. Mach. Learn. Res. **10**, 2013–2038 (2009)
19. Leeb, H., Potscher, B.M.: Sparse estimators and the oracle property, or the return of Hodges estimator. J. Econometrics **142**(1), 201–211 (2008)
20. Hastie, T.: Tibshirani R Generalized Additive Models. Chapman and Hall, London (1990)
21. Simon, N., Friedman, J., Hastie, T., Tibshirani, R.: Regularization Paths for Cox's Proportional Hazards Model via Coordinate Descent. J. Stat. Softw. **39**(5), 1–13 (2011)
22. Rosenwald, A., et al.: The use of molecular profiling to predict survival after chemotherapy for diffuse large-b-cell lymphoma. The New England J. Med. **25**, 1937–1947 (2002)

Data Mining in System Biology, Drug Discovery, and Medicine

Automatic Cell Tracking and Kinetic Feature Description of Cell Paths for Image Mining

Petra Perner[✉]

Institute of Computer Vision and Applied Computer Sciences,
IBaI, Leipzig, Germany
pperner@ibai-institut.de, http://www.ibai-institut.de

Abstract. Live-cell assays are used to study the dynamic functional cellular processes in High-Content Screening (HCA) of drug discovery processes or in computational biology experiments. The large amount of image data created during the screening requires automatic image-analysis procedures that can describe these dynamic processes. One class of tasks in this application is the tracking of cells. We describe in this paper a fast and robust cell tracking algorithm applied to High-Content Screening in drug discovery or computational biology experiments. We developed a similarity-based tracking algorithm that can track the cells without an initialization phase of the parameters of the tracker. The similarity-based detection algorithm is robust enough to find similar cells although small changes in the cell morphology have been occurred. The cell tracking algorithm can track normal cells as well as mitotic cells by classifying the cells based on our previously developed texture classifier. Results for the cell path are given on a test series from a real drug discovery process. We present the path of the cell and the low-level features that describe the path of the cell. This information can be used for further image mining of high-level descriptions of the kinetics of the cells.

Keywords: Cell tracking · Similarity-based cell tracking · Microscopic image analysis · Description of the kinetic of the cells · High-content analysis · Computational cell biology · Image mining

1 Introduction

The utilization of dynamic High-Content Analysis approaches during preclinical drug research will permit to gain a more specified and detailed insight into complex sub-cellular processes by the use of living cell culture systems. This will effectively support future drug discoveries leading to a great outcome of highly specific and most effective drugs that come along with a well improved compliance. Live-cell assays are therefore used to study the dynamic functional cellular processes in High-Content Screening of drug-discovery processes or in computational biology experiments. The large amount of image data created during the screening requires automatic image analysis procedures that can automatically describe these dynamic processes. One class of tasks in this application is the tracking of the cells, the description of the events and

© Springer International Publishing Switzerland 2015
P. Perner (Ed.): MLDM 2015, LNAI 9166, pp. 441–451, 2015.
DOI: 10.1007/978-3-319-21024-7_31

the changes in the cell characteristics, so that the desired information can be extracted from it based on data-mining and knowledge-discovery methods.

There are several tracking approaches known that track a cell by detection in each single frame and associate the detected cells in each frame by optimizing a certain probabilistic function [1]. Other approaches track cells by model evolution [2]. This approach seems to be able to handle touching and overlapping cells well but is computational expensive [3].

We propose in this paper a similarity-based approach for motion detection of cells that is robust and fast enough to be used in real world applications. The algorithm is based on a specific similarity measure that can detect cells although small changes in morphology appeared. The algorithm can track normal cells as well as mitotic cells. In Sect. 2 we review related work. The material is given in Sect. 3 and the basic constraints of the algorithm resulting from microscopic life-cell images are explained. The algorithm is described in Sect. 4. Results are given on a test series from a real drug-discovery process in Sect. 5. The features that describe the extracted path are explained in Sect. 6. Finally, we give conclusions in Sect. 7.

2 Related Work

Li et al. [4] proposes a system that combines bottom-up and top-down image analysis by integrating multiple collaborative modules, which exploit a fast geometric active contour tracker in conjunction with adaptive interacting multiple models motion filtering and spatiotemporal trajectory optimization. The system needs ten frames to initialize its parameters. If the number of frames in a video are high this is expectable otherwise it is not. Cells that have not been recognized in the initial frame are not considered anymore. The system cannot track new appearing cells and it has a high computation time. The evaluation of the categorical analysis shows that it has a high rate of false detection of mitosis and new cells. Swapping of cell identities occurred mostly in densely populated regions, where the boundaries between cells are highly blurred. This arises the question why to use snakes for tracking the cells.

In Sacan et al. [5], a software package is described for cell tracking and motility analysis based on active contours. Since snakes relies on the assumption that the movement and deformation of the tracked object is small between consecutive frames they propose like an ensemble of different tracking methods in a coarse to refined fashion in order to make their method more robust. Their combined method performs the following steps: the overall displacement and rotation of the object is first determined using a template matching method. The resulting contour is used as the initial state to pyramidal Lucas–Kanade optical flow-based deformation; statistical outlier detection and local interpolation is applied to achieve resistance against errors in the optical flow evaluation. They report that the ensemble method achieves accurate tracking even for large displacements or deformations of the objects between frames but an deep evaluation of the method is not given in the paper. The speed, area, deformation, trajectory and detailed tracking of the cells are computed and displayed for analysis. Besides automated cell detection and tracking capability, the interface also allows manual editing to initialize or modify the tracking data. The software can be used to work with movie or image files of a variety of file formats. The tracking results can be exported either as raw text data for further numerical analysis, or as movie or image files for visualization, sharing and publishing. Cell lineage is not reported in the paper.

In Wang et al. [6] is reported a system that can segment, track, and classify cells into cell types such as interphase, prophase, metaphase and anaphase. They use the distance transform to binarize the image and seeded watershed transformation to detect the cell area. The tracking path is determined by using local tree matching method. Graph matching approaches require large computation time. For each cell are calculated a large set of features such as texture, shape, gray level. The relevant feature set is determined by using a feature selection procedure. On-line support-vector machine is used to classify the nuclei. This approach is able to adapt the classifier to changing appearances of the objects.

Cohen et al. [7] propose an algorithmic information theoretic method for object-level summarization of meaningful changes in image sequences. Object extraction and tracking data are represented as an attributed tracking graph (ATG), whose connected subgraphs are compared using an adaptive information distance measure, aided by a closed-form multi-dimensional quantization. The summary is the clustering result and feature subset that maximize the gap statistic.

In Maška et al. [10] the coherence-enhancing diffusion filtering is applied on each frame to reduce the amount of noise and enhance flow-like structures. Then, the cell boundaries are detected by minimizing the Chan–Vese model in the fast level set-like and graph cut frameworks. To allow simultaneous tracking of multiple cells over time, both frameworks have been integrated with a topological prior exploiting the object indication function.

Magnusson et al. [11] propose a global track linking algorithm, which links cell outlines generated by a segmentation algorithm into tracks. The algorithm adds tracks to the image sequence one at a time, in a way which uses information from the complete image sequence in every linking decision. This is achieved by finding the tracks which give the largest possible increases to a probabilistically motivated scoring function, using the Viterbi algorithm.

Padfield et al. [12] present a general, consistent, and extensible tracking approach that explicitly models cell behaviors in a graph-theoretic framework. They introduce a way of extending the standard minimum-cost flow algorithm to account for mitosis and merging events through a coupling operation on particular edges. They show how the resulting graph can be efficiently solved using algorithms such as linear programming to choose the edges of the graph that observe the constraints while leading to the lowest overall cost.

Guo et al. [13] propose an optical flow method for automatic cell tracking. The key algorithm of the method is to align an image to its neighbors in a large image collection consisting of a variety of scenes. Considering the method cannot solve the problems in all cases of cell movement, another optical flow method, SIFT (Scale Invariant Feature Transform) flow, is also presented. The experimental results show that both methods can track the cells accurately.

We wanted to have a cell tracking method that can use an operator at the cell line without an initialization phase or heavy human interaction. The method should be fast and robust enough and should not have any subjective human interaction. Based on our observation, we concluded that it must be possible to track the cells by identifying the same cell from one image to the next image. Therefore, we introduced a similarity-based approach [20, 21] for cell tracking. This idea has been overtaken later on by Padfield et al. [8] and by Kan et al. [9]. Both approaches [8, 9] do not evaluate their approach with our approach.

Padfield et al. [8] use spatiotemporal volumes in order to keep track of moving cells over time. The methods uses a standard level set approach for segmentation to extract the G2 phase nuclei from the volume and a Euclidean distance metric for linking between the G2 phases across the other phases. The level-set approach allows to recognize cells at different granularity level but will not prevent the algorithm to detect changing morphology. The Euclidean distance is a simple similarity measure that does not take into account the specific characteristics of visual objects.

Given an estimate of the cell size only, the method of Kan et al. [9] is capable of ranking the trackers according to their performance on the given video without the need for ground truth. Using the cell size for tracking is a simple measure that is included in our application as well.

These two methods [8, 9] cannot track normal cells as well as mitotic cells as in our proposed approach. The Euclidean distance used as similarity measure in Kan et al. [10] is not robust enough to handle changing cell morphology as it is in our case. We use a very flexible similarity measure [17] that considers the similarity of the corresponding pixels as well as the similarity to its surrounding pixels and therefore can handle morphological changes from one time-frame to the following time-frame. Cells that appear new in the microscopic image window can be tracked as well.

3   Material and Investigation of the Material

The live cell movies are comprised of a sequence of 113 images. These images are taken in a time interval of 30 min. The whole experiment runs over two and a half days. The cells show no steady movement. They might suddenly jump from one direction to the opposite one. The cells might also turn around their own axis that will change their morphological appearance. They also might appear and disappear during the experiment as well as re-appear after having gone out of the focus of the microscope. It is also highly likely that the cells might touch or overlap. The tracking algorithm should run fast enough to produce the result in little computation time.

The following conditions have been decided for the algorithm: Only cells that are in focus should be tracked (background objects are not considered). Fragmented cells at the image borders are eliminated. Each detected cell gets an identity label and gets tracked. Disappearing cells are tracked until they disappear. Newly appearing cells are tracked upon their appearance and get a new identity label. Cells overlapping or touching each other are eliminated. They are considered as disappearing cells. Splitting cells are tracked after splitting and both get an identity label that refers to the mother cell. Cell fragments are eliminated by a heuristic less than 2 × Sigma of Cell Size. Note that we decided to exclude overlapping cells for the reason of the resulting higher computation time. If we want to consider this kind of cells, we can use our matching

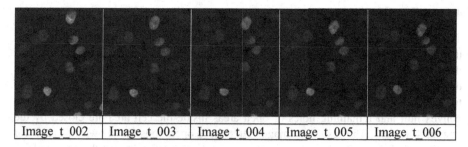

| Image_t_002 | Image_t_003 | Image_t_004 | Image_t_005 | Image_t_006 |

Fig. 1. Time series of image of a live cell project

algorithm in [15] to identify the portion of touching cells belonging to the cell under consideration (Fig. 1).

4   The Object-Tracking Algorithm

The unsteady movement of the cells and the movement of the cells around their own axis required special processing algorithms that are able to handle this. The successive estimation of the movement based on mean-shift or Kalman filters [14] would only make sense if we can expect the cells to have steady movement. Since it can happen that the cells jump from one side to another, we used a search window around each cell to check where the cell might be located. The size of the search window is estimated based on the maximal distance a cell can move between two time frames. The algorithm searches inside the window for similar cells. Since we cannot expect the cell to have the same shape after turning around its own axis and as also the texture inside the cell might change, we have chosen a similarity-based procedure for the detection of the cell inside the window. The overall procedure is shown in Fig. 2.

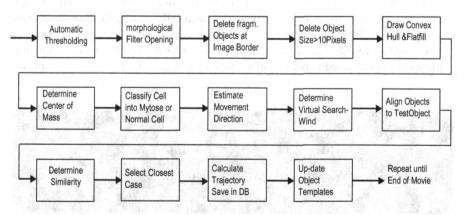

Fig. 2. Overall procedure of the algorithm

4.1   Overall Algorithm

The image gets thresholded by Otsu's well-known segmentation procedure. Afterwards the morphological filter opening by a 3 × 3 window is applied to close the contour and the inner holes. Fragmented cells at the image borders as well as small remaining objects of a size of ten pixels are deleted. The cells at the image borders are only tracked when they fully appear inside the image. Around the object is drawn the convex hull and remaining holes or open areas inside the cell area are closed by the operation flat-fill. The resulting images after these operations are shown in Fig. 3 for the six time frames. This resulting area is taken as the cell area and the area with its grey levels is temporarily stored as template in the data base.

Then the center of gravity of the object is determined and the search window is tentatively spanned around the object. A raw estimation of the cell's movement direction is calculated by the mean-shift filter over 3 frames. In the resulting direction is started the search for similar cells. Cells fragmented by the window border are not considered for further calculation. Each cell inside the window is compared by the similarity measure to the respective template of the cell under consideration. Before the similarity is determined the cells are aligned, so that they have the same orientation. The cell having the highest similarity score to the template is labeled as the same cell moved to the position x, y in the image $t + 1$. The template is updated with the detected cell area for the comparison in the next time frame. The position of the detected cell is stored into the data base under the label of the template. Mytotic cells are detected by classifying each cell based on the texture descriptor given in [16] and the decision tree classifier.

Let C_t be the cell at time-point t and C_{t+1B} the same cell at time point $t + 1$. Then the rule for labeling a cell as "disappeared" is: **IF** C_t has no C_{t+1B} **THEN** Disappearing_Cell.

4.2   Similarity Determination Between Two Cells

The algorithm [17] computes the similarity between two image areas A and B these images are in our case the bounding box around each cell (see Fig. 4). According to the specified distance function, the proximity matrix is calculated, for one pixel at position r, s in image A, to the pixel at the same position in image B and to the surrounding pixels within a predefined window. Then, the minimum distance between the compared

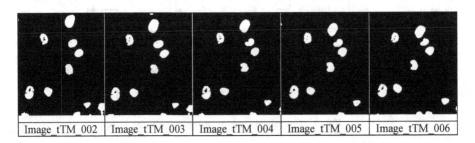

| Image_tTM_002 | Image_tTM_003 | Image_tTM_004 | Image_tTM_005 | Image_tTM_006 |

Fig. 3. Image after thresholding and morphological filtering

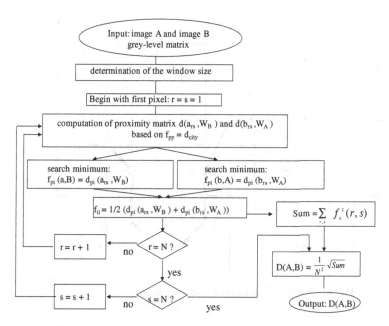

Fig. 4. Flowchart of the algorithm for computing the similarity measure.

pixels is computed. The same process is done for the pixel at position r, s in image B. Afterwards, the average of the two minimal values is calculated. This process is repeated until all the pixels of both images have been processed. The final dissimilarity for the whole image is calculated from the average minimal pixel distance. The use of an appropriate window size should make this measure invariant to scaling, rotation and translation (Fig. 5).

The resulting similarity value based on this similarity measure for the pairwise comparison of cell_1 to cell_1, cell_1 to cell_2, and cell_1 to cell_3 in the preceding time frame is given in Fig. 6. The lowest similarity value is obtained when comparing cell_1 to cell_1 in the proceeding time frame.

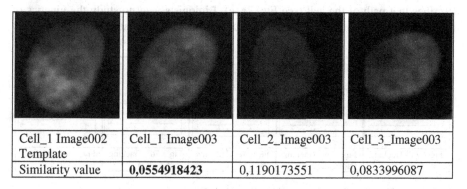

Cell_1 Image002 Template	Cell_1 Image003	Cell_2_Image003	Cell_3_Image003
Similarity value	**0,0554918423**	0,1190173551	0,0833996087

Fig. 5. Example cells used for comparison

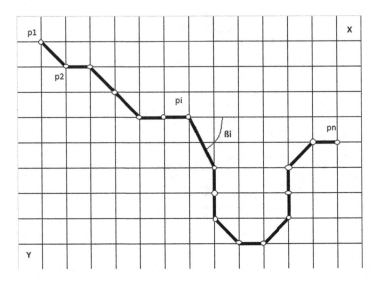

Fig. 6. Path of a cell marked with coordinate points

5 Results of Tracking Algorithm

Results in Fig. 7a–e show the tracking path of six different cells. We compared the manually determined path of a cell from an operator of the High-Content Screening Process-line with the automatic determined path by the tool *IBaI-Track* for 10 videos with 117 frames each. If both methods gave the same path we evaluated it as positive otherwise as negative. We observed a correspondence between these two descriptions of 98.2 %. The computation time for a sequence of 117 images of 674 × 516 pixels each and on average 80 cells per image is 11 min 42 s on PC with 1.8 GHz.

6 Information Extracted from the Path

The output of the cell-tracking algorithm is a tuple of coordinates for each cell that describes that path of the cell (see Fig. 7a–e). Biologists want to study the kinetics of the cells. Therefore, we have to extract features from this path that describe the motility

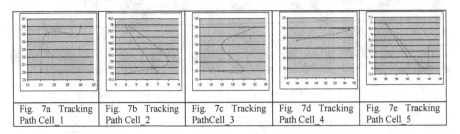

| Fig. 7a Tracking Path Cell_1 | Fig. 7b Tracking Path Cell_2 | Fig. 7c Tracking PathCell_3 | Fig. 7d Tracking Path Cell_4 | Fig. 7e Tracking Path Cell_5 |

Fig. 7 a Tracking Path Cell_1. b Tracking Path Cell_2. c Tracking Path Cell_3. d Tracking Path Cell_4. e Tracking Path Cell_5

Table 1. Measures for motility and velocity of a cell

Measure	Definition
Total distance traveled	$d_{tot} = \sum_{i=1}^{N-1} d(\mathbf{p}_i, \mathbf{p}_{i+1})$
Net distance traveled	$d_{net} = d(\mathbf{p}_1, \mathbf{p}_N)$
Maximum distance traveled	$d_{max} = \max_i d(\mathbf{p}_1, \mathbf{p}_i)$
Total trajectory time	$t_{tot} = (N-1)\Delta t$
Confinement ratio	$r_{con} = d_{net}/d_{tot}$
Instantaneous angle	$\alpha_i = \arctan(y_{i+1} - y_i)/(x_{i+1} - x_i)$
Directional change	$\gamma_i = \alpha_i - \alpha_{i-1}$
Instantaneous speed	$v_i = d(\mathbf{p}_i, \mathbf{p}_{i+1})/\Delta t$
Mean curvilinear speed	$\bar{v} = \frac{1}{N-1}\sum_{i=1}^{N-1} v_i$
Mean straight-line speed	$v_{lin} = d_{net}/t_{tot}$
Linearity of forward progression	$r_{lin} = v_{lin}/\bar{v}$
Mean squared displacement	$\text{MSD}(n) = \frac{1}{N-n}\sum_{i=1}^{N-n} d^2(\mathbf{p}_i, \mathbf{p}_{i+n})$

Table 2 Output of the celltracking tool

Cell number	Total distance traveled	Maximum distance traveled	...	...	...	...	Mean squared displacement
Cell_1	X11	X12	...	...	...	...	X112
...	...	...	...	...	...	...	
Cell_n	Xn1	Xn2	...	...	...	...	Xn12

and velocity of a cell. Table 1 shows features that are used to describe the path of a cell [18]. These features (see Table 2) are provided in a table to the biologist for further study. Please note, one image contains many cells. As result, we obtain a bunch of numerical values for one image that is hard to overlook for a human. More high-level descriptions are necessary that summarize these features and their feature values in a more compact information about the kinetics of the cells in one image. Recently, biologist use statistical tools to study these features. More complex image mining methods such as decision tree induction and conceptual clustering [19] can be of help in order to bring out the higher-level descriptions.

7 Conclusions

We have presented our new cell-tracking algorithm for tracking cells in dynamic drug discovery or computational biology experiments. The algorithm uses a window around the position of the cell in the preceding time frame for searching for the next position of the cell. This is necessary since the movement of the cells cannot be steady. The search inside the window is started in the expected direction of the cell. To calculate this direction, we use a mean-shift filter over 3 time frames. The detection of the cell is done based on a similarity determination of the grey-level profile between the cells in the preceding time-frame and the following time-frame. The cell giving the highest similarity value is selected as the same cell in the following time-frame. The template is updated with the new cell. A similarity measure that can handle small changes in the morphology of the cells is used. This similarity measure is robust enough to detect the cell with high accuracy. The tracking algorithm can track normal cells and mitotic cells by our formerly developed classifier. Eleven features describe the path of a cell. The resulting data are stored in a data file and can be used for further image-mining analysis. We propose that biologists and drug discovery experts think about more high-level terms to describe the events they want to discover since the large bunch of numerical values created during the tracking process is hard to overlook by a human. For that can be used proper image mining algorithm such as decision tree induction and conceptual clustering.

Acknowledgement. This work has been sponsored by the German Federal Ministry of Education and Research BMBF under the grant title "Quantitative Measurement of Dynamic Time Dependent Cellular Events, QuantPro" grant no. 0313831B.

References

1. Debeir, O., Ham, P.V., Kiss, R., Decaestecker, C.: Tracking of migrating cells under phase-contrast video microscopy with combined mean –shift processes. IEEE Trans. Med. Im. **24**, 697–711 (2005)
2. Padfield, D., Rittscher, J., Sebastian, T., Thomas, N., Roysam, B.: Spatiotemporal cell segmentation and tracking in automated screening. In: Proceedings of the IEEE International Symposium Biomedical Imaging, pp. 376–379 (2008)
3. Li, K., Miller, E.D., Chen, M., Kanade, T., Weiss, L.E., Campbell, PhG: Cell population tracking and lineage construction with spatiotemporal context. Med. Image Anal. **12**, 546–566 (2008)
4. Li, K., Miller, E.D., Chen, M., Kanade, T., Weiss, L.E., Campbell, P.G.: Cell population tracking and lineage construction with spatiotemporal context. Med. Image Anal. **12**, 546–566 (2008)
5. Sacan, A., Ferhatosmanoglu, H., Coskun, H.: Cell track: an open-source software for cell tracking and motility analysis. Bioinform. Appl. Note **24**(14), 1647–1649 (2008)
6. Wang, M., Zhou, X., Li, F., Huckins, J., King, R.W., Wong, St.T.C.: Novel cell segmentation and online SVM for cell cycle phase identification in automated microscopy. Bioinformatics **24**(1), 94–101 (2008)
7. Cohen, A.R. Bjornsson, C., Chen, Y., Banker, G., Ladi, E., Robey, E., Temple, S., Roysam, B.: Automatic summarization of changes in image sequences using algorithmic information theory. In: 5th IEEE International Symposium on Biomedical Imaging: From Nano to Macro, pp. 859–862 (2008)
8. Padfield, D., Rittscher, J., Thomas, N., Roysam, B.: Spatio-temporal cell cycle phase analysis using level sets and fast matching methods. In: Proceedings of the IEEE International Symposium Biomedical Imaging, pp. 376–379 (2008)

9. Kan, A., Markhamb, J., Chakravortyb, R., Baileya, J., Leckiea, C.: Ranking cell tracking systems without manual validation. Pattern Recogn. Lett. **53**(1), 38–43 (2015)

10. Maška, M., Daněk, O., Garasa, S., Rouzaut, A., Muñoz-Barrutia, A., Ortiz-de-Solorzano, C.: Segmentation and shape tracking of whole fluorescent cells based on the Chan-Vese model. IEEE Trans. Med. Imaging **32**(6), 995–1005 (2013)

11. Magnusson, K.E.G., Jalden, J., Gilbert, P.M., Blau, H.M.: Global linking of cell tracks using the viterbi algorithm. IEEE Trans. Med. Imaging **34**(4), 911–929 (2014)

12. Padfield, D., Rittscher, J., Roysam, B.: Coupled minimum-cost flow cell tracking for high-throughput quantitative analysis. Med. Image Anal. **15**(4), 650–668 (2011)

13. Guo, D., Van de Ven, A.L., Zhou, X.: Tracking and measurement of the motion of blood cells using optical flow methods. IEEE J. Biomed. Health Inform. **18**(3), 991–998 (2014)

14. Perner, P., Perner, H., Jänichen, S.: Recognition of airborne fungi spores in digital microscopic images. J. Artif. Intell. Med. **36**(2), 137–157 (2006)

15. Comaniciu, D., Ramesh, V., Meer, P.: Real-time tracking of non-rigid objects using mean shift. In: Proceedings of IEEE Conference on Computer Vision and Pattern Recognition, vol. 2, pp. 142–149, June 2000

16. Attig, A., Perner, P.: A comparison between Haralick's texture descriptor and the texture descriptor based on random sets for biological images. In: Perner, P. (ed.) MLDM 2011. LNCS, vol. 6871, pp. 524–538. Springer, Heidelberg (2011)

17. Zamperoni, P., Statovoitov, V.: How dissimilar are two gray-scale images. In: Sagerer, G., Posch, S., Kummert, F. (eds.) Proceedings of the 17th DAGM Symposium, pp. 448–455. Springer, Heidelberg (1995)

18. Beltman, J.B., Mar'ee, A.F.M., de Boer, R.J.: Analysing immune cell migration. Nat. Rev. Immunol. **9**, 789–798 (2009)

19. Perner, P. (ed.): Data Mining on Multimedia Data. LNCS, vol. 2558. Springer, Heidelberg (2002)

20. Perner, H., Perner, P.: Similarity-based motion tracking of cells in microscopic images. IBaI report, IB2006–11 (2006)

21. Perner, P.: Progress in cell image analysis and life cell tracking. QuantPro-BMWI Presentations and Reports May/2006, September/2007, March/2008 Grant no. 0313831B

Author Index

Printed in the United States
By Bookmasters